Praise for *Wine and Place*

"For me, the construct of *terroir* has always been the big pumping heartbeat of wine. Unhinged from it, wine would be dead—a hollow shell emotionally and intellectually. And yet do any of us spend enough time thinking about *terroir,* the very lifeblood of what we love? With *Wine and Place,* Tim Patterson and John Buechsenstein have given us a great gift—a fantastic book that explains why wine moves us and reminds us why wine has meaning. I could not put this book down."

KAREN MACNEIL, author of *The Wine Bible* and editor of *WineSpeed*

"Wine's magic appears to be closely tied to its place of birth. In this tasty volume, the subject inspires passionate writing from some of the best of our wine writers."

KERMIT LYNCH, author of *Adventures on the Wine Route: A Wine Buyer's Tour of France*

"Patterson and Buechsenstein's book presents a detailed compilation of some of the finest writing on *terroir,* the concept that's at the heart of fine wine. As such, it's a vital distillation of thinking on this important topic, thoughtfully arranged and interestingly presented. It's an important contribution to the wine literature."

JAMIE GOODE, author of *I Taste Red: The Science of Tasting Wine*

"What a bonus to find a book about the taste of place compiled by two actual winemakers— Buechsenstein, an accomplished professional, and Patterson, a passionate amateur—who made wines from scores of different *terroirs* in their careers. This scholarly and often witty compilation of viewpoints is the best there is."

JIM GORDON, editor of *Wines & Vines*

"Where something comes from is always intriguing. With wine, all the more so. Finding words to describe all that goes into that elusive bugaboo, *terroir,* is tough. Patterson and Buechenstein have worked every angle to help us understand—and hey, any book that includes magma is worth one's time. Site matters."

VIRGINIE BOONE, contributing editor for *Wine Enthusiast*

"I am so sad that I did not have hours and days and years to spend with Tim Patterson, talking about the subject dearest to us both, the mysterious, vexatious question of *terroir*. Thankfully, he—along with coauthor John Buechsenstein—has left us *Wine and Place*. This is a must-have volume for both terroirists and counter-terroirists alike, curious to understand how a wine might most profoundly express itself."

RANDALL GRAHM, author of *Been Doon So Long: A Randall Grahm Vinthology*

"In their chosen roles as compilers and contrarians, the experts behind *Wine and Place* have initiated a crucial dialogue about *terroir*. They have assembled, with erudition and wit, the perspectives of scholars, journalists, and winemakers, and they have created fruitful and engaging juxtapositions as to the definition, the construction, the meaning, the analysis, and the power of *terroir*. Everyone will learn something new, from wine aficionados to scientists to students of wine history and culture."

AMY TRUBEK, author of *The Taste of Place: A Cultural Journey into Terroir*

WINE AND PLACE

THE PUBLISHER AND THE UNIVERSITY OF CALIFORNIA PRESS
FOUNDATION GRATEFULLY ACKNOWLEDGE THE GENEROUS
SUPPORT OF THE AHMANSON FOUNDATION ENDOWMENT
FUND IN HUMANITIES.

Wine and Place

A Terroir Reader

Tim Patterson and John Buechsenstein

 UNIVERSITY OF CALIFORNIA PRESS

University of California Press, one of the most distinguished university presses in the United States, enriches lives around the world by advancing scholarship in the humanities, social sciences, and natural sciences. Its activities are supported by the UC Press Foundation and by philanthropic contributions from individuals and institutions. For more information, visit www.ucpress.edu.

University of California Press
Oakland, California

© 2018 by Tim Patterson and John Buechsenstein

Library of Congress Cataloging-in-Publication Data

Names: Patterson, Tim, author. | Buechsenstein, John, 1949- author.
Title: Wine and place : a terroir reader / Tim Patterson and John Buechsenstein.
Description: Oakland, California : University of California Press, [2018] | Includes bibliographical references and index. |
Identifiers: LCCN 2017014298 (print) | LCCN 2017016708 (ebook) | ISBN 9780520968226 (ebook) | ISBN 9780520277007 (cloth : alk. paper)
Subjects: LCSH: Terroir.
Classification: LCC SB387.7 (ebook) | LCC SB387.7 .P385 2018 (print) | DDC 634.8—dc23
LC record available at https://lccn.loc.gov/2017014298

Manufactured in the United States of America

27 26 25 24 23 22 21 20 19 18
10 9 8 7 6 5 4 3 2 1

To Nancy Freeman and Nancy Buechsenstein

CONTENTS

FOREWORD

Patrick J. Comiskey

This book begins, like so many do, with an epiphany, of the sort that thousands of wine lovers experience to one degree or another. You are *moved* by a wine, in precisely the same way that you are *moved* by a painting, a poem, a passage in a novel, a sunset, a mountain vista. Just taking a sip of this wine is transporting, humbling, it fills you with wonder. Like a great meal, great sex, a great musical performance, the feeling will be powerful but fleeting, rousing but ephemeral, grounded in the senses and yet cosmic all the same.

Inevitably, you start to wonder why. What makes this wine much more graceful, powerful, unique? What about this wine takes it out of the combinatory matrix of fruit, tannin, acid, and alcohol and into another, more interpretive realm? You wonder about that subtle spice, that consistently warm core of fruit, the grace and balance found here and here alone, distinct from vineyards just a few meters away. You marvel at the consistencies of texture, the tension, the power, the finesse, that seem to inhabit this wine no matter when you taste it and where—its constancy is uncanny, confounding, thrilling. Suddenly the experience of tasting this wine is (if I may coin a word) extra-vitreous: it takes you out of the glass, and into a speculative place.

For wine lovers, the discovery of *terroir* is a breakthrough that cannot be unbroken. Whether it's the somewhereness of a region like Chablis or the particulars of its prized Grand Cru vineyards, it is an irrevocable event in your wine consciousness: once you've found it you'll seek it in all wines for the rest of your life. *Terroir* will take over your understanding of wine, it will leave you gobsmacked by discoveries, and straining to grasp at things that aren't there.

Tim Patterson sought a practical explanation for that incredulity. As a wine writer he came up against the concept of *terroir* all the time, he became its student whenever he tasted and evaluated wines, when he walked vineyards with grapegrowers and winemakers, when he sniffed at the dirt and pocketed rocks from between vine rows, mementoes from hallowed ground.

But in addition to being a writer, Patterson was a home winemaker, using purchased fruit to make his own wine. This fact is critical: since he didn't grow the grapes, didn't live on the land, that sense of place wasn't something inherent in his interpretation of that fruit; its signature was something he had to discover, to isolate and express. He wondered constantly, as he sniffed his vats of bubbling grapes, just what did he have to do, or not do, to tease out a site's uniqueness—was there a skeleton key, a secret procedure, that would bring out the wine's *terroir*? Or conversely was there something that would inadvertently mask that character, and what could he do to avoid that fate? When did he need to step in, and when did he need to get out of the way?

John Buechsenstein had similar concerns over the course of his long career as a winemaker and wine educator in California and elsewhere, whether in the old vine fruit in the MacDowell Valley in Mendocino County, or the estate plantings at Fife, Phelps, and other places, and not least in his last ambitious winemaking venture, the Sauvignon Republic Wine Company, where with Paul Dolan and John Ash he parsed out the nuances of Sauvignon Blanc from three wildly differing regions, New Zealand, South Africa, and California. Had the project gone forward, they would have made Sauvignon Blanc from seven different global locales—*terroir* exploration on an unprecedented scale.

These ruminations, in many articles, over many vintages and many wines, grown, made, and drunk, are the germ of this book.

In 2008, at a wine conference in Portland, Patterson ran into Buechsenstein, and they got onto the topic of *terroir*. A few glasses of wine later they'd hatched the idea of a *Terroir Reader*, a compendium of texts that would get at these questions, allowing the two authors the chance to think about them deeply, systematically, and skeptically, and address the mysteries, strip the concept of its fairy dust, the cosmic claims, the dubious assertions, the siren song of marketing, the whiff of bullshit. The book was meant to be inclusive, to compile in a single volume the many contradictions inherent in the topic, encompassing both passionate belief and healthy skepticism.

Why skepticism? Because *terroir* as a concept is inevitably subjective and interpretive; its beholder is vulnerable to suggestion. As a wine writer Patterson was frequently exposed to marketing that exploited the subject, to suggestive interpretations of *terroir* by winemakers or spin agents eager to point out the uniqueness of the wine in front of him. As a globetrotting winemaker, Buechsenstein was subject to similar interpretive dances, literally on a global scale. His task was to decipher local *terroir* and minimize winemaking manipulation, so that his wines expressed their local flavors.

Not every wine, after all, expresses *terroir*. Some, like Yellow Tail or Two Buck Chuck, don't seek to make the distinction. But many more wines in the market that come from somewhere express little or nothing of the place they're grown; any vestige is obliterated by overripeness, indifferent winemaking, or adulteration—often all three. But that doesn't keep marketers from making claims of typicity. Skepticism is required in the *terroir* game, even if it's rarely employed. Patterson and Buechsenstein sought to avoid being seduced by their subject matter. They wanted to strip the concept of hype, to bring a sober eye to its examination.

John brought a familiarity with the scientific and technical literature. He foraged for relevant research material, scientific studies, conference papers, books, and articles. He knew where to find such pieces in the vast array of global sources. He brought, too, a familiarity with the technical and at times abstruse nature of these research efforts, and was a worthy translator and distiller of their research and ideas.

He and Tim spent long hours organizing and codifying the topical material in each chapter, but Tim brought the authorial and editorial thrust. It is his voice and his narrative skills that link the pieces of this compendium together, that allow them to flow and brush up against and inform one another, guided by his inquisitiveness and his abilities as a reporter and storyteller.

Tim Patterson lived much of his life in failing health. Born with extremely poor vision, he endured two autoimmune disorders, a non-functioning kidney, and a tubercular condition of the lymph nodes, for which he had his first surgery at age four. None of this stopped him, not from making wine, not from writing about it, not from tasting and enjoying it, and certainly not from having it consume him in the course of this project.

But in late March 2014, his health took a turn. He sent a worried email to Blake Edgar, then acquisitions editor at UC Press with some concern. "My vision," he wrote, "never much good, went into a tailspin—swarms of floaters, color vision way off, depth perception shot. So far, no diagnosis or explanation. Worst part is that I can't read anything, or type anything. I will keep you posted." By the time Edgar responded, Patterson was in the hospital; shortly after he was diagnosed with a glioblastoma—a cancerous growth in the connective tissue of his brain. It was untreatable, and less than two months later he passed away.

In his last days of consciousness, Tim lamented with Aaron Belkin, the cousin of his wife Nancy and a professor of political science at San Francisco State University, the fact that he would not live to finish the book. Belkin, who had spent many years helping authors with their manuscripts, promised Tim that he would shepherd the project to its conclusion. He has done a heroic job marshalling editorial forces (I became the developmental editor for the project), wading through a thicket of reprint and copyright permissions, pushing the project to its completion in keeping with Tim's final wish. With the guidance of Aaron and John, and as its final editor, I completed drafts, finished stalled chapters, brought some editorial polish to passages and chapters that Tim was too ill to address himself.

I've worked hard to preserve Tim Patterson's voice in these passages. Tim had a nimble and restless mind, a probing intellect, and a unique ability to inject a kind of dialectic tension into his prose as he grappled with an idea, an effect that was enhanced by the joy he had with language and in particular with wordplay: in these passages and commentary, he was often able to bring a touch of levity to a very demanding topic. Wherever possible I tried to preserve that feel to the prose here, even as he and John led me down pathways of inquiry I wouldn't have thought to explore.

As I said at the outset, *terroir* is a topic that few who love wine can escape, but what the authors have done here is a more complete and more exhaustive exploration than has ever been attempted before. They have made it all the more inescapable. For that, and for many other reasons, I'm grateful for having had the opportunity to work on this book. I hope you'll agree that this was a worthy project to bring to completion.

Patrick Comiskey
28 February 2017

INTRODUCTION
Why *Terroir* Matters

Does the world really need another *terroir* book? Of course it does.

The notion of *terroir* is at the heart of what makes wine special. No other foodstuff, no other agricultural commodity, grips the human imagination with such immeasurable force as a great wine from a great winegrowing area. When you taste a great wine it seems inevitable that a connection exists between those inimitable flavors and the particulars of that place—the soil, the climate, the elevation, the aspect, the parcel's unique position on the hill or in the vale. No other connection between food and place has inspired as extensive a body of literature as the earthly link in wine. Many agricultural products exhibit some degree of regional and sub-species variation, but since wine involves a dramatic transformation of raw grapes through fermentation, the lingering pedigree of origin is all the more remarkable. Wine is unique, and *terroir* is the reason. The Greeks and Romans had wine gods; there is no record of any deity responsible for, say, Vidalia onions, tasty as they are.

All over Europe, a rudimentary sense of *terroir*, a belief in the direct connection between soil, climate, and wine, was the conventional wisdom among growers for millennia. This instinctive, pre-scientific association of place and taste underwent considerable refinement before it emerged as the modern concept of *terroir* in the late nineteenth and early twentieth centuries. But in its earlier, inchoate form, the belief that dirt controlled wine's destiny was widespread, elemental, and obvious, no more remarkable than the daily rise of the sun in the east. There are books about winegrowing and winemaking dating back centuries, but in all that literature, only fleeting passages focus on *terroir*.

1

That changed dramatically in the middle of the twentieth century, when New World wine regions—Australia, California, Chile—began to produce wines of undisputed quality from vineyards that had no established *terroir* credentials. Modern winemaking seemed more important than the traditions derived from ancient terrain, and many New World winemakers, researchers, and writers dismissed *terroir* as a marketing ploy, a fuzzy French philosophical concept, even an excuse for poor hygiene. Once *terroir* came under attack, its defenders took up their pens, typewriters, computers, blog posts and decanters with a vengeance, both in the Old World and the New. The world of wine is now awash in talk of *terroir,* some of it inspiring and substantive, some of it fanciful or downright loopy.

In recent years, the notion of *terroir* has been applied far beyond wine (see chapter 2), with some justification. Certainly there are pronounced variations in cheeses and their molds, breads and their starters, even breeds of livestock and their favorite feeds or pasturage. But again, wine is special, in large part because wine grapes are special. No other raw fruit or vegetable has a flavor chemistry as complex as *Vitis vinifera* (though tea leaves may come close), with hundreds of aromatic compounds or their precursors present in varying degrees and limitless combinations. This cornucopia of chemistry means that wine grapes and wine are uniquely suited to register and reflect subtle, nuanced differences in the natural environments where they develop. It's why wine vintages vary far more widely than potato vintages, and why we can taste the difference between neighboring vineyards.

The dance of wine flavor chemistry is one of the many ways that the application of science intersects with the romance of *terroir.* It's true that a scientific approach casts doubt on certain *terroir* claims—like the idea of literally tasting the vineyard in the glass—and that, we think, is useful conceptual demolition work. But it's also true that many positive insights into the workings of *terroir* have come from science-based research projects. If *terroir* is real, and we think it is, then it is helpful that it exist not only as a vague entity in certain vineyards, or solely as a philosophical concept, but as a provable, measurable phenomenon.

At any rate, our goal in compiling this volume is to put all of these approaches between one set of covers. We think that placing a literary-minded Burgundophile like Matt Kramer and a climate prognosticator like Greg Jones side by side, so to speak, is all to the good; the observations of a wine-loving petroleum geologist like James Wilson are inevitably enhanced when read alongside those of a wine country *evocateur* like Gerald Asher.

We chose the vehicle of an annotated reader for several reasons. The first is humility: neither of us has done extensive original research, nor written classic essays on the subject. But we are confident that, having both spent years thinking about wine in general and this topic in particular, we know enough to identify the good stuff. Rather than simply giving footnote credit to the people who have set the standards for digging into *terroir,* we let them have their say directly. Some of the excerpts come from landmark publications, like Matt Kramer's advocacy of "somewhereness," Gerard Seguin's work on soil hydraulics, and Ann Noble's application of modern methods of sensory analysis to the wines of Bordeaux; some are lesser known, but equally on point.

As you will quickly see, the authors we reprint do not all agree with each other, and since they started these fights, we let them square off in their own words. You will also discover that we are not shy in offering our opinions of their opinions along the way.

We suspect that our readers' attitudes about the subject of wine *terroir* will fall into one of four categories. One group knows deep in their bones that *terroir* is real; they have experienced it and tasted it and reveled in it, and they aren't in need of more explanation. They just want the next glass to reveal the wonders of its origins. A second group thinks the *terroir* effect is real, but that it can and should be understood scientifically, by analyzing vineyard geology and rainfall patterns and grapevine clonal DNA and applying rigorous sensory analytical methods. (The authors are in this camp.) A third group is just curious, intrigued by the concept or maybe by a great *terroir* story and wondering what it's all about. Finally, there are *terroir* skeptics—*terroir* deniers?—who think the whole concept is shot through with romanticism, sophistry, and foolishness.

All of these viewpoints are represented in this book, and all are challenged as well.

ORGANIZATION AND USE OF THIS BOOK

This reader starts with an immersion in the emotional power of the concept of *terroir*, conveyed by a series of masterful evocations of its beauty and meaning. If you ever wondered why all the fuss about *terroir*, chapter 1 is your answer. Chapter 2 surveys the history and evolution of the concept, from its ancient prehistory through up-to-the-minute formulations, including the transformation of the term *terroir* from put-down to high praise. Chapter 2 also tracks the concept of *terroir* as it's used to elucidate regional foodways, broader cultural patterns, and changes in social organization.

Chapter 3 examines the role of soil, which lies at the heart of the concept of *terroir*. Chapter 4 examines climate with the same rigor and detail. Chapter 5 looks at how viticulture and grape physiology impact our perception of *terroir*, a crucial element often left out of the discussion. Chapter 6 turns to winemaking, where we'll delve into both the dramatic transformation of juice to wine by microbes and yeasts and the indispensable human element in the equation.

Chapter 7 scrutinizes the sensory evidence for the reality of *terroir*, and discusses what more might be done to validate the concept perceptually. Chapter 8 investigates the marketing of *terroir*, which is at once necessary (if anyone outside your village is ever to know how great your wines are) and perilous (wherein you believe your own hype). Chapter 9 examines the prospects for the future of *terroir* in the light of climate change, heavy-handed winemaking, and global expansion of vineyard land, as amazing new growing areas are discovered every year.

Chapter 10 gives us a chance to summarize what we think has been established to date—and what has not—about the mechanisms and effects of *terroir*, and what new directions are worth pursuing.

We do not expect many people to sit down and read through this book cover to cover in rapt concentration. More likely, the self-contained chapters should provide food for thought, a chance to meditate on discrete aspects of *terroir*, whether the minutiae of soil composition or the perils of promotional hyperbole. To make certain scientific concepts of *terroir* more accessible, most chapters include explanatory sidebars. A comprehensive bibliography appears at the end of the book, offering the reader the opportunity to follow up on any of the individual fragments presented in the text.

We want to acknowledge up front that this reader, for all its breadth, has certain limitations. For example, despite an international array of authorship, we've included only readings available in English or English translation, even though the concept of *terroir* did not originate among English speakers. We are also painfully aware that the use of excerpts and sections of whole works can fail to convey the flow of the originals, and we encourage readers to consult the complete originals, all listed in the bibliography, for further enlightenment. Because many of the readings are partial excerpts, we have done some minor editing to reduce confusion—for example, eliminating references to figures that are not included. Footnotes are included where they appear in the original texts.

A note on usage, and a warning. In both scholarly and everyday contexts, the term '*terroir*' is employed in both a descriptive and a prescriptive sense, conveying either that a particular wine is *distinctive* because of its origins or that it is *better* than other wines because of those same origins. At the same time, *terroir* gets applied on multiple scales: to broad wine regions, to particular districts within regions, to single vineyards, or portions of vineyards. Thus, if someone says a certain wine truly shows off its *terroir*, the speaker may be referring to a wine typical of a region (even if the wine is inferior) or to a spectacular expression of a tiny parcel. These claims are markedly different. We make every reasonable attempt to make clear which usage is in play, but readers are well advised to keep track for themselves.

THESES ON *TERROIR*

Finally, as another form of overview, we present an outline of our own ideas on the subject, what might be called our Ten Theses on *Terroir*:

I. We believe the effects of *terroir* are real and undeniable: wines made from grapes grown in different places smell and taste different, vintage after vintage, to multiple, experienced, trustworthy tasters. This is true for both macro-climate regions—Alsatian Riesling is different from Austrian, from German, from Finger Lakesian—and for specific vineyards and sites.

II. We believe that many if not most of the standard depictions of this phenomenon, however, are worthy of skepticism—at least from the standpoint of modern science. Wine flavors and aromas, for example, do not come directly from the soil through passive plants and into the glass intact; rather, they (or their precursors) are created inside the plants and berries and later during fermentation by yeast action. Vines remain mute as to their preference of whether they get their water through natural rainfall or drip irrigation, though they clearly respond to how much they get and when. And so it is unreasonable to argue that Old World dirt is somehow inherently superior to New World dirt.

III. We believe that for regional variations (macro-*terroir*s), climate and grape variety are both more important than soil; for vineyard variation, soil can be critical. New Zealand Sauvignon Blanc, for example, burst onto the international scene with a recognizable aromatic and flavor profile, clearly due to that country's cooler climate and winemaking proclivities. At least some of the perceived differences between bottlings within that climate zone reflect variations in soil structure and composition.

IV. Despite the obvious allure of that famous marketing trope, "Great wines are made in the vineyard," we wish to point out that great wine is in fact *never* made in the vineyard: it's made in the winery. As such, the human factor has to be included in any sensible view of the workings of *terroir*. Likewise, we feel it is necessary to factor in culture as part of the array of *terroir* elements a wine might possess, including the gastronomic and culinary milieu within which the wine has accrued its traditional cultural meaning and importance. Gaining an intimate knowledge of a wine's origin has a profound impact on how it tastes forever after.

V. Adding layers to the concept, however, inevitably creates tension and confusion as well. Limiting *terroir* to a discussion of dirt and place may be useful, even powerful, but it's clearly incomplete. When you add factors such as climate, traditional growing and winemaking practices, when you get humans into the act, you get a fuller picture—but you also generate controversies about *terroir*'s purest expression, and what the human element obscures. Maybe wine drinkers have to be included, too, since without a taster, there is no *goût du terroir*. It's easy to see that as components get included, *terroir* as a concept is in danger of becoming so broad it ends up meaning nothing at all. Balancing the power of the core concept with the complexities of actual winegrowing and winemaking is the challenge.

VI. A pair of observations that aren't strictly speaking, thesis statements, but need to be included in our initial salvo:

First, we note that two critical dimensions are almost entirely missing from standard discussions of *terroir*. First, despite millions of personal testimonials connecting a particular wine with a particular place, precious few rigorous sensory studies have been conducted, studies which, one would think, would validate the concept. If a sensory effect cannot be captured by careful sensory methods, we have a serious problem.

Second, we note that almost no attention has been paid to what the *vines* do, to the photosynthetic and physiological mechanisms that actually create the chemical compounds behind the distinctive flavors and aromas of *terroir*-driven wines. Grapevines are not neutral transmitters of metaphysical essences; they are living flavor factories. Without a rigorous examination of these two elements, *terroir* will remain more an article of faith than a true window into the natural world.

VII. We believe that there is a legitimate debate about whether the concept of *terroir* can or should be evaluated through the methods of science. Some hold that the concept is essentially spiritual, that reducing it to soil chemistry and climate charts is what's wrong with the modern approach to winemaking and wine appreciation. Others believe that the emotional connection between a wine drinker, a place and its wines is more important than anything else. The notion of *terroir* clearly arose in a pre-scientific context. We think that while science is not the only lens to employ, the environmental causes and sensory consequences of *terroir* expression are tangible, material, and often measurable. The *terroir* debate can benefit from rigorous investigation: claims advanced as scientifically valid but which, in fact, are not, have to be challenged.

VIII. We believe that because the sensory attributes of great wines are the result of so many dynamic, constantly interacting processes—and because wine tasting is an inherently subjective experience—we are unlikely ever to be able to say, definitively, that wine 'X' has the unique quality 'Y' because of the factor 'Z.' Wine will never be that simple. *Terroir* expression is not a single thing—in one wine, it may be found in the mouthfeel, in another the aromatics—and the origins of distinctiveness may be due to many different factors. Thus the goal of exploring *terroir* is not to "explain" the properties found in a single bottle of wine, the way we might explain a lunar eclipse or a skin rash, but rather to understand more thoroughly how even minute variations in soil, climate, viticulture and winemaking can have real, perceptible and, yes, magical consequences.

IX. As long as the concept is overgrown with mythology, as long as crucial aspects of it go untended, as long as it is allowed to mean anything to anybody, we believe that *terroir* will be vulnerable to exploitation by marketing departments happy to claim whatever appeals to consumers or reinforces a branding initiative. We have no moral quarrel with industrial, mass-produced beverage wine, or self-caricaturing fruit-and-oak bombs, but they are not and will never be wines of place, whatever their back labels claim.

X. In summary, we believe that the concept of *terroir* needs a good housecleaning in order to defend it. Without scrutiny, standards, and common understanding, *terroir* can be easily dismissed, challenged, or distorted beyond recognition. We think the most compelling, elusive, defining quality of fine wine should not be threatened with a fate no good winemaker would tolerate in his or her wines: severe dilution.

1

THE LURE AND PROMISE OF *TERROIR*

Wine *terroir*—the idea that certain wines uniquely express the special character of the places they come from—is no ordinary concept. Many important notions about wine—what makes a wine "balanced," for example, or when an age-worthy wine is at its peak—are thoroughly subjective, and thus the focus of endless, heated, highly enjoyable debates. But *terroir* encompasses a full-fledged belief system, perhaps several overlapping at once; it grips the vinous imagination like nothing else. *Terroir* is more than a mere attribute of wine: for its committed adherents, from winegrowers to writers to impassioned wine lovers, it is the conceptual gateway to wine's *meaning*.

Every well-made glass of wine worth putting in your mouth has *flavor* (or at least it should). It should have fruit, body, alcohol, tannins, and acidity for structure and texture; it ought to possess some facility with food. But a few wines, exceptional wines, have something extra: they have *meaning*. Part of what makes a wine exceptional is its connection to the earth, to the physical properties of a particular place, to the traditions of that place, to the generations of winegrowers who have nurtured the soil and the vines in that place. As industrial-technical wines overtake the global marketplace, with homogenized flavors and styles that blur the lines between Old World and New World, wines of place might seem headed for an endangered species list, but they are worth fighting to preserve.

The lure and promise of *terroir* has inspired a considerable body of great wine writing, and that is where our exploration starts. Before mining the more prosaic aspects of the concept—temperature gradients, soil drainage properties, nutrient uptake, commercial versus "natural" yeasts, the influence of marketing hype—we start by celebrating the very notion of *terroir:* the beauty it represents, the satisfaction it offers to those who approach wine in its thrall, the promise it holds out for experiencing something more than meets the glass.

And so we begin with Matt Kramer's classic essay on Burgundy and *terroir,* followed by pieces from prose masters Rod Smith and Gerald Asher, personal testimonies from eloquent Old World winegrowers, ruminations by winemaker/philosopher/*provocateur* Randall Grahm, and an ode to emotional connectedness by importer-evangelist Terry Theise. Bringing up the rear, we'll allow two skeptics the chance to have their say.

SOMEWHERENESS

Among American wine writers, Matt Kramer is perhaps the most consistent and self-conscious proponent of the importance of *terroir.* In a lucid series of books and in the pages of *Wine Spectator,* he has for decades championed the great wines of Burgundy and the tiny places they're sourced from. He has been tireless in advocating practices—low yields, gentle winemaking, restrained use of oak—that let the underlying *terroir* speak, and is critical of shortcuts—over-cropping, cellar manipulation, the injudicious use of oak—that take the soul out of wine.

Kramer coined the term "somewhereness" to describe the quality that great wines have to possess. His best-known treatment of the subject, "The Notion of Terroir," originally appeared in 1990 as a chapter in *Making Sense of Burgundy,* the second of several books he has written devoted to demystifying wine and the places it comes from. Even then, the concept of *terroir* was largely unexplored among American consumers, and underappreciated by those familiar with it. The following excerpts from that essay exemplify not only Kramer's passion for wine but also his intellectual rigor. He's not only willing to question current winemaking fads, but states unequivocally that understanding *terroir* requires a different mindset from the modern, linear, science-based worldview. And he also makes clear that an appreciation of *terroir* does not come easily, and is not for everyone.

FROM MATT KRAMER, "THE NOTION OF TERROIR"

Always the beautiful answer who asks a more beautiful question.

—E. E. CUMMINGS, *Collected Poems,* 1922–1938

The "more beautiful question" of wine is *terroir.* To the English speaker, *terroir* is an alien word, difficult to pronounce ("tair-wahr"). More frustrating yet, it is a foreign idea. The usual capsule definition is "site" or "vineyard" plot. Closer to its truth, it holds—like William Blake's grain of sand that contains a world—an evolution of thought about wine and the Earth. One cannot make sense of Burgundy without investigating the notion of *terroir.*

Although derived from soil or land (*terre*), *terroir* is not just an investigation of soil and subsoil. It is everything that contributes to the distinction of a vineyard plot. As such, it also embraces "microclimate": precipitation, air and water drainage, elevation, sunlight, and temperature.

But *terroir* holds yet another dimension: It sanctions what cannot be measured, yet still located and savored. *Terroir* prospects for differences. In this, it is at odds with science, which demands proof by replication rather than in a shining uniqueness.

Understanding *terroir* requires a recalibration of the modern mind. The original impulse has long since disappeared, buried by commerce and the scorn of science. It calls for a susceptibility to the natural world to a degree almost unfathomable today, as the French historian Marc Bloch evokes in his landmark work, *Feudal Society:*

> The men of the two feudal ages were close to nature—much closer than we are; and nature as they knew it was much less tamed and softened than as we see it today. . . . People continued to pick wild fruit and to gather honey as in the first ages of mankind. In the construction of implements and tools, wood played a predominant part. The nights, owing to wretched lighting, were darker; the cold, even in the living quarters of the castles, was more intense. In short, behind all social life there was a background of the primitive, of submission to uncontrollable forces, of unrelieved physical contrasts.

This world extended beyond the feudal ages, as rural life in Europe changed little for centuries afterward. Only the barest vestiges remain today, with the raw, preternatural sensitivity wiped clean. The viticultural needlepoint of the Côte d'Or, its thousands of named vineyards, is as much a relic of a bygone civilization as Stonehenge. We can decipher why and how they did it, but the impulse, the fervor, is beyond us now.

The glory of Burgundy is its exquisite delineation of sites, its preoccupation with *terroir*. What does this site have to say? Is it different from its neighbor? It is the source of Burgundian greatness, the informing ingredient. This is easily demonstrated. You need only imagine an ancient Burgundy planted to Pinot Noir and Chardonnay for the glory of producing—to use the modern jargon—a varietal wine. The thought is depressing, an anemic vision of wine hardly capable of inspiring the devotion of generations of wine lovers, let alone the discovery of such natural wonders as Montrachet or La Tâche. *Terroir* is as much a part of Burgundy wines as Pinot Noir or Chardonnay; the grape is as much vehicle as voice.

The mentality of *terroir* is not uniquely Burgundian, although it reaches its fullest expression there. It more rightly could be considered distinctively French, although not exclusively so. Other countries, notably Germany and Italy, can point to similar insights. But France, more than any other, viewed its landscape from the perspective of *terroir*. It charted its vineyard distinctions—often called *cru* or growth—with calligraphic care. Indeed, calligraphy and *cru* are sympathetic, both the result of emotional, yet disciplined, attentions to detail. Both flourished under monastic tutelage.

■　　■　　■

But in France there exists, to this day, a devotion to *terroir* that is not explained solely by this legacy of the Church. Instead, it is fueled by two forces in French life: a long-standing delight in differences and an acceptance of ambiguity.

The greatness of French wines in general—and Burgundy in particular—can be traced to the fact that the French do not ask of one site that it replicate the qualities of another site. They prize distinction. This leads not to discord—as it might in a country gripped by a marketing mentality—but consonance with what the French call *la France profonde,* elemental France.

This is the glory of France. It is not that France is the only spot on the planet with remarkable soils or that its climate is superior to all others for winegrowing. It is a matter of the values that are applied to the land. In this, *terroir* and its discoveries remind one of Chinese acupuncture. Centuries ago, Chinese practitioners chose to view the body from a perspective utterly different than that of the dissective, anatomical approach of Western medicine. Because of this different perspective, they discovered something about the body that Western practitioners, to this day, are unable to see independently for themselves: what the Chinese call "channels" or "collaterals," or more recently, "meridians." The terminology is unimportant. What is important is that these "meridians" cannot be found by dissection. Yet they exist; acupuncture works. Its effects, if not its causes, are demonstrable.

In the same way, seeking to divine the greatness of Burgundy only by dissecting its intricacies of climate, grape, soil, and winemaking is no more enlightening than learning how to knit by unraveling a sweater. Those who believe that great wines are made, rather than found, will deliver such wines only by the flimsiest chance, much in the same way that an alchemist, after exacting effort, produces gold simply by virtue of having worked with gold-bearing material all along.

Today, a surprising number of winegrowers and wine drinkers—at least in the United States—flatly deny the existence of *terroir,* like weekend sailors who reject as preposterous that Polynesians could have crossed the Pacific navigating only by sun, stars, wind, smell, and taste. *Terroir* is held to be little more than viticultural voodoo.

The inadmissibility of *terroir* to the high court of reason is due to ambiguity. *Terroir* can be presented, but it cannot be proven—except by the senses. Like Polynesian seafaring, it is too subjective to be reproducible and therefore credible. Yet any reasonably experienced wine drinker knows upon tasting a mature Corton-Charlemagne or Chablis "Vaudésir" or Volnay "Caillerets," that something is present that cannot be accounted for by winemaking technique. Infused in the wine is a *goût de terroir,* the savor of the site. It cannot be traced to the grape, if only because other wines made the same way from the same grape lack this certain something. If only by process of elimination, the source must be ascribed to *terroir.* But to acknowledge this requires a belief that the ambiguous— the unprovable and unmeasurable—can be real. Doubters are blocked by their own credulity in science and its confining definition of reality.

The supreme concern of Burgundy is—or should be—making *terroir* manifest. In outline, this is easily accomplished: small-berried clones; low yields; selective sorting of the grapes; and, trickiest of all, fermenting and cellaring the wine in such a way as to allow the *terroir* to come through with no distracting stylistic flourishes. This is where *terroir* comes smack up against ego, the modern demand for self-expression at any cost, which, too often, has come at the expense of *terroir.*

It is easier to see the old Burgundian enemies of greed and inept winemaking. The problem of greed, expressed in overcropped grapevines resulting in thin, diluted wines, has been chronic in

Burgundy, as are complaints about it. It is no less so today, but the resolution is easily at hand: Lower the yields.

But the matter of ego and *terroir* is new and peculiar to our time. It stems from two sources: the technology of modern winemaking and the psychology of its use. Technical control in winemaking is recent, dating only to the late 1960s. Never before have winemakers been able to control wine to such an extent as they can today. Through the use of temperature-controlled stainless-steel tanks, computer-controlled winepresses, heat exchangers, inert gases, centrifuges, all manner of filters, oak barrels from woods of different forests, and so forth, the modern winemaker can insert himself between the *terroir* and its wine to a degree never before achieved.

The psychology of its use is the more important feature. Self-expression is now considered the inalienable right of our time. It thus is no surprise that the desire for self-expression should make itself felt in winemaking. That winemakers have always sought to express themselves in their wines is indisputable. The difference is that today technology actually allows them to do so, to an extent unimagined by their grandparents.

Submerged in this is a force that, however abstract, has changed much of twentieth-century thinking: the transition from the literal to the subjective in how we perceive what is "real." Until recently, whatever was considered "real" was expressed in straightforward mechanical or linear linkages, such as a groove in a phonograph record or a lifelike painting of a vase of flowers. Accuracy was defined by exacting, literal representation.

But we have come to believe that the subjective can be more "real" than the representational. One of the earliest, and most famous, examples of this was Expressionism in art. Where prior to the advent of Expressionism in the early twentieth-century, the depiction of reality on a canvas was achieved through the creation of the most lifelike forms, Expressionists said otherwise. They maintained that the reality of a vase of flowers could be better expressed by breaking down its form and color into more symbolic representations of its reality than by straightforward depiction.

How this relates to wine is found in the issue of *terroir* versus ego. The Burgundian world that discovered *terroir* centuries ago drew no distinction between what they discovered and called Chambertin and the *idea* of a representation of Chambertin. Previously, there were only two parties involved: Chambertin itself and its self-effacing discoverer, the winegrower. In this deferential view of the natural world, Chambertin was Chambertin if for no other reason than it consistently did not taste like its neighbor Latricières. One is beefier and more resonantly flavorful (Chambertin) while the other offers a similar savor but somehow always is lacier in texture and less full-blown. It was a reality no more subject to doubt than was a nightingale's song from the screech of an owl. They knew what they tasted, just as they knew what they heard. These were natural forces, no more subject to alteration or challenge than a river.

. . . In seeking to establish the voice of a *terroir,* one has to concentrate—at least for the moment— not on determining which wines are best, but in finding the thread of distinction that runs through them. It could be a matter of structure: delicate or muscular; consistently lean or generous in fruit. It could be a distinctive *goût de terroir,* something minerally or stony; chalky or earthy. Almost always, it will be hard to determine at first, because the range of styles within the wines will be distracting.

And if the choices available are mostly second-rate, where the *terroir* is lost through overcropped vines or heavy-handed winemaking, the exercise will be frustrating and without reward. *Terroir* usually is discovered only after repeated attempts over a number of vintages. This is why such insight is largely the province only of Burgundians and a few obsessed outsiders.

Nevertheless, hearing the voice of the land is sweet and you will not easily forget it. Sometimes it only becomes apparent by contrast. You taste a number of Meursault "Perrières," for example, and in the good ones you find a pronounced mineraliness coupled with an invigorating, strong fruitiness. You don't realize how stony or fruity, how forceful, until you compare Perrières with, say, Charmes, which is contiguous. Then the distinction of Perrières clicks into place in your mind. It's never so exact or pronounced that you will spot it unerringly in a blind tasting of various Meursault *premiers crus.* That's not the point. The point is that there is no doubt that Perrières exists, that it is an entity unto itself, distinct from any other plot.

Such investigation—which is more rewarding than it might sound—has a built-in protocol. When faced with a lineup of wines, the immediate impact is of stylistic differences, a clamor of producer's voices. Once screened out, the lesser versions—the ones that clearly lack concentration and definition of flavor—are easily eliminated. Some are so insipid as to make them fraudulent in everything but the legal niceties. Then you are left with the wines that have something to say. At this moment you confront the issue of ego.

The ideal is to amplify *terroir* without distorting it. *Terroir* should be transmitted as free as possible of extraneous elements of style or taste. Ideally, one should not be able to find the hand of the winemaker. That said, it must be acknowledged that some signature always can be detected, although it can be very faint indeed....

Too often, signature substitutes for insufficient depth. It is easier, and more ego-gratifying, to fiddle with new oak barrels and winemaking techniques than to toil in the vineyard nursing old vines and pruning severely in order to keep yields low. Character in a white Burgundy, even in the most vocal of sites, does not come automatically. One need only taste an overcropped Montrachet— it is all too common—to realize how fragile is the voice of the land when transmitted by Chardonnay. As a grape, it is surprisingly neutral in flavor, which makes it an ideal vehicle for *terroir*—or for signature.

Character in a red Burgundy is just as hard-won as in a white, but its absence is not as immediately recognizable because of the greater intrinsic flavor of Pinot Noir. That said, it should be pointed out that flavor is not character, any more than a cough drop compares with a real wild cherry.

Where Chardonnay is manipulated to provide an illusion of depth and flavor, the pursuit with Pinot Noir is to make it more immediately accessible and easy down the gullet. An increasing number of red Burgundies are now seductively drinkable virtually upon release only two years after the vintage. Such wines can be misleading. Rather than improving with age, their bright, flashy fruitiness soon fades, like an enthusiasm that cools. The wine drinker is left stranded, stood up by a wine that offered cosmetics rather than character.

All of which underscores why *terroir* is the "more beautiful question" of wine. When the object is to reveal, amplify, and transmit *terroir* with clarity and resonance, there is no more "beautiful answer" than Burgundy. When it is ignored, wine may as well be grown hydroponically, rooted not in an unfathomable Earth that offers flashes of insight we call Richebourg or Corton, but in a manipulated medium of water and nutrients with no more meaning than an intravenous hook-up. Happily, the more beautiful question is being asked with renewed urgency by both growers and drinkers. A new care is being exercised. After all, without *terroir*—why Burgundy?

SOME WORDS FROM THE FRENCH

Terroir is, of course, a French word, with no exact, comprehensive, one-word equivalent in any other language. The term with its current meaning only emerged in the late nineteenth century (see chapter 2), but the association of wine and place has been a staple of French winegrowing at least since the Cistercian monks started paying attention to distinct vineyard plots in Burgundy in the Middle Ages.

One very useful compilation of Old World perspectives, mostly French, is *Terroir and the Winegrower,* published in 2006. Assembled, organized, and edited by Jacky Rigaux, an author and wine educator at the University of Burgundy, the book brings together presentations and short philosophical pieces by dozens of winegrowers, most of whom delivered their remarks at an annual event that Rigaux stages in Burgundy called "Les Rencontres Henri Jayer, Vignerons, Gourmets, et Terroirs du Monde" ("Henri Jayer Encounters: Winemakers, Wine Growers, and Terroirs of the World"). At this conference, winemakers describe and document with great passion their relationship with the vine and the plot of ground they till.

Terroir, writes Rigaux in a single magisterial sentence, "is a concept that cannot be reduced to folklore, that represents complex interactions between geology, climate, topography, 'stoniness,' drainage, slope, altitude, soil, subsoil, microbes, soil fauna, vines, grapes, yeasts, and indigenous bacteria, cellars . . . not forgetting the winegrower, of course, as it was he who decided to turn these slopes, which are unsuitable for cultivating cereals or for livestock farming, into terroirs giving birth to wines which are proud of their identity." (Rigaux 2008, 24–25)

This collection of work begins with a piece by Henri Jayer himself, a legendary Burgundian winegrower based in Vosne-Romanée who, starting in the 1950s, became one of *terroir's* great early champions, who was one of the earliest postwar growers to reject chemical farming in French viticulture, producing spectacular wines from several tiny parcels of land. It was Jayer who more or less coined the phrase "great wine is made in the vineyard." Jayer's comments are entirely matter-of-fact: no rants, no polemics, just the simple explanation of how good wine should respect its *terroir*.

FROM HENRI JAYER, "I HAVE A DREAM . . ."
TERROIR

Terroir is of capital importance in Burgundy since vineyards are planted in only one variety and the wine produced can therefore not be corrected by any blending. The generous nature of Burgundy has polished exceptional and very varied terroirs in the course of thousands of years, sometimes on very small surfaces. The Côte is very chaotic on the whole, cut in curves, with changing orientations and soils that abruptly change their character beyond a road or a footpath. This aspect is wont to provoke the joy of informed amateurs and the despair of new initiates who wonder how they will ever be able to understand this territorial complexity. A complexity one finds in the wines that are often clearly marked by their terroir—this is where man comes in, hence his importance for the future quality of the wine.

THE WINEGROWER

The varietal planting of a terroir means a future commitment of at least 50 years: the winegrower must select both the best rootstock and the best matching graft in order to produce a wine worth its A.O.C. and its reputation. Then the vines must be cultivated in conditions apt to produce grapes of outstanding quality. This requires respect for the soil and for the environment, allowing nature to take its course and intervening only when necessary. With the time for harvesting well chosen, he would be able to deliver a product of quality, which would be further refined through a severe process of selection at the winery. At this time, he will be able to manage the vinification according to the grapes and to make a wine in the way he likes and that suits his image: his philosophy allows him to create a product of pleasure, good to drink young as well as old, and that will always develop in a positive way, varying from year to year and leaving the consumer with the impression that he is always tasting a great wine although it never tastes quite the same.

THE WINE

A good bottle should exude a certain mystery when it is opened, since nothing is more frustrating than knowing in advance what kind of wine one is going to taste. Let us avoid making the same standard wine from year to year. Clients are becoming more and more demanding and the reign of appellation is starting to give way to the winemaker. At the moment, many wine drinkers prefer a wine of character that respects terroir and vintage while presenting a personal touch from their creator's imagination. It is excellent for Burgundy's reputation that more and more young winemakers are trying to distinguish themselves from their neighbours. Nature welcomes all tastes and there is a place for everyone on this earth. It is up to the winemaker to communicate with the client he receives and to explain the message he would like to transmit through his wine, so that his work will be appreciated at its just value at a tasting. Dear winegrowing friends and wine lovers, this is, in a few sentences, my personal conception with which I would like to open the debate regarding terroirs, their importance for quality and also the bigger or smaller space left for the winegrowing winemaker and his philosophy on wine. ∎

HERE IS Marc Kreydenweiss, a biodynamic grower and winemaker based in Alsace, but with projects in the Rhône Valley also. In describing his role as winegrower, he situates that role within the larger framework of culture, flavor, and sensory experience.

FROM MARC KREYDENWEISS, "MY PHILOSOPHY AS A MAN OF THE VINES"

Why am I a man of the vines? . . . Thanks, without doubt, to the fact that I was born into a family where the lineage of vintners goes back more than three centuries. Above all, I have become a vintner because of a true passion for this wonderful vocation where each day is one of discovery and investigation, a life divided between the contemplation of nature and hidden moments in the depths of our cellar. My last reason . . . has to do with meeting a woman who loves and respects this passion. It is to her that I owe whatever is feminine in our wines.

Terroir is the most important element, showing off a wine's worth and adding grandeur. It is directly reflected in a wine. From elements caused by the decomposition of rocks, wines obtain dry extracts, characterized by a long persistency in the mouth. The variety is responsible for the style. A Riesling would be straight, nervy and deep, often closed-up when the bottle is opened. Pinot Blanc is more accessible and forthcoming. Muscat is floral and ethereal. Pinot Gris is ample and voluminous and Gewürztraminer aromatic and robust. If the year allows it, we can introduce variants according to levels of ripeness, such as late harvests or noble late harvests.

The winegrower adapts his philosophy and his knowledge according to multiple elements. He is like a conductor who must use all these elements as wisely as possible to create a work of art with each wine. By planting those varieties that are best suited to each terroir, by regulating vineyard yields, working organically, harvesting at optimal ripeness and, of course, looking carefully after the ageing process, the winegrower imbues the wine with his style.

Our wines slake thirst and nourish the spirit. They form part of culture and are the showpiece of our very way of life. We may all delight in tuning in to our inner being, discovering ancient secrets in each sip. Wine is in itself a part of history and has so many mysteries to reveal. But it reveals itself only when we give it enough of our time and attention.

I am talking about those Noble Wines that make me quiver, those whose spirit sings in the bottle, those wines that delight us even as we fetch them from the cellar. Comfortably settled, I meditate and try to understand their language. A Great Wine can be compared to an intelligent creature that does not reveal all its attractions at once: that is what makes it appealing.

Allow yourself to be lulled by the discovery of the senses and the well-being that wine brings, an enchanted melody of colors, perfumes and flavours, subtle and refined. ∎

FROM BORDEAUX, winegrower and consultant Stephan Derenoncourt comments on some of the many ways *terroir* expression—the term, the concept, the connotation—is used and misused.

FROM STEPHAN DERENONCOURT, *"TERROIR"*

Faced with the complexities of a definition, the *"terroir"* term is shrouded in a veil of mystery and adapted to any purpose. It is readily used as a conclusion to certain likely questions when trying to identify an aroma or a taste, coming like a bolt from the blue: "It's *terroir.*" However, it can also sound like a marketing argument and ring like a cash register. Lastly, the concept is mentioned when discussing the inability to subscribe to the values of the modern agricultural society, dedicated to specialization. Today, one reads scientific reports emanating from well-known researchers proclaiming that no link can be shown between a wine and its origin. A modicum of observation suffices to realize that these publications are the result of laboratory work. We are far from the knowledge accumulated over generations by a peasant population inspired by their observations and experiences in the field. Similarly, there are a few famous and megalomaniac winegrowers who publicly express their annoyance when the word "minerality" is pronounced, as they have difficulty in accepting the idea that they were not involved in what is yielded by *terroir!*

The *terroir* term is starting to take on an almost mystical connotation: should one become a believer? Unquestionably, this spiritual dimension has always been the driving force of my work as a winegrower and a consultant. Observing and classifying plots, looking after the soil in an adapted way, isolating patches of land to allow a simultaneous wine-making process (as the same features are carried through by the grapes), or creating "cuvées" from plots which seem to be complementary: that is the source of our quest for authenticity in wine. Beyond the satisfaction of improving one's knowledge on a daily basis, this approach constantly reminds us to stay humble before the complex work of nature. And even if we often make mistakes, our joy is that much greater when we are able to isolate an original character in an estate's "*cuvée*" for the first time. It may be a light petroleum note on a Cabernet Sauvignon from a schistous soil in Tuscany, a floral character on a "*tempranillo*" from Ribera del duero's gravelly limestone, or a noble and beautiful black truffle on the "*astéries*" limestone of the Saint-Émilion region. The quest for the *terroir* wine is a quest for an inimitable product. ■

AND FINALLY from the Loire Valley, the late Didier Dagueneau, winegrower at Pouilly-Sur-Loire, champions the ability of Sauvignon Blanc to express the flinty soils of Pouilly-Fumé. In doing so, Dagueneau paints a vivid picture of how grower and place create a powerful bond that, with care and diligence on the part of the winemaker, finds its way into the glass.

FROM DIDIER DAGUENEAU, "MAN AND TERROIR"

I have always thought that the most wonderful gift was a patch of land . . . Both feet on the earth and your head in the stars, such is the best position for man!

Paradoxically, I am not going to tell you about the flint stones of the Saint-Andelain hill, nor the Kimmeridgian marls or Portland limestone of Pouilly, nor the particular climatology caused by the mists from the Loire at its feet; other winegrowers could do it just as well as I can. No, I would like to speak to you about an element that is just as important in the idea of *terroir* as the soil or the climate and that is too often forgotten: man. Not more important than but just as important as the two others. I consider *terroir* to be the result of the interaction between these three elements. One cannot dissociate them. *Terroir* is conceived from these three elements together. Soil and climate are elements that have no set parameters, that have to be taken into account but cannot be changed and that are going to bring about the typicity of the vintage. The winegrower must understand and reveal them, perhaps even transcend them. Man is the detonator that reveals the identity of a *terroir*. Some men may massacre and even destroy the *terroir*, proof that a *terroir* does not exist without man.

Keeping your feet on the earth . . . Before becoming a winegrower, I saw myself as a farmer. I am deeply attached to the earth. There must be osmosis between man and his *terroir*; to express this well, he must know how to tune into his soil, be sensitive to it, train all the fibers of his being to sniff the air, to see if the vine is suffering, if the soil is breathing. For everything revolves around air, water, sun and a man who faces these elements and understands them, interprets them, expresses them and transforms them into wine. Although I do not have many certainties, experience has given me at least one: wine is made in the vineyard. A good wine can only be made with good primary matter and a good grape is the result of a multitude of small parameters, small details, daily decisions made in the vineyard in the course of working with the vines. Certainties about these decisions? I have none. I think that there is not only "one" truth when it comes to the vineyard, but several. There is no unique truth and I belong to no "sect," neither that of the "organics," not any other. Totalitarian fundamentalism does as much harm in this domain as in any other! Each must find "his" truth on his *terroir*, a truth that suits him and his land. I claim to be a free being, free in my choices but also in my mistakes: they are also part of my freedom.

One of my greatest pleasures is to see a vineyard in good condition, nice straight plants, well aligned and well pruned. It gives me great pleasure, like when I have completed a good day's work and look back on it: I have the impression of a job well done! I know the vines will return the favor. One must love the soil and experience the artisanal satisfaction of a job well done in order to understand this sentiment. I try every day to fashion my vineyard like a work of art. I see it as a work of art, moving and moreover, consumable. This has an important impact on my wine. The wine I would like to make on this Pouilly *terroir* must be cultural; in other words, not only a simple and basic product that may respond to a certain demand for alcohol and induce intoxication but, on the contrary, a wine that calls out, that demands a certain attention from the drinker, that makes him reflect, provoking not only pleasure but also emotion.

Apart from the soil, the climate and the personality of the winegrower, the wine should also reflect the culture of its *terroir*, in other words, its tradition and its know-how. This is why I have insisted for several years now on the appellation "Blanc Fumé de Pouilly" for my wines. The difference between this name and Sauvignon? None. It is indeed the same variety. But at Pouilly, Sauvignon is called "Blanc Fumé" and this is no coincidence. Certain vintages express the wine like a little caricature of its name with a slight but ineffable smoky taste. *Terroir* also means respecting this name, this local particularity. "Purity" is often mentioned concerning my wines. I am above all interested in a fragile equilibrium: intensity on a fine and elegant structure. Concentration, strength and finesse at the same time, in wines that should be drinkable in the first place and that should not remain in the bottle having saturated the drinkers too rapidly. But make no mistake; there are few wines of which I am proud. I like austerity, rigor, acidity but I have to compose the wine with the elements that nature gives me every year. The vintage years must be accepted as is with its qualities and its weaknesses. I do my best with what I receive, but I cannot always do what I want to do.

FEELING *TERROIR'S* TOUCH

Rod Smith, one of the most graceful and sensual of American wine writers, began exploring the concept of *terroir* as early as the 1970s, years long before it was fashionable in California, which was then fully in thrall with the pre-eminence of the winemaker. In articles for the *Los Angeles Times, Wines & Spirits,* and other publications, Smith championed the *terroirs* of several prime California growing areas, and helped educate a generation of wine lovers about the interaction of soil, climate, and vines in the U.S.

This evocative piece from the *Los Angeles Times* ran in early May, 2002, just as the California wine country was waking up from its annual winter dormancy. It won a James Beard award for distinguished wine writing. The scene is set on the Sonoma Coast, but its focus is the state as a whole—what might be called macro-macro-*terroir*. Even on this grand geographical and topographical scale, *terroir* translates into direct personal experience.

FROM ROD SMITH, "WAITING FOR THE VALLEY TO INHALE"

California's winegrowing season started the other day. I had to put on a jacket. It felt like a bucket of cold water down the back of my neck. As a friend said through chattering teeth, "Somebody just opened the icebox door."

It might seem odd that a blast of frigid air would herald the opening of fruit-growing season, but that's what makes California distinctive.

The day certainly didn't start out in the refrigerator. Just before sunset we had arrived at a high ridge near the coast in western Sonoma County. We'd brought along grilled pork sandwiches and a bottle of Russian River Valley Pinot Noir, built a small fire and settled in to watch the immense full moon drift up out of the redwoods in dead-calm air, turning from orange to blazing white as it cleared the lowland haze over distant Santa Rosa.

It was T-shirt and shorts weather. We lounged comfortably in wildflowers and grass that still had weeks of green left before it would turn brown for the summer. The campfire smoke curled lazily straight up to the stars.

The wine was the '99 Dehlinger "Goldridge" Pinot Noir. It was perfect for the evening, a gracefully exuberant Pinot with piercing red berry and rose scents and bold red and black fruit flavor. Its clear flavors and bracing acidity spoke of vines rooted in Gold Ridge soil, the light sandy loam found primarily in the cool Green Valley appellation near Sebastopol, in the southern Russian River Valley.

The botanist Luther Burbank established a farm in Sebastopol because he felt this soil was a magical medium for growing luscious fruit, especially apples. This Pinot corroborated that vision. But more than that, it spoke of a special grape-growing climate—not just in the Russian River Valley but in all of coastal California.

Winemakers take a lot of credit for creating fine wines. Growers do, too, with considerably more justification. But it's really the Pacific Ocean that makes California's wines outstanding.

California is blessed with a cold coastwise current. Frigid to begin with, it also causes an upwelling of even colder water from great depths. In combination with onshore winds, these currents have a significant cooling effect on vineyards in coastal wine districts which otherwise would be much hotter.

Without the ocean's influence, California would be too hot for grapes such as Pinot Noir, Chardonnay and Cabernet Sauvignon. Extremely ripe fruit produces heavy, lackluster wines. The ocean's cooling effect allows the grapes to ripen fully without losing the precious acidity that creates intensity and fine structure. It emulates the effects of altitude and distance from the equator. In viticultural terms, vineyards near the coast seem to be at a higher latitude than the map would indicate.

Here's how it works: The northern and southern extremities of California's Central Valley (with its two great lobes, the Sacramento and the San Joaquin) reach very high temperatures during the summer because of the way the terrain concentrates solar radiation. As the hot air rises, it creates a vacuum that sucks cold marine air inland via the San Francisco Bay/Sacramento River system, the Russian River Valley and other gaps in the coastal mountain ranges south to Santa Barbara County.

The cool air typically flows inland during the afternoon, and it's often accompanied by fog that lingers until dissipated by the sun next morning. In the height of summer, the marine incursion is often large enough to cool the Central Valley significantly. As for the coastal valleys, the fog may cover them well into the afternoon for days on end, until enough heat accumulates inland to drive it offshore again. At that point the Central Valley begins super-heating again, and the cycle is repeated.

From the grape's perspective, the pace of sugar accumulation and physiological ripening in the latter part of the season are influenced by the way these fog events affect photosynthesis. Thus the character of a vintage is marked by the rhythm of marine incursions.

This regular five- to seven-day cycle is perhaps best understood using an organic model: Looking at a topographical map, you can visualize California as a living, breathing organism, with the Sacramento and San Joaquin valleys operating as a pair of lungs, the Golden Gate and San Francisco Bay as a mouth and the delta as trachea, and Lodi somewhere around the tonsils. The ongoing respiration, as the Central Valley inhales and exhales throughout the growing season, is what makes California wine grapes so good. The situation is similar to the southern Mediterranean coast of France—in fact, Lodi could be called the Languedoc of California.

That's just what we were experiencing on that ridge in western Sonoma County. It had been warm for a few days, and the interior had super-heated (well, for that time of year) to the point where the ocean began to breathe on the land. We felt the first stirrings as it came ashore a few miles west. Its briny tang mingled with the warm scents of redwood and pinecone, of green grass and wildflowers and woodland under-story. The air had body, a texture and volume.

And then it was just cold. Under the blasting ocean breeze, the smoke from the campfire ran down the slope. It reminded me of standing on Russian Hill in San Francisco, watching the great fog serpent slithering in through the Golden Gate. We drank the last of the Dehlinger Pinot Noir and then scrambled for fleece and long pants, and without any discussion began packing things up for the moonlight hike down the hill.

Far below us on the plains and hillsides of the Russian River Valley, hundreds of thousands of grape vines were feeling the change. They were feeling it in Dry Creek Valley, in Alexander Valley and in Chalk Hill. They were feeling it in Carneros and in Lodi. Similar marine incursions would be happening farther south—in the Salinas Valley, at San Luis Obispo, in the Santa Maria and Santa Ynez valleys.

The next morning was cool but clear. It will be another month or so before the great weather machine cranks into high gear, before the vine-tempering cold cycle begins in earnest and the fog comes into play.

But for grape growers, wine producers and a few wine fans who happened to be out and about when the change occurred, the signal had been unmistakable. The 2002 growing season is underway.

TRAVELING INTO *TERROIR*

In the normal order of things, wine writers write about wine, and travel writers write about going places. For several decades, British-American author Gerald Asher has entranced readers by writing about wines by describing the places they come from, capturing the feel of a wine's origins and its terrestrial place in a way no standard tasting note ever could.

Asher's leisurely pieces for *Gourmet* magazine over three decades, many of them now reprinted in three collected editions, covered the globe, Old World and New, from legendary parts of Bordeaux and Burgundy to lesser-known parts of Italy and California. Asher's great gift as a writer is to provide layers of context—geological, climatological, but also culinary, cultural, and personal—for the wines and the regions he covers. By the end of the essay readers can almost taste the wines, without having left home or popped a single cork. In fact the word *terroir* rarely appears in his writing; it doesn't have to.

In this selection, Asher explores Galicia, the Albariño-producing region of Spain, and a region that would have been relatively unknown to *Gourmet* readers in 1999. These engaging white wines came onto the world stage out of nowhere in the 1990s—except that, as Asher explains, they came out of *somewhere*. Asher's tale may also dispel any illusion that *terroir* just up and announces itself, hanging out a sign saying "Bottle Me!" Without determined effort in the vineyard—from building terraces to restricting yields and training vines into complex trellising systems (the pergolas in the piece), not to mention modern winemaking equipment and practices—*terroir* wouldn't have the opportunity to express itself.

FROM GERALD ASHER, "RÍAS BAIXAS—ALBARIÑO: A FRAGRANT WINE OF THE SEA"

Has a totally unfamiliar white wine ever given us such a jolt? Such pleasure? Albariño from Rías Baixas, in northwest Spain, is still no more than an eddy in a puddle compared to the ocean of Chardonnay we consume every year. But in the United States, sales of this seductively aromatic wine have bounded from just two thousand cases in 1992 to twenty-two thousand in 1998. If production could have supported it, sales would have risen even faster.

In Spain, the wines from this small corner of the province of Galicia have won praise on all sides—they are now on the list of every serious restaurant from Seville to San Sebastián—and for at least the past decade they have had a place at the royal table.

Albariño is actually part of a much larger secret. In summer, when northern Europe invades Spain's Mediterranean beaches, the Spanish themselves disappear to Galicia, a private green refuge tucked between Portugal's northern border and the Atlantic. Santiago de Compostela, the region's capital, ranks with Rome and Jerusalem as a place of pilgrimage for the world's Catholics, of course. But for most Americans it's usually just a stop on a wider European tour. Having paid their respects to the cathedral—its altar ablaze with gold and silver—and allowed themselves to be beguiled for a day or two by the medieval charm of narrow, winding streets that open abruptly onto vast plazas of an austere splendor, they are on their way. Rarely do they venture the short distance to the coast, with its countless bays and inlets—the *rías*—or discover, farther inland, the rivers that long ago carved out deep canyons in Galicia's ancient mountains (now mostly protected as natural parks), or walk in woods where paths are banked with creamy rock-roses and clearings edged with beds of tiny scarlet wild strawberries.

Galicia is hardly the Spain of popular imagination: of strumming guitars, stamping heels, and carnations between the teeth. The Moors never established themselves here, so Galicia dances in slippers to the Celtic drum and bagpipe, and its pilgrims walk in sneakers and parkas (but still with cockleshells pinned to their hats) on roads punctuated by tall crucifixes directing them to the tomb of St. James. For curious travelers there are mysterious cave drawings and prehistoric dolmens, feudal castles and Romanesque churches, Roman bridges (still in daily use) and monasteries built near remote passes where pilgrims in centuries past could find a night's shelter and protection from brigands unimpressed by their piety. Elsewhere are sober, seventeenth-century stone *pazos*—manor houses presiding over villages that still live on what they can wrest from a wild land and an even wilder sea. Like Brittany and the west coast of Ireland, Galicia is a place of sudden storms and drowned fishermen, of lighthouses, ghost stories, wee folk, and things that go bump in the night.

Above all, however, it's a place where one eats and drinks well. There are mussel and oyster beds in the *rías,* and every village up and down the coast has its own line of fishing boats. Only Japan consumes more fish per capita than Spain, and in this region, which boasts two of Europe's most important fishing ports—Vigo and La Coruña—meat is rarely more than a footnote on the menu. Even the scruffiest of the bars and taverns that line Santiago's twisting Rua do Franco serve just-caught fish of unimaginable variety cooked with a confidence that would put any restaurant in Paris or New York on the defensive.

No one knows when the vine was introduced to this part of Spain. It can be assumed that the Romans, who settled in the area two thousand years ago to mine the hills near Orense for precious metals, would have provided for themselves somehow. The first reliable record we have that links past to present, however, is of vineyards planted in the twelfth century by Cistercian monks at Armenteira in the Salnés Valley near the fishing port of Cambados. (According to local legend, Albariño is descended from cuttings of Riesling brought from Kloster Eberbach in the Rheingau by some of these Cistercians. It's a pretty story, but recent DNA research has shown beyond a doubt that there is no such connection; the variety's origin is, therefore, still a matter of speculation.)

More vines were planted later on lands granted to the monastery of Santa Maria de Oia, forty or so miles south within the sharp angle formed by the estuary of the Miño—the river that establishes the frontier with Portugal—and the Atlantic. From these early beginnings evolved the vineyards of Val do Salnés to the north and those of O Rosal and Condado do Tea in the south. (A fourth defined zone roughly midway between them, Soutomaior, was created recently, but it produces such a small quantity of wine that, commercially, it has had no impact as yet.)

There are all kinds of shadings and subtleties to be explored—with wine there always are—but an explanation of the differences between Salnés on the one hand and Rosal and Condado on the other is enough to illustrate most of what one needs to know about Albariño. The vineyards of Salnés form an open bowl facing west to the ocean. From the terrace of the Martín Códax winery on Burgans Hill, above Cambados, there is a limitless view of the bay and the Atlantic beyond. The individual shelves and terraces of vines face this way and that to catch the sun, but all are exposed to whatever blows in from the sea. The vineyards of Rosal and Condado also turn about, to accommodate the rise and fall of a terrain shaped by the streams and rivulets that feed the Miño, but their broad direction is always to the south, to the river. Rosal is affected by ocean less directly, and Condado—because it is farther upstream—even less. Their vineyards are drier and warmer than those of Salnés; they get less rain and have more hours of sun.

There are other differences, too. Vines in Salnés are on a fairly homogeneous granitic sand, while those of Rosal and Condado also contend with crumbled schist, clay, and the rolled pebbles typical of any place where the water's flow has shifted. These differences of soil and prevailing weather do not make any one zone better than the others (though you can be sure that's not the way the growers see it), but they do impose distinct and varied characteristics on their wines. A Salnés Albariño is bolder than one from either Rosal or Condado. It has good acidity, a pungent aroma and flavor (some say of pineapple), and it gives a powerful, fleshy impression. Its focus is intensely varietal. Rosal wines—and to an even greater degree those from Condado—are more graceful and more supple. Their flavor steals across the palate and lingers there. If a Salnés wine tends to express the varietal more than the site, a Rosal or Condado wine does the opposite....

The aroma and bite for which Albariño is celebrated owe much to the cooling proximity of the ocean and to the stress-free growth guaranteed by ample spring rain. In Spain, at least, these are conditions unique to the Rías Baixas. Ground humidity brings potential problems, of course, even though Albariño has a thick skin. As protection from rot, the vines are trained over high, horizontal pergolas. The hefty granite pillars used are not the most elegant supports, but they fit in with the local custom of scattering vines about in handkerchief-size plots like untidy afterthoughts. A vineyard is sometimes no more than an arbor attached to side of a house, an arrangement of granite and wire in an awkward bend of the road, or a backyard shared with a patch of turnips or cabbages. For years, most families in Rías Baixas have had vines to make wine only for their own consumption. It was a crop grown for the household—like peppers, corn, and potatoes—that sometimes provided a surplus that could be sold....

The Martín Códax winery, one of the first to come onto the scene in Salnés, was founded in 1986, funded by a group of small growers who knew they could achieve more together than if each tried to make and market wine on his own. With modern equipment—good presses, refrigeration for cool fermentation, easy-to-clean stainless-steel tanks—they were able to reveal Albariño's forgotten qualities. "And who is Martín Códax?" I asked Pablo Bujan, the firm's sales manager. "The prime mover of the project? The partner with the biggest holding?"

"He was a thirteenth-century Galician poet and troubadour from this very place," Bujan replied. "He drank a lot of wine."

The thirsty Códax would doubtless be pleased to know that the winery named for him is probably the most successful in Rías Baixas, producing almost 1.5 million bottles of wine in a normal year. To understand the difficulty of this feat, one must remember that the grapes are grown by more than two hundred partners who together own nearly four hundred acres of vines divided over twelve hundred separate plots.

Over the years, land in Galicia has been divided and divided again through successive inheritance. Vineyard plots on the terraced hillsides got smaller and smaller, and in the end those at a distance from the villages were hardly worth the effort needed to cultivate them. In the 1940s and 1950s, under General Franco, most were given over to stands of eucalyptus trees to provide pulp for a paper mill in Pontevedra. In the rain their drooping leaves and peeling bark lend a wistful melancholy to the landscape. But where the trees have been cleared, the original terracing can still be seen.

"Our natural trees are oak and chestnut. Eucalyptus trees give a poor return and are bad for the soil," says Javier Luca de Tena, director of Granja Fillaboa, with a dismissive wave of the hand. Angel Suárez, manager of the Lagar de Cervera (now owned by the Rioja Alta winery), endured several years of painstaking negotiations with disparate owners to put together enough land to restore a rational, workable vineyard. Often a single row of trees belongs to one family and the next one, to another. Each of four brothers will cling to his clump of half a dozen eucalyptus trees when their entire holding, taken together, makes up a block no bigger than a kitchen garden.

A program encouraging owners to swap land in order to build up the size of individual holdings has not been a great success. Apparently, the smaller the patch, the greater its mystical importance to someone unwilling to part with it. To put together an economically viable vineyard, therefore, takes a great deal of forbearance and much money. (One doesn't buy land in Galicia, one bribes the owner to part with it.) . . .

While in Galicia recently, I lunched with Angel Suárez at a restaurant near the old fishing harbor of La Guardia. We chatted over a bottle of his 1998 Albariño and a dish of assorted fish—hake, turbot, and monkfish—simmered with potatoes and served in a sauce of garlic and *pimentón,* a very small pepper, dried over a slow wood fire and ground to a fine powder. It has a hauntingly smoky, sweet-sharp taste, and no Extremaduran or Galician kitchen is ever without it.

I enjoyed the wine, but Suárez said 1998 had been a difficult year for Rías Baixas. Spring came early: The vines sent out tender shoots in February, but the weather then turned very cold, with a predictable effect. Hailstorms in April did further damage, and rain in June meant that the fruit set poorly. As a result, the crop was barely 40 percent of the previous year's and never did reach the level of quality hoped for.

"A red wine region copes with this kind of difficulty more easily," he continued. "Red wine is usually aged for a year or two, and differences of crop size from one year to another can be blurred, at least, by delaying or bringing forward the release date. Albariño is best when bottled and consumed young. We have no stocks to fall back on. Production is increasing a little every year, but at present, even when the crop is of a good size, it seems there is never quite enough to meet demand."

He poured the last of the 1998 into my glass, leaving none for himself.

THE QUEST FOR NEW WORLD *TERROIR*

The concept of *terroir* remains closely associated with Old World winemaking and wine thinking. But some of the most articulate exponents of the concept hail from the New World, where *terroir* is more a quest than an everyday assumption. Europe has a few thousand years' head start in learning the ways of place, but Californians and Chileans and New Zealanders are eager to catch up.

One of the leading voices on behalf of *terroir* in the United States is Randall Grahm, a pioneer in working with Rhône varieties in California, the "President for Life" of Bonny Doon Vineyard, and a flamboyant advocate for out-of-the-box wine causes. In the last decade, as a writer and a *vigneron*, Grahm has taken up the elusive quest for true *terroir* in California or elsewhere in the New World.

In several early issues of *The World of Fine Wine,* Grahm explored this quest in very personal terms. We draw from two here. In the first, titled "A Meditation on Terroir: The Return," Grahm captures nicely the difference between the wines that are all around us and the wines he dreams of making.

FROM RANDALL GRAHM, "A MEDITATION ON TERROIR: THE RETURN"

The French make a salient distinction between *vins d'effort* and *vins de terroir*—wines that are notably marked by the imprint of human efforts, as opposed to wines whose character primarily reflects their place of origin. Ultimately, *vins d'effort* are wines easy to like—presumably they are constructed with precisely that in mind—but difficult to love, at least in the thoroughly obsessive, *I've just seen a beautiful face and I will go mad if I cannot see it again* kind of way. *Vins d'effort,* especially those of the New World, attempt to hit the stylistic parameters of "great" wine—concentration, check; new wood, check; soft tannins, check. And yet the net result is like a picture of a composite, computer-generated "beautiful" person: it is never as compelling as the picture of an aesthetically "flawed" but unambiguously real person. ■

AND IN this excerpt from 2007, Grahm lays out just how hard the quest can be.

FROM RANDALL GRAHM: "THE SEARCH FOR A GREAT GROWTH IN THE NEW WORLD"

For the longest time, I supposed myself to be an agnostic on the question of finding *terroir* in the New World. It was not that I was a member of the anti-*terroirist* brigade. Quite the opposite. For me, a wine that conveyed a sense of place was the only wine that was truly "necessary," that incrementally made the world more interesting, beautiful, and more meaningful, in the sense that it created more connection between ourselves and the natural world. And yet there was a plethora of fatal impediments to creating or discovering a true *terroir* wine in the New World; the appearance of something like a New World Great Growth would be nothing less than a miracle.

This is the central paradox, or perhaps the existential dilemma, facing anyone who is "serious" about making great or meaningful wine in the New World. You can try to learn from what others have done successfully or unsuccessfully—which is the logical way to proceed. In the New World we do something very well, maybe diabolically well, but despite our vociferous protestations to the contrary, what we do is, I gently suggest, not about *terroir*. Were one to be serious about producing an authentic *vin de terroir* in these parts, one should be filled with a great and overpowering sense of existential angst, and there should be much gnashing of teeth and rending of garments. The very cornerstone of the proposition to produce great, authentic, and original wine—the identification of an outstanding site and the planting and cultivation of a vineyard thoroughly appropriate to that site—is tenuous; it is the construction of an elaborate edifice on shifty sand, gravel, or perhaps clay loam.

The belief in *terroir* holds that a *vigneron* might produce a wine that exposes both the inherent distinguishing qualities of a particular vineyard site and the unique qualities of the vintage year. The differentiating qualities of the site and ultimately of the wine, its originality, are most certainly linked with the site's ability to solve particular environmental challenges—its ability to drain water in the event of excess precipitation and, conversely, its ability to provide moisture to the plant in a thrifty and measured way in a vintage that is hydrologically challenged, a condition that obtains essentially every year in most places in California where grapes are grown. There are small but telling differences that distinguish the site from neighboring ones, that create the clarionlike distinctiveness of the more favored sites. *Terroir* is about solving environmental challenges, and certainly about solving them *elegantly*. In a warm and challenging environment, a great site allows vines to keep their cool in a suave, Cary Grant–like way.

Put another way, in the Old World, there has evolved a sort of homeostasis, a "learning" between vine and soil under the watchful eye of a human being. Vines have in some sense "taught" soils—through the mediation of their symbiotic microbial demiurges, mycorrhizae, and the like—what their specific needs are. Soils (and microclimates) have in some sense taught vines what they might expect over a growing season. Over many generations a sort of stately, rhythmic dance has emerged. The most interesting wines that arise from this dance are the ones that have captured this distinctive rhythm or waveform, this unmistakable signature, which we call *terroir*. My great dilemma, the dilemma of anyone interested in producing a truly distinctive wine is: How are you going to be so clever or lucky to work out this exquisite balance, this elegant harmonic in the one lamentably brief lifetime that you are given? How can you identify a priori a site capable of expressing *terroir* and real distinction? There is so much to be done and, at the same time, so much not to be done.

■   ■   ■

New World viticulturists tend to live in the culture of control, and this is the single major stumbling block to the discovery of the truly original. It is a non sequitur to talk about the *terroir* of an irrigated vineyard, rather like trying to accurately describe the natural history of an animal by observing its behavior in a zoo. So, if you sincerely seek to produce a wine that expresses *terroir,* where do you begin? The lessons learned from growing clonally selected, virus-free, irrigated grapes of selected popular varietals on modern trellis systems have little applicability to the establishment of a truly original vineyard.

■　■　■

The question remains: how to find something that is truly authentic, something that truly matters and that makes the world richer and happier for its existence? If you are lucky or skilled enough to find it, how do you know you have found it, especially if its virtues are opaque to the relevant opinion-makers?

I love the fact that I do not know where the project will ultimately go, and that the road there will be a very, very slow and meandering one.

KNOWING A PLACE, KNOWING ITS WINES

In its purest form, the concept of *terroir* says that the natural, physical properties of a place indelibly shape the wines that come from it. But there's another form of cause-and-effect at work as well, one which every *terroir* lover would subscribe to: that knowing a place forever changes your appreciation of and personal bond to the wines of that place. It's the emotional and psychological side of *terroir.* First-time travelers to France or Italy always fall in love with the wines of whatever region they happen to visit, not because of the local geology, but because they're excited to be in Europe, and things taste better there.

Terry Theise has done as much as anyone in the recent past to introduce American wine drinkers to the wonderful, sometimes magical wines of Germany and Austria, particularly those coming from small, lesser known growers and producers. He has been tireless in instructing tasters on the subtle differences in Germanic sites, from vineyard to vineyard and vintage to vintage in these peripatetic regions, spreading the gospel of *terroir* with wit, showmanship, and passion.

The annual catalogs of Theise's German and Austrian selections have become required reading for wine buffs, including many who drink little Riesling but lap up Theise's expressive, opinionated prose. In 2010, he published a small book, *Reading Between the Wines,* laying out his philosophical stance, his guilty pleasures and pet peeves, and making a forceful case for wines that mean something. The selection below, from a section entitled "Connectedness," captures the web of relationships that ties it all together: Theise and the winemaking families, the families and the land, the wines themselves and the inner reaches of the soul.

FROM TERRY THEISE, *READING BETWEEN THE WINES*

I woke up this morning thinking about Germany's Mosel valley, and about a vintner family I know well, the Selbachs of Zeltingen. I first met them in 1985; it was Hans and his wife, Sigrid, in those days, with eldest son Johannes waiting in the wings. Hans died recently, and suddenly, and when I visit the Mosel each March to taste the new vintage, I pay a visit to his grave, which has a view of the silvery river and the village of Zeltingen, where he and Sigrid raised their family. Indeed, if the steep hill behind the St. Stephanus church weren't a cemetery, it would certainly be planted with vines, and Hans and his vines alike rest deep within the slate.

He died at home, surrounded by family. His body was carried through the house, through the bottle cellar (one of his sons told me, "Terry, it was as if you could see and hear the bottles stand and applaud Papa"), before it was placed at last in the ground, perhaps three hundred meters from the house. It is not only his spirit that lingers genially among his survivors; his body itself is near at hand.

My own father died abruptly. I was about to be a senior in high school. I came home one afternoon from my summer job and he was slumped over the kitchen table. He died six hours later in a hospital room while I waited at home with my small sister. He is buried in an enormous cemetery in Queens, New York; I doubt if I could even find the gravesite.

My story may not be typical, but neither is it all that unusual. We were suburban folk, and a certain existential disconnect was a defining parameter of our experience. Nor do I claim this is necessarily tragic. Disconnection has its silver linings if you're a lone wolf.

But when I contemplate the connectedness the Selbachs nurture and presume upon, it becomes clear that their wines are also connected, that *they* are a defining parameter in a complex of connections. This is as invisible and vital as oxygen to the Selbachs and people like them.

Johannes speaks nearly perfect English. In fact, he gets along in French and for all I know can mumble articulately in Chinese. What I didn't know, or had forgotten, is that along with his native German he also speaks *Platt,* or regional dialect. I heard him speak it when we visited another grower together. It was the Merkelbachs, two bachelor brothers now about seventy, who have barely ever left their village and who make a living from a scant five acres producing some thirty different casks of Riesling, each of which they bottle separately. As I heard Johannes lapse into dialect it struck me what a piece of social glue this was; it was Johannes's way of reassuring Rolf and Alfred, *We are brethren,* another marker of connection and identity. One might almost claim that Mosel Riesling is what it uniquely is because of the dialect *it* speaks.

I find I am *satisfied* in some essential way by connected wines. It doesn't even matter whether I like them. I happen never to have met a Priorat I enjoyed, but I respect Priorat for its authenticity—it is manifestly the wine of a place, speaking the dialect of the sere, barren terraces in northeastern Spain. I may not like it—I have issues with high-alcohol wines—but I'm glad it exists.

I can't summon up anything but weariness for the so-called "international" style of wine (ripe, "sweet" fruit, loads of toasty oak, a spurious seductiveness), since it's either connected to things I don't care about or connected to nothing at all. I've had more than enough disconnect in my life. Many of us have. When I consider a Mosel family like the Selbachs—like any of the people with whom I work—everything I see expresses identity rooted in connection; they themselves, their wines. You could not disconnect these things even if you tried.

And it salves a kind of loneliness. Though it isn't my home it is at least *a* home, and the people are particular people, and the wines are particular wines. I spend too much of my life driving among strip malls and their numbing detritus, and so when I descend the final hill over the Eifel and the village of Zeltingen comes into view, sitting peacefully along the Mosel, I have a momentary thrill of *arriving.* Here is *somewhere.* I see it, I know it, I will soon embrace people who embody it—and I also get to *taste* it.

I will not settle for less from any wine. Nor need you.

When I'm there I stay at home with the Selbachs, and since the family likes to eat and knows how to cook it is an ongoing challenge to maintain my trim, boyish profile. I need to tramp, ideally every day, and the surrounding vineyards are ideal for tramping, steep and beautiful. One morning I set off into the misty, moist freshness, with a high fog riding about five hundred feet above the valley. I stamped up to a trio of wild cherry trees blooming halfway up the Himmelreich vineyard. I pushed at top speed to get warm. Kept climbing. Got up into the woods too high for vines and listened to the birds fluting away, new birds with unfamiliar voices.

The Himmelreich hill leads back into a small combe that gives way to the next hill, corresponding to the Schlossberg vineyard, and this in turn leads southeast to the great Sonnenuhr. I was on a high path with the Mosel vertically beneath me through the vines, and only the woods above. Some workers were pruning and binding here and there, and it seemed lovely to be out on such a sweet, cool morning working with the vines in such pretty surroundings. I know very well it isn't always like this—these vineyards get hot in the summer, or grapes wouldn't ripen—but I seemed to have passed through a membrane, and everything was suddenly and clearly *divine.* The small teams of people working, the birds noisily peeping, the languorous Mosel below, the smells of slate and wet trees. These fugue states are so sudden; you just take a small step through what you thought was yourself, and you're in some silent, airy space that's strangely durable while you're there. I passed a group of workers replanting in the Sonnenuhr and bade them *Guten Morgen,* glowing and goofy with joy and certain we all were as giddy as I was. But of course to them I was probably just another crazy tourist blown away by the view.

I turned to head back down and walked past Hans Selbach's grave, and I wanted to stop and talk to him, to tell him, It's still like it always was, old friend; it's a beautiful foggy morning and the workers are working and the birds are birding and it's all as it should be, and you were right, it is divine and full of love and patience, this little bit of the world. I got back late and my colleagues were waiting impatiently, but I didn't feel too bad; I'd burned a bunch of cals and had a mystical reverie—all before 10:00 A.M.

Visiting Hans isn't a duty; I need to do it. I love that he lies in the slate, the soil where his Riesling grew. I love that he views the village and the river. I love thinking of harvesttime, when the air will be full of the voices of the pickers and the thrum of the tractors, all nearby. Later the first snow will sift and settle over the graves where Hans and his neighbors lie in the slate above the river.

I think that we who love Mosel wine do so with a special tenderness. That is partly because of the wines' particular sparrowy charm, but if you have ever been there you find in these wines a taproot from which you can drink from your soul's purest waters. These wines do not merely hail from a culture; they're so deeply embedded in that culture you can't tell anymore where one ends and the other begins. The cohesion is both stirring and unnerving. Looking at the mourners at Hans Selbach's funeral, many of their faces could have been carved on Roman coins. They were the people of *this* place in the world. It's no accident that there are almost no international consultants, the "flying winemakers," from here. The Mosel gives its vintners all the stimulus they need.

Yet as much as I love this culture, I recognize its shadow side. It is not exclusively lyrical. If its air is rarefied, that's partly because it isn't always as fresh as it might be. There are all the petty jealousies and Hatfield-McCoy chicaneries that afflict small-village life around the world. But there is more.

I represent two Mosel producers who are neighbors on the same site; their parcels are contiguous. One producer hadn't quite finished picking grapes when his Polish harvest workers' work visas expired, meaning the crew had to return to Poland. No problem, said the neighbor; we'll pick for you. *We'll pick for you!* It really is another world. People may know one another for twenty years and still address each other as *Herr*-this and *Frau*-that. They have all the ratty bullshit that can possibly exist among people, but—"We'll pick for you."

Sigrid Selbach (Hans's widow and the matriarch of the family) told me a story once. "We picked our Eiswein last year on Christmas Day," she began. "The day before, when we saw it might be cold enough on Christmas morning, we hesitated to call and ask for help with the picking. But you know, we called twelve people, and they all agreed to help, and they were all cheerful to do it. We went out into the vineyard before dawn to check the temperature, then phoned them at 6:00 on Christmas morning, and they all came and all in a good mood. Afterward they gathered here at the house for soup and Christmas cookies. And when they left they were all singing out Merry Christmas as they went home to their families. Isn't that wonderful?"

I ask you! I too am amazed that people would cheerfully agree to get out of their warm beds before dawn on Christmas morning while their loved ones slept, to go out into the frigid vineyards and gather enough fruit for a few hundred bottles of wine that nobody makes any money on. This is more than mere neighborliness. It is simply assumed that certain traditions are ennobled by observing them with a hale kindness. When nature gives you a chance at an Eiswein, you *celebrate* the opportunity Your grapes might have rotted or been eaten by wild boars. But this time the gods smiled.

Being a Mosel vintner signifies membership in a human culture much deeper than mere occupation. This is true of every vintner, whether his wines are great, good, or poor. This may seem abstruse to the "consumer," but there are many ways to consume and many things to *be* consumed in a glass of wine. You can see it merely as an object and assess it against its competitors using some arbitrary scale. Or you can drink something that tells you it was made by human beings who want to show you the beauty and meaning they have found in their lives. *You* decide.

ON THE OTHER HAND . . .

After luxuriating in such impassioned prose, it's hard not to be swept away with a sense of the beauty of *terroir*. But not everyone is a believer. Skeptics abound. While few deny *terroir*'s existence altogether, there are plenty of articulate wine lovers who believe it doesn't tell the whole story, or even the main story, behind fine wine. The battle is usually joined over the relative importance of natural assets versus human intervention.

In the 1820s and 1830s, when the modern concept of *terroir* was just beginning to germinate, Australian wine pioneer James Busby journeyed repeatedly to France for the plant material that would serve as the foundation for that country's vine-stocks. Along the way, he solicited insights that might help develop a similar industry in the New World. He compiled many useful observations in his *Treatise on the Cultivation of the Vine and the Art of Making Wine* and his 1833 *Journal of a Tour Through some of the Vineyards of Spain and France* (from which the following is taken). It was there that he expressed some skepticism about the value constantly placed on place.

FROM JAMES BUSBY, *JOURNAL OF A TOUR THROUGH SOME OF THE VINEYARDS OF SPAIN AND FRANCE*

The limited extent of the first rate vineyards is proverbial, and writers upon the subject have almost universally concluded that it is in vain to attempt accounting for the amazing differences which are frequently observed in the produce of vineyards similar in soil and in every other respect, and separated from each other only by a fence, or a footpath. My own observations have led me to believe, that there is more of quackery than of truth in this. In all those districts which produce wines of reputation, some few individuals have seen the advantage of selecting a variety of grape, and of managing its culture so as to bring it to the highest state of perfection of which It is capable. The same care has been extended to the making, and subsequent management of their wine, by seizing the most favourable moment for the vintage—by the rapidity with which the grapes are gathered and pressed, so that the whole contents of each vat may be exactly in the same state, and a simultaneous and equal fermentation be secured throughout,—by exercising equal discrimination and care in the time and manner of drawing off the wine, and in its subsequent treatment in the vats or casks where it is kept—and lastly, by not selling the wine till it should have acquired all the perfection which it could acquire from age, and by selling, as the produce of their own vineyards, only such vintages as were calculated to acquire or maintain its celebrity. By these means have the vineyards of a few individuals acquired a reputation which has enabled the proprietors to command almost their own prices for their wines; and it was evidently the interest of such persons, that the excellence of their wines should be imputed to a peculiarity in the soil, rather than a system of management which others might imitate. It is evident, however, that for all this a command of capital is required which is not often found among proprietors of vineyards; and to this cause, more than to any other, it is undoubtedly to be traced, that a few celebrated properties have acquired, and maintained, almost a monopoly in the production of fine wines. ■

And for a final dash of defiance, here are excerpts from an op-ed posted at Wine Review Online by Bay Area writer W. Blake Gray in 2010.

FROM W. BLAKE GRAY, "DEMOTING *TERROIR*"

Terroir isn't the most important factor in wine quality. The producer is.

This is what California wineries used to preach in the 1980s: That they could overcome their lack of limestone with superior science. That oak treatment in the winery mattered more than the plot of ground where the grapes were planted. That a superstar winemaker was more important than a superstar vineyard.

We moved away from all that in the 1990s, led by winemakers themselves, who recognized a consistent difference in the quality of grapes from different areas.

Today, among wine aficionados, *terroir* is king again, as it was before Robert Parker launched what was essentially a career-long broadside against it with his *Wine Advocate* newsletter.

But you know what? Parker was and is right. And so were those blasphemous, non-dirt-worshipping Californians. And it's about time we all acknowledge it.

You, dear reader—yes, you with the skeptical look—agree with me. Don't shake your head. You agree with me, and I'm going to prove it to you.

Which would you rather drink: A Ridge Zinfandel from Amador County, or a wine made from Amador's best vineyard by a winery you've never heard of?

Which would you rather drink: A California-appellation Moscato made by Heidi Peterson Barrett, or a home winemaker's Cabernet made from grapes from To Kalon vineyard? Be honest, because I have judged homemade wines and I know what the correct answer is.

■   ■   ■

Last March I attended a Clos Vougeot tasting at the chateau. All the wines came from the same piece of land; some came from neighboring rows of vines. The whole *clos* is a Burgundy Grand Cru: If there are differences in climate and soils between vines, they are miniscule.

The wines varied tremendously. Many were, frankly, ordinary. But a few were wonderful, and they were without exception made by people who also made great wines from other vineyards.

I found this to be the case throughout Burgundy, supposedly the place where *terroir* matters more than anywhere else. Premier Cru wines from great producers are better than Grand Cru wines from mediocre producers. In fact, village-level wines from great producers are better than the worst examples of Grand Cru wines.

If *terroir* were truly king, that wouldn't be true: A Grand Cru wine would always stand out. But indifferent winemaking can minimize a vineyard's greatness. Poor winemaking can obliterate it.

■   ■   ■

Sure, the science-based winemakers of the 1980s took it too far. I don't want to drink a wine that's purely a product of technology. I don't want oak overpowering grape. And I don't usually choose to drink multi-region blends, even if they're delicious, because I want my wine to taste like it came from somewhere. If I can't pinpoint the individual vineyard, I want it at least to show typicity of region.

But let's face it: There is no such thing as a wine made with no human intervention. The single most important decision in winemaking is when to pick the grapes. A Donkey and Goat Winery is one of my favorites in California, and they talk a lot about their non-intrusive, natural techniques in the winery. But the fact is their wines have that "Donkey and Goat" character because they pick earlier than everybody else who shares the same vineyards. If Kent Rosenblum made wine from the same vineyards, using the same equipment and same "natural" yeast, it would taste different because he would pick weeks later.

Terroir is queen. Winemaking is king. Q.E.D. Gentlemen, start your overreactions.

2

HISTORY AND DEFINITIONS

As with most human concepts born of observation and recognition, *terroir* does not technically exist unless it is *perceived*. Just like the proverbial tree falling in the woods, whose sound is questioned if no one hears it, wild vines undoubtedly produced flavorful fruit even though no one was there to taste it (and rhapsodize.) *Terroir*, in other words, must be communicated: its message is meaningless if it is not received and appreciated.

From the earliest days of organized agriculture, when sites for food crops were chosen by trial and error, methods evolved and were refined through successes and failures, as Neolithic humans moved from hunting and gathering to farming in locations that proved prolific and consistent. The history of wine, meanwhile, almost certainly begins with a chance occurrence, an assortment of grapes left on their own to ferment in whatever vessel they were stored in, resulting in a pleasing, perishable beverage bearing the wondrous, mysterious properties of intoxication. Other mysteries, like *terroir* expression, would have been a low priority in an epoch when survival was still paramount and wine could not be conserved beyond immediate consumption.

However, this changes. Grapes could not make themselves into wine without human assistance and intervention. Agricultural sites are developed and refined over seasons, and we can infer that certain sites and conditions were favored over others. Before long, there is graphic evidence of wine's important role in the culture: pictographs of trellised grapes on Egyptian walls and amphorae mark the beginnings of human appreciation for this precious commodity.

In his remarkable book of viticultural anthropology, *Uncorking the Past: The Quest for Beer, Wine, and other Beverages,* Patrick McGovern offers not only a vivid and engaging picture of early winemaking, but a rationale for its role in Paleolithic culture.

FROM PATRICK MCGOVERN, "THE PALEOLITHIC HYPOTHESIS"

How far back in the mists of archaeological time have humans savored alcohol, and how has it shaped us as a biological species and contributed to our cultures? The drunken monkey hypothesis explores the biological side of the question. The Neolithic period, beginning around 8000 BC in the Near East and China, provides a rich trove of archaeological material to mine answers to the cultural questions. Humans were then settling down into the first permanent settlements, and they left abundant traces of their architecture, jewelry, painted frescoes, and newly invented pottery filled with fermented beverages. For the Paleolithic period, beginning hundreds of thousands of years earlier, we are on much shakier ground. Yet this is undoubtedly the time when humans first experimented with alcoholic beverages, as they relished their fermented fruit juices and came to apprehend their ecstasies and dangers.

The evidence from Paleolithic archaeology is scant, and it is easy to overinterpret and read modern notions into this fragmented past. Archaeologists once thought that early humans were meat eaters on a grand order because their encampments were littered with animal bones. Then it dawned on someone that the remains of any fruits or vegetables simply had not survived, and that the abundance of bones, which were infinitely better preserved, indicated only that meat constituted some portion, possibly minor, of the early human diet.

In [my previous book] *Ancient Wine* I outlined a plausible scenario, which I refer to as the Paleolithic hypothesis, explaining how Paleolithic humans might have discovered how to make grape wine. In brief, this hypothesis posits that at some point in early human prehistory, a creature not so different from ourselves—with an eye for brightly colored fruit, a taste for sugar and alcohol, and a brain attuned to alcohol's psychotropic effects—would have moved beyond the unconscious craving of a slug or a drunken monkey for fermented fruit to the much more conscious, intentional production and consumption of a fermented beverage.

In an upland climate where the wild Eurasian Grape (*Vitis vinifera* ssp. *sylvestris*) has thrived for millions of years, such as eastern Turkey or the Caucasus, we might imagine early humans moving through a luxuriant river bottom. Using roughly hollowed-out wooden containers, gourds, or bags made of leather or woven grasses, they gather up the right grapes and carry them back to a nearby cave or temporary shelter. Depending on their ripeness, the skins of some grapes at the bottom of the containers are crushed, rupture, and exude their juice. If the grapes are left in their containers, the juice will begin to froth or even violently bubble up. Owing to natural yeast on the skins, it gradually ferments into a low-alcohol wine—a kind of Stone Age Beaujolais Nouveau.

Eventually, the turbulence subsides, and one of the more daring members of the human clan takes a tentative taste of the concoction. He reports that the final product is noticeably smoother, warmer, and more varied in taste than the starting mass of grapes. The liquid is aromatic and full of flavor. It goes down easily and leaves a lingering sense of tranquility. It frees the mind of the dangers that lurk all around. Feeling happy and carefree, this individual invites the others to partake. Soon everyone's mood is elevated, leading to animated exchanges. Perhaps some people sing and dance. As the day turns into night and the humans keep imbibing, their behavior gets out of hand. Some members of the clan become belligerent, others engage in wild sex, others simply pass out from intoxication.

Once having discovered how to make such a beverage, early humans would likely have returned year after year to the wild grapevines, harvesting the fruit at the peak of ripeness and even devising ways to process it—perhaps stomping the grapes with their feet or encouraging better anaerobic fermentation of the juice and pulp by covering the primitive container with a lid. The actual domestication of the Eurasian grapevine, according to the so-called Noah hypothesis, as well as the development of a reliable method of preventing the wine from turning swiftly to vinegar (acetic acid), was still far in the future. A tree resin with anti-oxidant effects might have been discovered accidentally early on, but it would only have delayed the inevitable by a few days. The wiser course was to gorge on the delectable beverage before it went bad. ∎

THROUGH THE Neolithic, Bronze and Iron periods of human civilization, wine evolved as a beverage possessing transformative powers, with mystical and religious properties and the capacity to induce euphoria and abandon. It was also a safe alternative to water and a commodity so widely desired that early civilizations—Phoenicians, Greeks, Romans—produced it and transported it for sale around the Mediterranean Rim. Pliny the Elder in his *Historia Naturalis* and Columella in his *De Re Rustica* detailed Roman agricultural practices, including those of vineyards, in the first century AD.

Roman armies took vines to every distant conquered outpost, where successes inspired further exploration and development. Wines were aesthetically potable for a short time, though many needed flavor amelioration; wine's tendency to spoil motivated producers to find ways of preserving it, at least until the seasonal cycle replenished the stock: each new vintage, in more ways than one, was a cause for celebration. Meanwhile the seasonal cycle of wine, with its inherently salubrious and celebratory properties, worked its way into the cultural fabric of European communities as a harbinger of replenishment and renewal.

As Rome expanded its empire, vine growing and winemaking spread throughout Western Europe. Early wine descriptions exist, but since travel and trade were limited, so were comparative tasting records. Exclamations of local quality no doubt sprang from chauvinistic pride rather than anything that would pass for a critical acumen.

Notions of quality are all but omitted in early accounts, until the Middle Ages, when winemaking and viticulture were raised to an art form owing to the efforts of Benedictine and Cistercian monks; without them, perhaps the very notion of *terroir* would have gone unnoticed for hundreds of years. Both orders were responsible for the development of wine regions—Champagne, Bordeaux, Burgundy, to name but three—whose vineyards have been in continuous use since their development, regions that use systems of classification and quality control established by these talented monks. The Dark and Middle

Ages are marked by spectacular breakthroughs in viticulture and husbandry, soil amendment, cellar practices, winemaking and, ultimately, wine quality. Texas Tech University physics professor Stefan Estreicher discusses this early monastic husbandry.

FROM STEFAN K. ESTREICHER, "DARK AGES, LIGHT WINES"

The quality of wines and the opportunities associated with the wine trade gradually increase during the second half of the Middle Ages and several key wine producing regions of France emerge. This begins in the late 11th century in Burgundy, because of independent and hard-working monks. Then, the region of Bordeaux returns on the wine scene, an unforeseen consequence of the marriage of Eleanor of Aquitaine to the future King Henry II of England. Côtes du Rhône comes next, as the papacy moves from Rome to Avignon in the early 14th century. Popes demand luxury, including good wine, and new vineyards are planted. Burgundy, Bordeaux, and, to a lesser extent, the Côtes du Rhône acquire a reputation of excellence that continues today.

This is not to say that these are the only wine regions. In fact, around the year 1000, wine is produced almost everywhere in Europe and around the Mediterranean basin. However, in the Iberian Peninsula, North Africa, and the Near East, Islam limits, and at places prohibits, its production and trade. Italy and much of central Europe are divided and politically weak. The wine production remains local. Thus, as far as wines with a reputation of quality are concerned, Burgundy and Bordeaux have little competition, and their reputation will remain for centuries. In the 1300s, a major climatic change in Europe, the "Little Ice Age," followed by the first waves of Black Plague, shake the social structure to its core, mark the beginning of the end of feudality, and change the geographical distribution of vineyards. The first Golden Age of Burgundy occurs under the reign of four powerful dukes.

CISTERCIAN WINES

Until the 11th century, all bishops are appointed by the King, a principle imposed by Charlemagne. Pope Leon IX (1049–1054) and his successor Gregory VII (1054–1085) initiate a policy of emancipation from the temporal power. In 1075, Gregory decrees that only the Pontiff of Rome can be called Universal. Only he can appoint and depose bishops. Further, since the Pope crowns kings and emperors, he can also depose them. Excommunication is his weapon: an excommunicated ruler loses legitimacy. But monasteries also fit in the secular feudal system. Kings, dukes, or other powerful rulers establish them, give them land, and then appoint their abbots. This is also about to change.

In 909, William, Duke of Aquitaine and the monk Berno establish a Benedictine monastery at Cluny, near the city of Mâcon in Burgundy, which benefits from a "pontifical exemption." This means that it is under the direct authority of the pope, free from secular obligations. The influence and power of Cluny rapidly grows. In order to counteract the influence of the Benedictines (who wear black habits), Cistercian monks (who wear white habits) establish the monasteries of Molesme (1075) and then Citeaux (1098) just east of Nuits-St.-George, also in Burgundy. They follow the same rules, those of St. Benedict, but apply them to the letter, with principles of penitence. They are also under the direct authority of the pope. These independent monasteries encourage reform within the Church. Monastic life—pray, work, rest—quickly becomes popular. The appeal of religious life increases. One reason is the new fervor associated with the Crusades. Another reason is the insecurity of cities caused by frequent Viking raids. Yet another reason is that monastic life is overall safer than life in cities or small rural communities.

As the population of an abbey outgrows its capacity, a group of monks leaves and establishes a new abbey elsewhere. The number of Cistercian abbeys grows enormously until the onset of the Little Ice Age in the early 1300s. This begins with the daughter-abbeys of Citeau, Pontigny, Clairvaux, and Morimond. From a handful in 1100, the number of Cistercian abbeys grows to about 300 in 1150, 500 in 1200, 650 in 1250, and 697 in 1300. These numbers exclude hundreds of nunneries. Cistercian abbeys are built from Portugal to Russia and from Sicily to Scotland. Most of them are impressive structures, with magnificent churches, buildings, and fantastic cellars for keeping wine, the first extensive wine cellars.

The Cistercians are quality fanatics. They have plenty of time, manpower and financial resources. In Burgundy, it is claimed that they even taste soil samples before deciding where to plant new vineyards. They bring the Chardonnay grape to the region of Chablis. They introduce the notions of *terroir* and *cru*. *Terroir* refers to the soil, exposure, slope, and other physical characteristics of a vineyard. *Cru* relates a wine to the grapes that grow in a specific geographical location. The Cistercians vastly improve winemaking techniques, mostly by systematic trial and error. This extends from the pruning of vines to winemaking techniques. The quality of the wines they produce in Burgundy, including red wines, increases substantially. Cistercian wines are far better than those produced elsewhere at the time. ∎

IN HIS highly readable popular history of wine, Paul Lukacs takes the story of the Cistercian monks and contextualizes it, positing that the monks' practices established a level of vine-growing that had never before been achieved, perhaps with any agricultural product. That quantum leap begins, he says, with the Cistercians' careful, meticulous farming and their fine-tuned improvements in viticulture and vine-growing. Their practices provide the underpinnings of our modern understanding of *terroir*.

FROM PAUL LUKACS, *INVENTING WINE*

Some superior dry wines did exist at the start of the Renaissance, but only a handful ever came to the market. Over the next three hundred years an awareness of these wines developed slowly. They would not become valued broadly in Europe until the 1800s, when new technologies and new scientific knowledge allowed vintners to produce a far wider range. Radical changes then transformed wine's production as well as its consumption, enabling it to become newly respected as an object of cultural connoisseurship. Yet the seeds of change had been planted much earlier–initially in the swales and slopes of viticultural Burgundy, then on the steep hillsides along the Rhine and Mosel rivers in Germany, and finally in gravelly vineyards near the city of Bordeaux. These were the three areas in which a special or fine wine's identity first became linked with the particularities of place, and in which people began to understand the complex interplay of soil, climate, and culture, which constitutes what contemporary enthusiasts call a wine's *terroir*.

Terroir itself is a modern term. Derived from the Latin *terratorium,* it entered the French lexicon during the Renaissance, but at that point it simply meant "territory." During the nineteenth century it acquired a second meaning as an area of land valued specifically for agricultural properties, but still rarely was employed in connection with wine. Only in the 1900s did it come to be used specifically to designate a vineyard's natural environment, including geology, soil type, topography, climate, and more. Then it also began to signify a particular feature of wines made with grapes grown in that environment. The presence of this feature, sensed physically but recognized intellectually, explains why discerning drinkers today claim to be able to taste it in fine wines coming from particularly fine vineyards.

As something smelled and tasted, *terroir* has a mental as well as a physical aspect. When applied to wine the word means more than just locale. It designates the human recognition of locale and so indicates both what may make a wine from one place taste unlike a wine from another place, and what helps that first wine taste like itself. Put another way, vin fin, unlike vin ordinaire, needs to be self-referential. That is, it has to taste like previous renditions indicate it should taste, meaning above all it must be true to its origins. A taste, or *goût de terroir,* thus invariably recognizes particularity as a necessary aspect of quality, one understood to be a property of the wine itself.

The first wines to be prized for their ability to display such individualized aromas and flavors were made in Burgundy by white-habited Cistercian monks in the late Middle Ages. Though these devout clerics did not use the word *terroir*, they clearly thought of their vineyards and their wines in this new way. Of course the notion that grapes from one place can produce better wines than grapes from another was not new. As the inscriptions on amphorae in the pharaohs' tombs demonstrated, the ancient Egyptians recognized as much. So did the Greeks, who repeatedly expressed a preference for wines from certain Aegean islands, as well as the Romans who venerated Falernian and other dried-grape wines from Campania. In antiquity, however, distinguishing between vines simply meant recognizing general merit. There was as yet no sense of a wine's individuality. The Burgundian monks also differentiated between wines in terms of merits or class, but their understanding of their work involved something else as well–identity. When they singled out particular parcels of land for special recognition, they did not necessarily judge the wines from one to be better than the wines from another. They simply contended that those wines tasted different, with particular characteristics that repeated themselves vintage after vintage. Thus they insisted that both the wines and the vineyards needed to be identified as distinct entities. And by doing so, they invented what over the following centuries became a crucial part of our modern concept of high-quality, or fine, wine, a concept that ultimately enabled wine in general to overcome the many challenges that it would soon have to face.

■　■　■

The Cistercians tended their vines with near-fanatical zeal, and the renown their wines came to enjoy surely owed as much to the care the monks devoted to their work as to the *terroirs* they discovered. Unlike growers in other regions, who cultivated vines interspersed with different crops, theirs was a monoculture. At the time, most farmers raised grapes alongside other fruits, vegetables, and cereal crops. But the newly plowed swaths of land on the well-drained hills above Citeaux contained only vines. The monks meticulously recorded which plots produced healthy grapes and which produced sickly ones. They noted where bud break came early and where late, as well as where the fruit matured evenly and where ripening proved irregular. Somewhat surprisingly, they found that significant differences in such matters could come in short measure, with only a few paces sometime separating an excellent grape growing spot from a merely good one. When the monks were certain—and it took a great many vintages to become certain—they then marked each separate locale. No matter that vines might be growing in an unbroken swath all the way up the hill, they would distinguish between what they understood to be individual plots on that hill. At times, especially when they prized a specific spot, they would build stone walls around it and create a *clos*—a cloistered or enclosed vineyard. Though most of these walls long since have fallen down, the boundaries between many of the sites that the Cistercians first delineated remain in place today, solidified not by stone but by the rules of the modern French appellation system, as those same borders identify many of contemporary Burgundy's most famous crus.

In Burgundy today, the word "cru" designates both a wine and a vineyard, a celebrated wine from a celebrated vineyard. The Cistercians certainly made celebrated wines, but their significance in the history of wine comes much more from their vineyards. That's because their winemaking techniques were just as rudimentary as those anywhere else. Grapes typically were foot-trodden. With whites, the liquid would be strained in open casks, where it would be left to ferment; with reds, the treading would take place in deep vats, so that the juice and skins could stay mingled together during fermentation. Sometimes, especially in large communities, presses were used, yielding potentially brighter whites and fuller reds. But, regardless of the method, speed always was of the essence. Though no one knew why, everyone understood that excessive exposure to air hastened a wine's souring. Thus all wine was fermented quickly and rushed into barrels as swiftly as possible. As a result, medieval and Renaissance Burgundies were lighter in body, with less alcohol and, in the case of reds, less color, than Burgundies today. The Cistercians made the most invigorating wines, wines that tasted not only unlike other wines but also subtly unlike one another. The monks distinguished them as crus, with names like Chambolle, Corton, Pommard, and Volnay, because they thought those more nuanced to differences reflected different terroirs.

·   ·   ·

The Cistercian focus on the particularities of grape and place marked a significant advance in the history of wine. While these monks certainly were not the first people to care about their vineyards or their wines, they were the first to care about their terroirs. Not surprisingly, ambitious lay farmers who saw the prices that these wines fetched began before long to employ equally meticulous viticultural practices. Not only in the Côte d'Or but also in select locales in the Loire and Rhône Valleys, as well as in Alsace and, most notably, in Germany, they labored long and hard to produce vin fin rather than vin ordinaire. Much as with the monks cloistered behind the walls of Vougeot, their goal was to capture an admittedly short-lived taste of individuality. ■

LUKACS GOES on to echo the sentiment expressed at the outset of this chapter, namely that the history of *terroir* is a cultural history, one in which the concept itself goes unnoticed without others to discern and appreciate it:

During the Renaissance and Reformation, production of the new terroir-driven wines always remained small. Thus an appreciation for particularity linked to a place developed slowly. For a long time very few wines demonstrated anything of the sort, so very few consumers had the opportunity to experience it. Those who did invariably lived in or near cities, as the markets for these special wines were always primarily urban. The iconic image of a farmer drinking the unspoiled fruit of the earth in his own vineyard has a certain pastoral appeal, but it is largely a romantic fantasy. Hardly any individual grape growers had the knowledge or the financial means to make vin fin. Even those few who did possess the necessary resources characteristically found an audience for their wines in the city (or at court), not in the country. Thus the paradox of terroir-driven wines: their distinctive taste originates in nature, but the human recognition and appreciation of that taste always have come more in man-made, usually urban settings. ∎

EVALUATION, THEN, was an essential component of a wine's *terroir* expression, and there's ample documentary evidence that such assessments were an essential part of European courtly life. It even, at times, takes on a literary form: in 1224 poet Henri d'Andeli set to verse an account of a competition in the court of Philippe Augustus which he called "La Bataille de Vins" (the Battle of the Wines), wherein dozens of wines from across Europe were placed before and judged by an English priest. The judge pronounced the good ones "Celebrated," and the rejected were "the Excommunicated." Those wines that remained in good graces were prized, at least in part, for their fidelity to place.

Indeed, many of the characters of Rabelais (1494–1553) display a working knowledge of the fact that quality is dependent on a vineyard's locale and circumstance; and the sorts of wisdom attributed to Bacchus in countless literary accounts establish not only an appreciation of wine's finer qualities, but its many regional expressions. Still other commentators from Gohory (1520–1576) to Montaigne (1533–1592) established relationships between place, plant, and man that signified a tacit knowledge and recognition of *terroir* centuries before the term would be in broad use.

It is Olivier de Serres, however, in his book, *The Theatre of Agriculture and the Management of Fields,* published in 1600, who elevates and codifies *terroir* as a fundamental tenet of French culture. According to Thomas Parker's accounting in his book *Tasting French Terroir: The History of an Idea,* the word *'terroir'* is used 87 times in de Serres book—and that book, he notes, began its life with a 16,000 copy print run, was reprinted eight times, and distributed to every parish in France. Parker provides in cogent terms an overview of the work and its importance to modern French thought.

FROM THOMAS PARKER, *TASTING FRENCH TERROIR: THE HISTORY OF AN IDEA*

... de Serres makes clear that the foremost matter at hand is the question of its determining force, which shapes food and drink—and wine in particular: "The climate and terroir provide wine with its taste and force in accordance with their properties, so that it is completely impossible to account for the diversity of wine by the species of grapes. Accordingly, the same vine put in different places will produce different kinds of wine as diverse as the soils where it is planted." . . . de Serres separates macro- and microclimatic differences, crediting terroir and not the grape for giving wine different tastes and qualities. ∎

THE 'THEATRE OF AGRICULTURE,' for de Serres, is organized according to *terroir;* indeed, writes Parker, the first line of the book "calls on the reader to understand the unique qualities of the *terroir* intended for planting: 'Agriculture's foundation is in the understanding of the nature of the terroir that one wishes to cultivate.'" The rational farmer, explains de Serres, should distribute his crops according to the diverse growing potentials of soils.

French agronomists and theorists move closer to a firm definition of *terroir* in the eighteenth and nineteenth centuries, with cultural observers like Pierre Jean-Baptiste Legrand d'Aussy and Brillat-Savarin keenly recognizing that many of the wines and comestibles in France were unique to the place where they were grown and made. Not only could they not be made anywhere else, but their very uniqueness depended on the particularities of place.

Eventually a geographer, Paul Vidal de la Blache, devises a theory which codifies this relationship still further. Vidal de la Blache studied and lectured on a concept he called 'genre de vie,' or the way in which people impact the landscape, and vice versa; this later came to be known as human geography. "What one hopes to explain in these pages," wrote Vidal de la Blache in his seminal work, the *Tableau de la Geographie de la France,* "concerns how can the history of a people be (or must be) incorporated in the soil of France? The rapport between the soil and the people is imprinted with an ancient character that continues through today."[1]

By the nineteenth century, the French have embraced not only the regional underpinnings of their gastronomical diversity, (maps of the country begin to appear with the nation's foodstuffs hovering over their respective locales) but have also established how place, tradition, and communities synchronize to create agricultural products—including wine, of course.

Terroir's modern meaning seems to have been codified in the mid-nineteenth century. This definition, culled from a mid-twentieth century reference work, refers to a passage from agronomist Jules Guyot, from his 1861 work, *Culture de la vigne et vinification.* Guyot's definition includes some demonstrably pejorative connotations.

FROM YVES RENOUIL, DIRECTOR, PAUL DE TRAVERSAY, COLLABORATOR, *DICTIONNAIRE DU VIN* (1962)

TERROIR (Goût de).   This flavor refers to specific characteristics due to/rendered by the soil/earth itself. Sometimes the "terroir flavor" which characterizes a wine contributes flavor notes which make it particularly esteemed. But, more often, it shows up as a disagreeable character that can be described as excessive manure, as a type of fertilizer, the flavor of too-long stem contact, or some

TERROIR BY LAW: THE ROLE OF THE APPELLATION D'ORIGINE CONTRÔLÉE (AOC)

As early as 1905 the French government began coordinating rules and laws to protect the reputations of certain regional products. By 1935, they'd put laws in place to establish the Appellation d'Origine Contrôlée or AOC laws, a system of certifying certain French regional food and beverage products, nearly all of which were determined to have a significant geographical component. The system, which includes meats, herbs, spirits, cheeses, honey, and wine, represents nearly 400 products, none more numerous, visible, or weighty than the wine AOCs.

Appellations are granted and maintained by another august French governing body, the Institut National des Appellations de l'Origine (INAO—its name was recently changed to Institut National de l'Origine et de la Qualité, though the acronym remains unchanged). They are in effect the arbiters of French *terroir.*

AOCs are never not controversial. They are designed, in the first place, to be a mark of quality, since the INAO forms committees and peer groups which taste and evaluate the wines in question and determine if they merit the certification of place. Thus their economic impact is indisputable—traditionally, if one's wine can be labeled with an AOC it will command a better price in the market than one that can't make the claim. But in every AOC there are producers whose products barely meet the minimum standard of quality—or there are arbiters, bureaucrats, whose notions of quality are questionable. Depending on whom you talk to, the part may tarnish the reputation of the whole.

Furthermore, with their limitations on varieties and yields, and with the strict guidelines on how a wine can be made, the AOC system inevitably dictates a region's range of expression, which may feel restrictive to certain winemakers. In the Loire Valley, for example, an intriguing indigenous variety Pineau d'Aunis can no longer be given the AOC designation "Touraine," even though it has been growing in that region for centuries, and hundreds of acres of old vines remain, capable of producing wines of tremendous character and, naturally, exhibiting the *terroir* character of the Touraine.

Which brings up another point: Who's to say that a wine made outside of the AOC strictures doesn't express *terroir* in its own right? A sensitive winemaker seeking to make a Pineau d'Aunis, for example, would likely make a wine that expresses *terroir.* And yet he would be obliged by law to use the words "Vin de France" on the label. It's a designation of last resort for winemakers who choose, or who are forced, to make wine outside the appellation rulebook. And while this designation was once thought pejorative, many winemakers now flaunt it, either as an act of defiance or of independence from rules they feel are restrictive.

And one other footnote from the annals of officialdom: in 2010 the OIV, or the Organization Internationale de la Vigne et Vin, asked for a resolution defining *terroir* and defined it, officially, by way of its General Assembly:

> Vitivinicultural "terroir" is a concept which refers to an area in which collective knowledge of the interactions between the identifiable physical and biological environment and applied vitivinicultural practices develops, providing distinctive characteristics for the products originating from this area. "Terroir" includes specific soil, topography, climate, landscape characteristics, and biodiversity features.

The final word, perhaps. Say no more!

uncertain cause. When the "terroir flavor" is not too noticeable it may diminish and almost disappear with age, as does a young wine's fermentation bouquet.

The particular odors which give rise to these flavors are absorbed on the grape skins and stems, and develop their disagreeable odor during tank fermentation. Dr. Guyot recorded a case where one vine cane grew along a wall and passed in front of a stable window through which odors emanated, and the grapes harvested from in-front of this window were inedible, due to the manure flavor which was quite evident, whereas the other grapes from the vine tasted normal.

Then there was the case of a classified wine which had acquired the flavor of tar following the tarring of a road which bordered the vineyard during the period of flowering. It is very difficult to lessen a flavor so conspicuous.

Our advice for prevention is to have primary fermentation as short as possible, a rapid settling, numerous rackings, to avoid problems from sediment. To correct these problems, fine with egg whites and a half-liter of olive oil per barrel, this, when it rises to the wine surface, absorbs essential oils that cause bad flavor. Unless this procedure is used at the time of racking and fining it will not achieve the result. Occasionally, and only with white wines, carbon treatment may be used.

Of course, the elements of this terroir flavor may constitute the charming appeal of a distillation, although in some brandies they show characteristics more or less important and more or less agreeable.

[John Buechsenstein, translation] ∎

IT IS difficult to pinpoint how the word *terroir* evolved away from such undesirable connotations. But it did soon become a desirable concept. An early twentieth century piece by French author Colette expresses the sentiment of *terroir,* even though she doesn't use the word itself.

FROM COLETTE (1873–1954), *TENDRILS OF THE VINE,* "SECOND MOVEMENT" (1908)

Alone among the world of plants it is the vine that distinctly shows us the true flavors of the earth. How faithfully she renders this translation! She shows it, expressing in its clusters the secrets of the soil. The flint, through her, makes us understand that it is alive, liquefied, nourishing. The barren chalk weeps wine, as tears of gold . . .

[John Buechsenstein, translation] ∎

IT WAS during this period that Burgundy emerged as the seat, conceptually, of French *terroir,* a place where *climats, lieu-dits, clos,* and *crus* had served as models for the concept for centuries. Soon, in the rest of the country, winegrowers sought out *climats,* villages, and vineyard plots already producing *terroir* wines that were as yet unrecognized and unheralded.

One thing is certain: The word *terroir* achieved household status in the twentieth and twenty-first centuries, its use and referral skyrocketing to be among the most popular, debated, controversial, and celebrated concepts in the world of wine. One way to note this

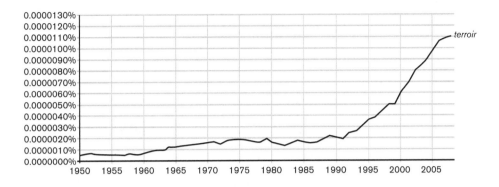

FIGURE 2.1

Ngram depicting the use of the word *terroir* in American English, 1960–2008.

is to search on Google Ngrams, a logarithm that tracks the frequency of usage for the word "terroir" in books written in American English for the last thirty years.

Nearly every reference book we consult offers a definition for *terroir*. While many are similar, some are limited and simplistic, others quite comprehensive. What follows here is a sampling of definitions extracted from numerous texts, encyclopedias, dictionaries, and articles. They tend to fall into distinct categories. Some relate *terroir* to merely soil and rock; others allow varying degrees of other parameters such as topography and climate. A surprising number say nothing about human involvement or culture.

DICTIONARIES AND ENCYCLOPEDIAS

FROM GERARD DEBUIGNE'S *DICTIONNAIRE DES VINS* (1969)

terroir. This word—which designates the earth, the soil—takes on, when it comes to wine, a special meaning in the expression '*goût de terroir.*' In this case it designates a flavor characteristic, a particular one, nearly indescribable, that is acquired by all wines that come from certain soils.

[John Buechsenstein, translation]

FROM PAUL CADIAU'S *LEXIVIN,* A DICTIONARY THAT TRANSLATES FRENCH WINE TERMS INTO ENGLISH AND OTHER LANGUAGES (1982)

terreux (goût). earthy (taste)

terroir. soil, local soil

(goût de) terroir. local specific taste (tang)/taste from soils of a certain region

FROM AN OIV LEXICON, *LEXIQUE DE LA VIGNE ET DU VIN* (1963)

B 1 *écologie viticole.* **(viticultural ecology)** The study of the inter-relationship between the vine and its environment (climate and soil) and the quality of the wine produced.

B 3 *lieudit, « climat ».* **(vineyard site)** A site with soil type, exposure, etc., that make it especially suited for the production of quality grapes.

B 4 *cru, finage, « climat ».* **(no translation)** Wine produced by a particular vineyard renowned for the quality of its products. The term is also applied to vineyards (France).

U 166 *gout de terroir.* **(earthy taste)** Special taste resulting from the type of soil on which the grapes were grown. *Italian: sapore de terroso; Spanish: gusto a terreno ó a terruño; German: Bodenge-schmack; Portuguese: sabor ao solo d'origem, gosto a rocha-mãe.*

FROM FRANK SCHOONMAKER, *ENCYCLOPEDIA OF WINE*, A LATE EDITION (1973)

TERROIR—(Tair-wahr)—Soil or earth, used in a very special sense in French in connection with wine, as *goût de terroir.* Certain wines produced on heavy soil have a characteristic, unmistakable, almost indescribable, earthy flavor, somewhat unpleasant, common, persistent. This is a *goût de terroir,* and the German equivalent is *Bodenton* or *Bodengeschmack.* Superior wines rarely if ever have much of this, which, if once recognized, will not easily be forgotten.

FROM CASSELL'S *FRENCH DICTIONARY* (1962)

terroir, n.m. Soil, ground. *Goût de terroir,* raciness (of style), native tang (of wine); *sentir le terroir,* to smack of the soil. [Ed note: this definition doesn't have its own entry, but is found under the definition of the word "territoire."]

FROM *THE OXFORD COMPANION TO WINE,* JANCIS ROBINSON, ED., 3RD ED. (2005)

terroir. much-discussed term for the total natural environment of any viticultural site. No precise English equivalent exists for this quintessentially French term and concept. Dubos and Laville describe it fully, and how it underlies and defines the French *APPELLATION CONTRÔLÉE* system. A definition is given in van Leeuwen *et al.* Discussion of terroir is central to philosophical and commercial differences between OLD WORLD and NEW WORLD approaches to wine.

Major components of terroir are soil, (as the word suggests) and local TOPOGRAPHY, together with their interactions with each other and with MACROCLIMATE to determine MESOCLIMATE and vine MICROCLIMATE. The holistic combination of all these is held to give each site its own unique terroir, which is reflected in its wines more or less consistently from year to year, to some degree regardless of variations in methods of VITICULTURE and WINE-MAKING. Thus every small plot, and in generic terms every larger area, and ultimately region, may have distinctive wine-style characteristics which cannot be precisely replicated elsewhere. The extent to which terroir effects are unique is, however, debatable, and of course commercially important, which makes the subject controversial.

FROM *THE LAROUSSE ENCYCLOPEDIA OF WINE* (1994)

Choosing the Site: Man can manipulate the basic growing conditions—by irrigating in dry areas, for instance—but his biggest contribution to wine quality is to choose the actual vineyard site. Ancient Rome's farmers knew that vines loved a sunny slope, a nearby river. Aspect to the sun, altitude, soil and underlying rock, taken together with climate both general and local, determine why wine from some vineyards is consistently better than that from others. France has developed the concept of *terroir*—the sum of all the factors pertaining to a particular plot of land.

The terroir approach is a traditional one and is challenged to some extent by modernists, who say that science can master nature's foibles. Few experienced vinegrowers, however, dispute that site does matter.

FROM LAROUSSE, *WINES AND VINEYARDS OF FRANCE* (1991)

Wine draws its greatness from the *terroir* that has given it birth. The alchemy of light, temperature, water and earth determines a wine's qualities and particular features. Science has begun to penetrate these mysteries, but as yet offers no more than the beginning of an explanation. Wine-growing *terroirs* are jealous of their secrets . . .

A *terroir* may be defined as a stretch of land with specific agricultural attributes resulting from the combination of local climate factors and the nature of the soil. A viticultural *terroir,* however, is rather more than this. There is a good deal of man's handiwork in it. Through a whole series of interventions, in the methods of nurturing the vines and the soil, the wine grower is in fact able to change the natural environment into a managed one that is generally favourable to the quality of its products. The reputation of a particular *terroir* may depend in the first instance on its natural potential, but it also depends to a considerable extent on the skill of the wine grower, and then on that of the wine maker.

The idea of a wine-growing *terroir* necessarily implies a degree of affinity between the characteristics of the environment and the potential of the grape varieties grown there. The grapes should not only be able to reach a satisfactory degree of ripeness, but should also exhibit pleasing organoleptic characteristics, subtle, nuanced, and varying according to the *terroir* they are grown in. The study of a *terroir* requires analysis and knowledge of both climate and soil.

FROM THOMAS PINNEY, *A HISTORY OF WINE IN AMERICA: FROM THE BEGINNINGS TO PROHIBITION* (VOLUME I, APPENDIX 2, 1989)
THE LANGUAGE OF WINE IN ENGLISH

One cannot talk or write long about wine in English without discovering that the language is weak in words for the activities of vine growing and winemaking. The solution is either to Frenchify one's language, since French is rich in just those words that English lacks, or to invent English equivalents, or, most often, to strike some sort of compromise.

■   ■   ■

Terroir is another unfamiliar concept; it refers, literally, to the contribution of the soil to the character of the wine, but in application it sometimes takes on almost mystical attributes. The French take terroir seriously; Americans so far remain skeptical.

FROM THOMAS PINNEY, *A HISTORY OF WINE IN AMERICA: FROM PROHIBITION TO THE PRESENT* (VOLUME 2, 2005)

Where a grape is grown and *how* it is grown appear to be quite as essential as *what* grape is grown. And so do the methods by which wines are made from grapes of the same name. So the modern wine drinker, looking at the hundreds of Cabernets now offered for his choice, can have only the most elementary guarantee from the fact that all the wines are certainly from Cabernet grapes. The simple uniformity of the name conceals an unpredictable variety of styles, characters, and qualities. In recent years the winemakers of California have increasingly recognized this fact and have turned, inevitably, to promoting the idea of location, usually under the fashionable term of *terroir,* as a way of establishing an identity and a claim to inimitable character. There is no doubt that different regions have different characters, but it is slow work discovering just what that character is and then getting the public to recognize and appreciate it. ■

IN THE decades following World War II, California researchers and vintners had the confidence that its winemakers could "do it all," i.e., accomplish quickly what had taken European vintners centuries to achieve.

Since then, there has been a reaction against such exuberant confidence—hubris, as some would have it. One sometimes feels now that an almost penitential renunciation of the old self-assurance has overcome American winemakers. Now we hear that wines are made in the vineyard, not the winery; that the old gravity-flow design is the best design; that wild yeasts make better wines than cultured ones; that wood is better than steel for fermentation; and so on through a long list of anti-technological reversions. The vogue for exalting the virtues of *terroir,* so conspicuous in recent years, is another form of this reaction, since the elements of *terroir,* however one may define them, are precisely those things about which nobody can do anything: soil, exposure, temperature, rainfall, and so on. Instead of boasting about his power to manipulate his materials to any desired end, today's winemaker is likely to adopt a position of reverent fatalism: he can alter what he is given only for the worse.

No doubt a simple-minded conflict between the "scientific" and the "traditional" is the wrong way to think about the relations of old and new; there must be a complex and reciprocal adjustment instead. But there was a time when the scientists were sure that the future of wine was in their hands, and perhaps they had good reason to think so. ■

AND HERE in *Wayward Tendrils Quarterly,* Pinney revisits the discussion of English-language wine words shown above, and provides an update (2011).

More than twenty years after I wrote a short appendix on the language of wine in English, our editor [of Wayward Tendrils] is giving me the chance to comment on what I said then. What would I add? Nothing, really. My prophecy about the "lively growth of novelty" in our stock of wine words looks like most prophecies—quite false. Or perhaps I have not been paying attention. At any rate, I can't think of anything to add, and I would be glad to hear of eligible nominations.

What would I change? A couple of things. The word "terroir," which I classified back in 1989 as one of those French words "still felt to be alien" is now, alas, part of every wine writer's word hoard. I say "alas" because I think, when it is properly understood, it means no more than "place," and why should one not simply say that instead of "terroir," which no native English speaker can pronounce anyway? Most of the time, the word is used to mean anything and therefore nothing. The plague of all writing on esthetic matters—such as all non-technical wine writing is—is the plague of nonsense. "Terroir" is a word that carries the plague. ∎

DURING THE last decade a number of very fine texts fully define *terroir* in all its dimensions. Whereas some earlier definitions focused solely on soil or geology, and others included climate, most did not explore the human element, the cultural niche. Here is a sample . . .

FROM PATRICK ILAND'S SERIES, *A TASTE OF THE WORLD OF WINE* (2009)
TERROIR

Terroir is a concept that connects the site (location) of a vineyard to the style and/or quality of the wine made from those vines. Although a French term, the concept of terroir can be applied to every site where vines are grown. The word terroir describes the characteristics of the site and refers to the interaction of the vine with its nearby environment—that space above and below the ground that affects the growth of the vine and how it ripens its fruit. The concept of terroir is best understood when sites are compared within a small geographical area. The comparison might be over just a few metres, or it might be over one or two kilometres. Comparisons made over large distances are best made using macroclimate differences rather than the terroir concept.

The Components of Terroir

Terroir can be thought of as the overall effect of several components on vine performance. We have chosen to think of terroir as having four components: aspect, canopy, soil and a human component.

The aspect component: Whether the vines are located on flat or sloping land—determines the amount of direct radiation that the site receives, and shelter from wind.

The canopy component: The degree of exposure of leaves and berries to sunlight and, as well, their nutrient status—determines how efficiently sugar is produced in the leaves and flavour compounds in the berries. Often some form of trellis is used to ensure that the leaves and bunches are placed in the most favourable positions for that site.

The soil component: Structure and fertility—determines how well the vine is supplied with water and nutrients, and this influences how efficiently sugar is produced in the leaves and flavour compounds in the berries.

The human component: Within a region, vineyard management practices, eg pruning, trellising and irrigation, may need to be modified to match different aspects and soil conditions. Over time, vineyard management practices evolve to achieve the best result at any one site. Therefore, it could be argued that because management decisions (made by humans) influence how the vine grows and ripens its fruit, a human component should be added to the terroir equation. One compelling argument for inclusion of a human component in the concept of terroir is that the decision to harvest is made by a person, often the maker of the wine. Because grape composition changes during ripening, harvesting at different times can have a large impact on wine style. The role of the human component remains a much-debated issue.

At some sites, all four components play a role in setting the terroir, while at others one particular component may be more important than the others.

Terroir can be thought of as a description of the characteristics of a site—its aspect, its soil properties and the shape and function of the canopy of its vines, as influenced by nature and the action of humans . . .

■ ■ ■

Every site has a terroir, but the terroir of some sites is better than others.
The terroir of each site gives the vines and their wine a sense of place.

In any one region, the best terroirs are considered to be those sites where the terroir components come together to create a pattern of ripening such that the grapes have, at harvest, the potential to produce the best wines from that region. This is why the owners of these special sites speak so passionately of their site's terroir, and why so much reverence is attached to sites which consistently produce great wines.

FROM JACQUES FANET, *GREAT WINE TERROIRS* (2004)
TERROIR: MYTH OR REALITY?
Enhancing the Natural Environment

What holds it all together [ref. to the French system of Appellations d'Origine Contrôlée] is that uniquely French notion of *terroir:* an umbrella term for a subtle interaction of natural factors and human skills that define the characteristics of each wine-growing area. There is no equivalent term in any other language, which may explain why it is occasionally misapplied. New World producers for their part strongly dispute any link between a wine and the soil that produces it. They say there is not a shred of evidence for such a claim and that the whole idea was dreamed up by French producers to keep out foreign competition. By restricting production to so-called reference areas, the AOC concept makes a wine more exclusive and more expensive as a result. You can see why New World producers think the way they do. When vines were first planted in the Southern Hemisphere, the rootstock was European and so was the know-how. The only thing colonial planters could not bring with them from the Old World were the soils. The settlers' first priority in any case was to find climate conditions compatible with vine cultivation rather than worry about the type of soil . . .

FROM DON GAYTON, *OKANAGAN ODYSSEY: JOURNEYS THROUGH TERRAIN, TERROIR & CULTURE* (2010)

Terroir . . . The French term refers to a group of wines made from grapes that share the same local ecology, climate and winemaking practices, all of which contribute to give them a specific local personality.

FROM ROBERT E. WHITE'S *SOILS FOR FINE WINES* (2003)
1.1 THE SOIL AND *TERROIR*

English has no exact translation for the French word *terroir.* But *terroir* is one of the few words to evoke passion in any discussion about soils. One reason may be that wine is one product of the land where the consumer can ascribe a direct link between subtle variations in the character of the product and the soil on which it was grown. Wine writers and commentators now use the term *terroir* routinely, as they might such words as rendezvous, liaison, and café, which are completely at home in the English language.

French vignerons and scientists have been more passionate than most in promoting the concept of *terroir* (although some such as Pinchon [1996] believe that the word *terroir* has been abused for marketing, sentimental, and political purposes). Their views range from the metaphysical—that "alone, in the plant kingdom, does the vine make known to us the true taste of the earth" (quoted by Hancock 1999, p. 43)—to the factual: "*terroir viticole* is a complex notion which integrates several factors . . . of the natural environment (soil, climate, topography), biological (variety, rootstock), and human (of wine, wine-making, and history)" (translated from van Leeuwen 1996, p.1). Others recognize *terroir* as a dynamic concept of site characterization that comprises permanent factors (e.g., geology, soil, environment) and temporary factors (variety, cultural methods, wine-making techniques). Iacano et al. (2000) point out that if the temporary factors vary too much, the expression of the permanent factors in the wine (the essence of *terroir*) can be masked. The difference between wines from particular vineyards cannot be detected above the "background noise" (Martin 2000). A basic aim of good vineyard management is not to disguise, but to amplify, the natural *terroir* of a site.

Terroir therefore denotes more than simply the relationship between soil and wine. Most scientists admit they cannot express quantitatively the relationship between a particular *terroir* and the characteristics of wine produced from that *terroir.* Nevertheless, the concept of *terroir* underpins the geographical demarcation of French viticultural areas: the Appellation d'Origine Contrôlée (AOC) system, which is based on many years' experience of the character and quality of individual wines from specific areas.

FROM ROBERT J. HARRINGTON, *FOOD AND WINE PAIRING: A SENSORY EXPERIENCE* (2008)

The idea of terroir is uniquely French in origin and a relatively new concept for wine makers in the New World growing areas. It generally reflects the unique interaction of natural factors (climate, soil, water, wind, etc.) and human skills that create definable characteristics in a specific wine-growing location.

FROM JAMIE GOODE AND SAM HARROP MW, *AUTHENTIC WINE* (2011)

The vine is more affected by the difference of soils than any other fruit tree. From some it derives a flavour which no culture or management can equal, it is supposed, upon any other. This flavour—real or imaginary—is sometimes peculiar to the produce of a few vineyards; sometimes it extends through the greater part of a small district and sometimes through a considerable part of a large province.

ADAM SMITH, *An Inquiry into the Nature and Causes of the Wealth of Nations,* 1776

We begin this pivotal chapter with an abrupt, provocative comment: most writing on terroir is simply nonsense, and many winegrowers take terroir far too seriously. But at the same time we believe that it is one of the most important concepts in wine. In fact, we go so far as to call it the unifying theory of fine wine. Our attitude seems a bit paradoxical: on the one hand, we are claiming that terroir isn't important; on the other, we are saying that it is. Here we'll try to explain why we are taking this seemingly absurd stance.

First, though, we must deal with definitions of terroir, a French term without an equivalent in English. On one level, terroir is a truism. Grapes grown in different places result in wines that taste different. Sometimes the effect is subtle; sometimes it is pronounced. It's hard to imagine anyone disagreeing with this rather simple concept. So terroir exists. But this is terroir defined in a very broad way; when wine writers and winemakers alike wield the term, they are frequently referring to somewhat different notions. Also, a concept framed so broadly is pretty much useless in practice. For discussions of terroir to make sense, we first need to constrain the definition to something a bit more specific. Then we need to try to reconcile the several overlapping definitions of the term terroir in common use.

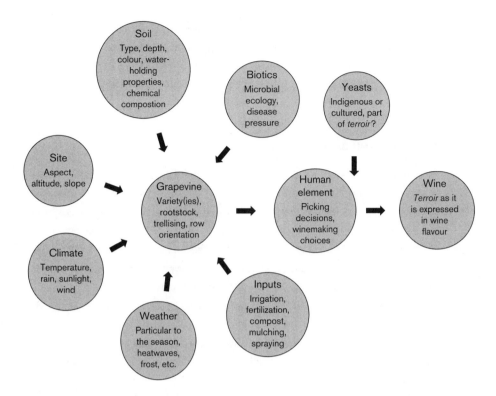

FIGURE 2.2
Influences that shape *terroir*.

The universal premise underlying the concept of terroir—the simple truism we've already described—is that vineyard differences can affect the flavour of wine. Take a single grape variety and plant it in three different spots. Handle the harvested grapes in the same way, and the wines will taste different. The differences are apparently attributable to the soil and subsoil properties of the vineyard site and to its local climate.

THE DEFINITIVE DEFINITION

It is time, at long last, for what may be the most comprehensive definition we've found, one that explores in minute detail the depth and reach of this remarkable concept. For this we turn to Warren Moran, a Professor in the Geography and Environmental Sciences Department of the University of Auckland in New Zealand, who in 2006 at the University of California Davis delivered a keynote address that addressed the concept in masterful detail.

It is a lengthy, comprehensive work. We'll be exploring just a fraction of it here and will save some facets of his definition for other chapters (his section on winemaking *terroir,* for example, appears in chapter 6; his discussion of marketing *terroir* appears in chapter 8).

FROM WARREN MORAN, "YOU SAID *TERROIR*? APPROACHES, SCIENCES, AND EXPLANATIONS"

In this presentation I interpret how and why *terroir* has, in the last decade, become the 'in' word within the wine industry. In the process it has often become banalised by attributing the distinctiveness and quality of different wines almost exclusively to the qualities of the natural environment where they are grown. This situation results from many commentators and even scientists using a very weak form of explanation of *terroir*—the oversimplified spatial association of wines with attributes of the physical environment. These attributes are often very coarsely defined, using broad terms such as climate, or soils, or geology. In many such discussions the functional (causal?) relationship between the natural environment and the species *Vitis vinifera,* let alone the wine made from its particular variety or varieties, is ignored. More scientifically reliable and sophisticated functional explanations are available. They derive from disciplines such as plant physiology, or, increasingly, molecular cell biology. But even these explanations constitute only a part of the concept of *terroir.*

When the characteristics of some wine are associated with some type of soil, or even worse the mapped geology of an area, and treated as if it is a valid explanation for its style or quality, the missing ingredient is almost invariably the complex intervening connections between the ecophysiology (*agro-terroir*) of the varieties of *Vitis vinifera* and the wine that results from them—including the empirical learning required to understand the environment. The word *terroir* is itself a social construction—its meaning has changed through time and will continue to change, especially as cellular biologists, chemists, and plant physiologists discover more about the real relationships between the processes in the environment that result in wines having a seductively subtle and intense association with place.

I argue that two important sources of such superficial approaches to *terroir* are the professional publicity for some French wine regions and the interpretations of the foundations of their appellation system by some wine writers. In many emerging (often loosely identified as 'New World') wine countries, these oversimplifications are perpetuated by being too readily adopted in the legislation defining the criteria used to delimit their wine localities and regions, now known internationally as Geographic Indications.

Running through the process is more than a touch of disciplinary hegemony. In some regions of France, for instance, geologists have captured the disciplinary high ground and tend to dominate the debate on *terroir* by claiming the influence of geology on differentiating the wine of one parcel of land, or commune, or locality from another. This is particularly the case in Burgundy, although I must also note how this tide is turning with the excellent work on the contemporary history of the vine and wine now emerging from the group working with Serge Wolikow at the Université de

Bourgogne, Dijon. Much less well known than the promotional approach to *terroir* is a rich French, and increasingly international, academic literature, deriving mainly from the humanities and social sciences, that stresses the influence of people, their political power, investment, and invention in crafting their wine regions by learning their environments and capitalising on their qualities. Its majestic figure is the French geographer and member of the *Collège de France,* Roger Dion, who wrote a series of provocative papers in the 1940s before integrating his ideas in the book first published in 1959 and entitled *Histoire de la vigne et du vin en France des origines au dix-neuvième siècle* (*History of the vine and wine in France from its origins to the nineteenth century*). Dion's work builds on a powerful thread of French scholarship that can be traced back to Vidal de la Blache who, in the first decades of the twentieth century, developed an approach to geography that emphasised the distinctiveness of the localities of France—*les pays*—and the manner in which locals had developed distinctive *genres de vie*—ways of life—that resulted in differentiated landscapes, from rural practices to housing styles and villages. Vidalian approaches have had a profound influence on the French social sciences and humanities from Braudel in history to recent (post-modern even!) theories of social interaction in economic development. Actor network theory, for instance, demonstrates a clear Vidalian connection.

People are responsible for the idea of *terroir.* Without people wanting to grow and manipulate some plant, in this case the vine, to ripen its fruit and to make wine from it, the word would not have the traction that it has achieved in the last decade. But the people who craft the ideas around *terroir*—and here I include both the practitioners and the researchers who grapple with understanding the vine and wine—tend to be a neglected part of the story. Unless we understand the importance of these two groups, and the way they work, we are in a very weak position to distinguish the validity of the arguments around the concept of *terroir* or even the possibilities of using it to add value to the wine that is being produced.

Successful wine enterprises, localities, regions, and countries result from patient and persistent work and learning by people in their environment. Although tempting, it is both practically and theoretically limiting to consider the people growing the grapes and making the wine as separate from the environment that they are learning to understand. It is their experience, day-by-day, year-by-year, decade-by-decade, and generation-by-generation that realises quality grapes and distinctive, fine wines. These people are not passive observers of the soil or the atmospheric environment. They are part of it. They work in it and with it. They select the cultivars, shape the canopy, pluck and trim the leaves, decide what fertiliser and sprays to apply, and help to balance the crop carried in the particular environment of each year by each vine. They do this to ensure a harvest of grapes that has qualities that they will be able to capture each year in their wine. And every year is distinctive. ■

MORAN THEN explores *terroir* from six interrelated facets: *Agro-terroir,* or the way in which *terroir* is expressed by the plant (and managed by man); *vini-terroir,* or the ways in which winemaking influences our perception of place; *territorial terroir,* where the associations with a specific territory can yield obvious commercial benefits; *identity terroir,* where the notions of territory are extended into the realm of national pride or privilege; *promotional terroir,* where the concept is exploited for its obvious promotional advantages; and *legal terroir,* where the notion of *terroir* is extended into legal terrain, as in the AOC system. We'll summarize, with excerpts, some of these ideas below.

AGRO-TERROIR

In describing this first facet, *agro-terroir,* Moran takes pains to point out that despite the obviousness of claims that a plant is the product of its natural environment, and that that environment is somewhat natural and unprocessed, nevertheless that environment is inevitably heavily managed by man—it is, in fact, a man-made environment: "The problem with taking the direct explanatory path from the natural environment to wine is that it omits the viticultural and winemaking practices that intervene," he writes.

> The vine is at the centre of all of these practices but they all involve decisions by people because the vine is a managed plant. These practices are themselves social constructions, as is the science on which they depend. And they both vary from place to place.

But among such relationships, Moran cites 'water balance' (*bilan hydrique*) as the one factor that best facilitates a plant's innate ability to express *terroir.*

> *Bilan hydrique* translates almost into 'water balance' but the French phrase captures more of the idea of the seasonality of water availability in relation to the phenology of the plant, in this case the vine. Its effectiveness lies in the combination of the qualities of soil (not geology) with the atmospheric environment in a manner that allows us to understand and monitor the progress of the vine towards producing grapes that reach physiological ripeness.

Moran refers to a "deceptively simple but powerful paper" by D. Martin, delivered at a recent symposium on cool climate viticulture, to illustrate how this balance is measured and achieved.

> The principle common denominator . . . between all great Chateaux is soils with physical properties that permit a *regular* but *gradually* limiting of water (and nutrient) supply to the vines in drier years.
>
> Site expression is when individual fermentation lots consistently demonstrate a certain specificity or originality . . . Terroir appears when site expression overrides varietal expression and interannual climatic variability.

Martin goes on to say that cooler climates are generally more liable to express *terroir* because they extend the optimum ripeness window; and that this might be the single most important factor in the making of a great wine.

VINI-TERROIR

As Denis Dubourdieu is fond of saying, *terroir* is about overcoming the disadvantages of the natural environment where you find yourself growing grapes for wine. Vinification and *assemblage* are one means of ameliorating, or even capitalizing on, such disadvantages, especially those of climate. The region Champagne is the ultimate cool climate success story and extreme example of this point.

Champagne, especially in its real home in France, is an innovative method of vinifying and assembling Pinot Noir, Chardonnay, and Pinot Meunier to make a distinctive wine. It capitalizes on the high acidity and aromas of these grapes that in many years would make relatively ordinary still wine. Part of its success lies in local people and enterprises selecting and developing over centuries cultivars, pruning methods, and canopy management that suit the Champagne style. Vines in Champagne yield at levels over four times higher than the same varieties grown in regions of France where still wines of quality are made from them (although often from different clones of Chardonnay and Pinot Noir).

■   ■   ■

Sauternes is another obvious example where an apparent disadvantage (moist atmospheric conditions prior to vintage) has been transformed into a wine of great distinction. Like everywhere else, the atmospheric conditions vary from vintage to vintage in Barsac-Sauternes (Roudié 1994). That vignerons are able to achieve reasonable quality in many seasons, and outstanding in some, is a considerable achievement in understanding how to manage the vine and its crop to ensure the highest probability that the berries are in fine physiological health when the weather conditions are ideal for the so-called "rot" to be noble.

TERRITORIAL *TERROIR*

Prior to the middle of the nineteenth century, when wines were formally classified, more overtly commercial methods were used to give advantages to those within the territory of Bordeaux. From as early as 1214 the first delimitation of territory occurred, when the *Bourgeois de Bordeaux* were exempt from customs duty on wine. Their wines were also given a commercial advantage over wines produced from vineyards more distant from the city. Such privileges lasted until the French Revolution. At the turn of the twentieth century, when the institution of a system of naming was being seriously considered to combat fraud, producers and negociants were again in conflict over the right to use the name "Bordeaux" for their wines. Not surprisingly, the appellation system that was first brought into law in 1919 ensured that, to some extent, the *rente d'appellation* could be shared among the land owners of each small communal territory.

In Burgundy the association of territory, wine, and the identity of settlements is most clearly seen in the naming of villages after vineyards. It began in the middle of the nineteenth century when most of the villages of the Côte de Beaune and Côte de Nuit took the names of their most prestigious vineyards. Nuit became Nuit-St-George, Aloxe added Corton to its name, and Gevrey became Gevrey-

Chambertin. When the formal AOC system was put in place, many of these names were designated as Grand Cru. Land owners within them have been privileged with a *rente d'appellation.*

The second strand of evidence for the association of territory and *terroir* is the development of the appellation system itself. The *Institut National des Appellations d'Origine* was formally founded in 1935, although its origins go back much further and deeper. Although the word *terroir* is also much older, it became more formally described in conjunction with words such as *typicité,* as the members and officials of the organization grappled with establishing their approach. Territorial control to combat fraud was widely discussed and implemented during and after the Phylloxera epidemic in the last three decades of the nineteenth Century, when the industry was in crisis.

Thirty four years later INAO was discussing *terroir* in much more sophisticated terms. These reflected the changing science of the period. *Terroir* now becomes:

> *A social construction within a natural space gifted with homogeneous characteristics, delimited on the cadastra, and characterised by a set of values—esthetic landscape values, cultural values of historical significance, patrimonial values of social significance, and values related to its reputation. (INAO 2000 as quoted in Hinnewinkel 2004)*

Evidence for more informal but similar processes from the emerging wine countries and their cultures is not difficult to find. People, in this case members of the grape growing and winemaking community, will go to considerable lengths to protect their territory. Some examples of the political use of *terroir* in the "New World" illustrate the point. The New Zealand trademark Gimblett Gravels, for instance, is not primarily about gravels. Its origin is a group of producers owning (or about to own) land in a locality, finding a way of publicizing their territory by defining it and attempting to convince journalists and consumers that it has the eco-physiological (*agro-terroir*) qualities to make great wines. The history of the definition of the Martinborough territory north of Wellington, New Zealand shows a diversity of reasoning to delimit its territory at different times, from lines joining trig points, to particular terraces, to the 800 mm rainfall isohyet, to everything within a 10 km circle of the town of Martinborough.

What are all of these groups defending? Not merely the natural environment where these vines are grown and the wine made, but in the European cases, often the centuries of effort that have gone into understanding how to grow grapes of outstanding quality in some years and make outstanding wine in some years. The "New World" producers are, like the Europeans, attempting to capture the increased income that comes from the reputation of wine that is grown in particular regions or on particular parcels of land.

IDENTITY *TERROIR*

The idea, concept even, of identity is difficult to separate from the idea of territory. In both cases the central association is between place and wine. The distinction between the two is land as an economic asset and place as a cultural experience. Roger Dion is undoubtedly the doyen of identity for the vine and wine in France. His words can equally well be read to illustrate territoriality, or indeed the whole humanist philosophy that increasingly underpins the French appellation system.

It suits us to see in the qualities of our wine regions, the effect of a natural privilege, of a particular grace accorded to the land of France, as if there were greater honour for our country to receive from the heavens than from the struggles of people this renowned wine industry in which our ancestors found a collective pride even before the feeling of a French nation stirred in them. (Dion, 1959, 8)

To geographers, an essential element of identity is the resolution or scale of its expression. In the above quotation Dion is talking about national identity, but he and many other writers also consider it at the local and regional level. When identity comes into play in defining territory it takes many different forms—some without fixed boundaries—such as neighbourhoods, rural communities or groups, or ethnicities, to provide a few examples. As discussed in the next chapter, enterprises (businesses) also set their own territorial boundaries, even though numerous national, state or provincial, and local authorities have precise delimitations of territory with which enterprises must comply.

ON THE OTHER HAND . . .

We give the last word to one of the wine world's more gifted satirists, Ron Washam, known by his nom de plume, the HoseMaster of Wine. No one weeds through bullshit better, and he does it here with aplomb, giving us a passably good definition in the process.

FROM RON WASHAM, *THE HOSEMASTER'S COMPREHENSIVE GUIDE TO WINE,* CHAPTER 8: "THE PROPER USE OF WINE TERMINOLOGY"
TERROIR

You can be forgiven for thinking "terroir" was Harry Waugh's older brother Terry, who certainly now smells of the soil, but you'd be horribly wrong. Terroir is a French word, used by wine connoisseurs, that has no meaning, and is interchangeable with the words "I have no idea what I'm talking about." For example, a wine lover might say, "This Chinon certainly shows fabulous terroir." Now that you have the insider information, you know that he's just remarked, "This Chinon certainly shows *I have no idea what I'm talking about."* Many people will imply that terroir is an expression that takes into consideration where the wine was grown, what soil it was grown in, the microclimate, the regional characteristics of the wine, the techniques used to produce the wine, and even the influence of the winemaker—as though there could be one word to express all that and have it make sense. Yeah, right. Well, there is one word for all that, and that word is bullshit. So, as another example, you read a winery marketing brochure and it reads, "Our winemaker's goal is to express our vineyard's unique terroir." Every lazy winery says this, as though other wineries are trying to express some other vineyard's terroir. This is why most wine marketing people are charter members of the Go !%#$ Yourself Club™. An advanced student of wine language understands this sentence to read, "Our bullshit winery script's goal is to express our vineyard's *I have no idea what I'm talking about."* Now you know that when you have no idea what you're talking about when it comes to a wine, yet you want to sound knowledgeable and educated about said wine, simply adopt your faux French accent and say confidently, "Wow, smell that terroir!" If someone asks you to define "terroir," do what wine experts do, look at them disdainfully, shake your head, and walk away.

NOTE

1. Taken from Amy Trubek, *The Taste of Place: A Cultural Journey into Terroir* (Berkeley: University of California Press, 2008) p. 23.

3

SOIL
The *Terre* in *Terroir*

The concept of *terroir* revolves around the sensory expression of a particular place, and the heart of that place is the land, the soil. Weather comes and goes, as do the vines and the winemakers, but except in the broad expanses of geologic time, the dirt stays put. The French word *terre*—earth, ground—is the root of the term *terroir,* but more importantly, soil serves as the basis for the entire conceptual framework linking wine and Nature.

Since plants are tethered to the earth, it's easy to understand how early thinkers assumed the existence of an almost mystical relationship between a plant and its host soil. For millennia, everyone from the local peasantry to a society's most esteemed authorities on agriculture viewed plants essentially as extensions of the soil they grew in, as leafy, upright manifestations of the earth below. Only at the beginning of the nineteenth century did scientists come to understand that plants grew by their own internal mechanism—photosynthesis—and not by way of a host of mysterious properties issuing from beneath the surface of the earth. (More on how vines work in chapter 5.)

But even without the preternatural connotations, the power of dirt is undeniable. Neighboring vineyards seem to produce recognizably different wines; vines growing in one soil type behave differently than vines in another, depending on type, depth, and composition. In a single vineyard plot, vines at the rocky crest of a hill will often produce fruit that tastes and feels completely different than fruit from vines grown at the bottom of the hill, where the soil is deep and gravelly. Even when climate, winemaking, harvest dates, and other factors are held constant—or as constant as reality permits—different patches of land make different batches of wine. Why in the world would this be true—isn't dirt just dirt?

By no means. The most orthodox advocates of *terroir* maintain that the soil and the parent rock beneath it are the source of the aromatic and flavor distinctions between fine wines. Wine importer Terry Theise, for one, is fond of likening a wine that stands out year after year to a clock: the land is the hour hand, and the vintage variation (owing mainly to climate) merely the minute hand.

Over the centuries, many explanations have been offered for how soil expresses itself in wine. Some believe that the soil literally comes through in the taste, the *goût de terroir* of special wines. Some emphasize the match of a particular soil type and a grape variety— that Riesling grown on slate, for example, is consistently different from Riesling grown on limestone or marl or loess. Some argue that it's not the chemistry of the soil, but the structure and drainage of it, the ways in which the soil provides just enough water at just the right time in the growth cycle; others point to the way heat is absorbed and released, depending on the soil composition.

Our survey begins with a classic French perspective on the role of soil from the *Larousse Encyclopedia of Wine,* and then moves to a richly detailed section from James Wilson's *Terroir,* an influential early foray into fine wine soil classification. Jonathan Swinchatt and David Howell walk us through the importance of the bedrock beneath the topsoil in *The Winemaker's Dance,* and geologist Kevin Pogue examines how the basalt undergirding the Columbia Basin in Washington state affects vineyard performance. In a seminal study, researcher Gerard Seguin describes the hydraulics of vineyard soils in Bordeaux. Unconventional but highly regarded soil scientist and *terroir* campaigner Claude Bourguignon makes the case for living soil. Finally, we'll conclude with some well-reasoned skepticism from Harold McGee and Daniel Patterson on whether soil character can find its way into the glass at all.

Since we assume that most of our readers, like us, are not soil scientists, the boxes and sidebars on soil topics are designed to help clarify the terminology in the selections below.

The traditional belief in the primacy of soil for the character of wine and the questions that confront that belief are captured nicely in this entry on "Soil and the Concept of Terroir" in Christopher Foulkes's *Larousse Encyclopedia of Wine.*

FROM CHRISTOPHER FOULKES, "SOIL AND THE CONCEPT OF TERROIR"

Does soil do anything more than hold the vine up? Even to ask such a question in France is utter sacrilege: the whole edifice of the *appellation d'origine contrôlée* system is founded on delineating the best soils. A Burgundy *Grand Cru* is better than a Burgundy *Premier Cru* because its soil is superior.

This view went unchallenged until California and Australia began to produce seriously good wines—from every kind of soil imaginable. French experts visited the New World and found that soil was not at the top of a winemaker's priority list. And there were no lines on maps to say that a certain kind of vine could be grown on this hillside, but not on the next.

New World winemakers look at a vineyard's water supply and temperature. If there is not enough
water from rainfall or the water table, they irrigate. If there is not enough warmth, or too much, they
plant a vineyard somewhere else. Many New World vineyards were recently sheep pasture or corn-
fields, and many will be again. It is hard to imagine the hill of Corton, or the gravel of the Médoc,
growing anything other than grapes, but in the New World there are no certainties.

Soil is thought to confer special flavours: every French or Italian textbook describes the way in
which the various minerals, clays, or limestones in a given soil affect the taste of the wine. This theory
has yet to be proved by analysis. Tasters can distinguish between wines from different vineyards. But
is it the soil they are tasting, or the myriad other factors that influence the grapes and the wine?

MAPPING THE GREAT VINEYARDS

The character of the soils underlying great vineyards has been a subject of discussion for
centuries, and careful geological analysis of important winegrowing regions dates back
several decades. But the scientific linkage of soil and fine wine received an enormous
boost with the publication in 1999 of petroleum geologist James E. Wilson's *Terroir*.
Though the subtitle—*The Role of Geology, Climate, and Culture in the Making of French
Wines*—portends a broad view of the factors that determine *terroir*, the focus of the
book rests on rocks, gravel, and dirt. For one major French wine region after another,
Wilson charts the geological history, the parent rock and the soils, and the correlation
between soil type and wine quality, all illustrated by extremely detailed maps, plates, and
drawings.

Wilson's unique geological insight and expansive knowledge of wine history made for fascinating reading, and the reception of *Terroir* upon publication was somewhere between highly positive and ecstatic. One reviewer after another declared that *terroir* had finally been put on a foundation as firm as the Earth itself.

Wilson's work greatly advanced the collective understanding of French regional vineyard geology. However, it did not directly address, let alone answer, the more important questions surrounding the concept. Did we mainly want to know the full story of the formation and nature of the soils of Burgundy, or did we want to know *how* and *why* those soils shape the aromas and flavors of those wines? If geological history had taken another course and swapped the soils of Burgundy and Bordeaux, what would those wines taste like today—and why? Clearly, knowing the *what* of vineyard soils is essential for understanding *terroir;* the task of connecting soil types to sensory experience, parsing out the mechanisms that make the character of one distinct from the character of another, is crucial.

The following excerpts come from Wilson's book, and examine a well-known streak of limestone running from the coast of England through some of the most famous wine regions of France—parts of Champagne, Chablis, and the Loire Valley—known as Kimmeridgian soils.

FROM JAMES WILSON, "THE KIMMERIDGIAN CHAIN: A BAND OF CHALKY SCARPS"

Like a set of nested dishes, the strata of the Paris Basin dip toward the center from all directions. The rim of one of these "dishes" along the southeastern part of the basin, the Kimmeridgian, is a remarkably uniform band of chalky marl, capped by a hard limestone called the Portlandian. For 200 miles (320 km), this classic caprock slope supports one wine area after another.

The wine areas are "islands" that are separated from the major regions with which they are traditionally associated: the Aube, 75 miles (120 km) southeast of the Marne Champagne; Chablis, the same distance north of the Cote d'Or; Pouilly-sur-Loire and Sancerre 80 miles (130 km) cross-country from Touraine-Anjou. Figure 1 (*Figure 3.1 here—eds.*) outlines the geology and geography of this arc of wine islands.

ENGLISH NAMES FOR FRENCH WINE ROCKS

The British Isles are a part of the Euro-Asian landmass, and southeastern England was at one time an arm of the Paris Basin. Downwarping during the Ice Age allowed the North Sea to flood the land bridge that joined England and France. The English Channel thus severed the English arm of the Paris Basin.

The French geologist Alcide d'Orbigny, working in southern England in the mid-18th century, named the massive Jurassic limestone on the Isle of Portland, Dorset, the Portlandian. The Isle is not in fact an island but a pendant-like peninsula where limestones have been quarried since the Middle Ages. Further eastward along the coast near Swanage, d'Orbigny named a dark marl below the Portlandian the Kimmeridgian *[named for Kimmeridge, a small town on England's Dorset coast—eds.].*

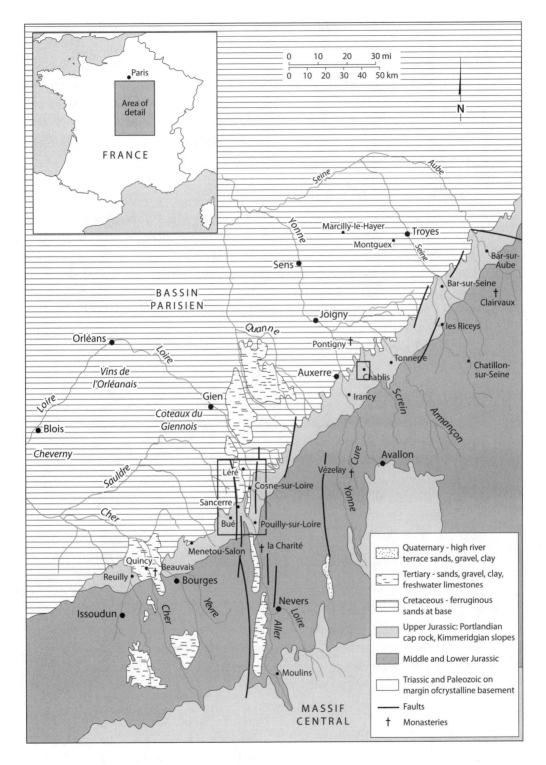

FIGURE 3.1

Map: Geology and geography of French wine regions, detail.

The marls in this area are petroliferous and when they were set on fire by lightning or spontaneous combustion were known as the "burning cliffs."

The Kimmeridgian of France is relatively uniform chalky marl and thin marly limestone containing many lenses or banks of sea shells. The fossils and fragments of frost-shattered Portlandian help aerate the slopes and aid drainage. The marly soils develop good structure and water retention characteristics and are easily cultivated. One of the fossils found in abundance is a small, comma-shaped oyster, *Exogyra virgule, virgule* being French for comma. In the Aube region, the abundance of the oyster in the upper Kimmeridgian gives it the name "Virgulien."

THE MAKING OF THE ARCHIPELAGO

The Cretaceous and Jurassic strata were deposited in widespread seas. Sagging of the central Paris Basin during the late Tertiary and Quaternary allowed erosion to fashion the concentric bands of ridges and plains illustrated in the Champagne chapter. The stream courses of the Seine, Aube, Yonne, and Loire were well established before the Paris Basin began to sag. The tilting was sufficiently slow for the rivers to be able to downcut through the rising ridges, much the way a buzz saw "eats" into a board. The rivers thus cut the Kimmeridgian-Portlandian outcrop band into an archipelago of wine areas.

■   ■   ■

CHABLIS

Chablis is the "big island" of the Kimmeridgian Chain. Situated about halfway between Paris and Beaune, a few miles east of the Autoroute du Soleil, the village is rather drab as wine villages go. But it is chief town of one of the most popular and most imitated wines in the world.

What makes Chablis so well known and highly regarded? To begin with, the Chardonnay, the grape of Chablis, is a perfect fit with the chalky marls of the Kimmeridgian. The crisp, greenish-gold wine is characterized by the distinct flavor of that grape.

. . . There are seven Grands Crus "climats," about a dozen and a half Premiers Crus, and very many commune grade Chablis. (*Climat* is a term used especially in Burgundy referring to a specific vineyard.) The roughly circular appellation is bisected by the Serein, a lazy, gentle stream, perfect for picnics and fly fishing. Drainage, largely by intermittent waterways, has eroded the low-dipping strata into an intricate pattern of slopes and ridges.

Facing the town of Chablis on the north side of the Serein is a classic caprock cuesta slightly over a mile long. This is the Grand Cru slope of Chablis. The inset map on Figure 1 (*Figure 3.2 here—eds.*) identifies the seven climats of this Grand Cru slope. The total area of the Grand Cru slope is 245 acres (100 ha), with Les Clos the largest at about 55 acres (22 ha). The Chablis climats are erroneously referred to as "Grands Crus," but technically it was the geologic slope that was classified Grand Cru, not the individual terroirs.

■   ■   ■

Geologic conditions identical to the Grand Cru slope extend both northwest and southeast, but the vineyards are classified only Premiers Crus. Experts are of the opinion that Premiers Crus such as

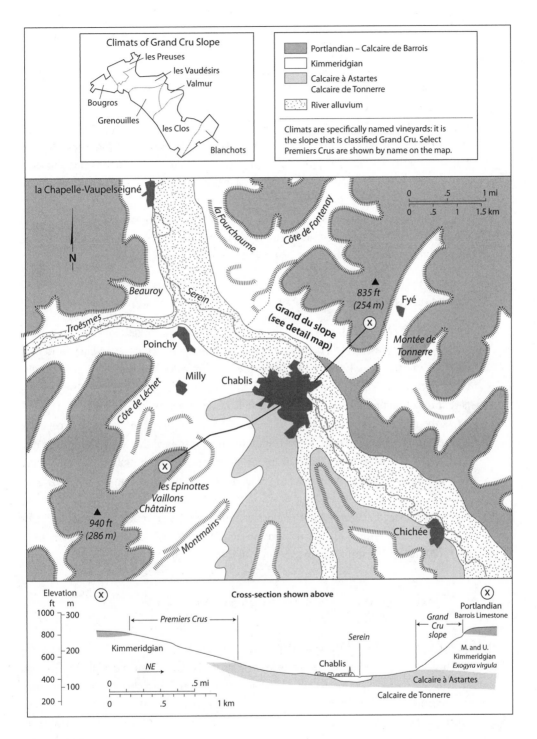

FIGURE 3.2
Map: General geology and cross-section of Chablis.

Montée de Tonnerre, Mont de Milieu, and Fourchaume are every bit as good as those of the Grand Cru slope.

On the opposite side of the Serein, long, Portlandian-capped finger-ridges extend northwestward. On the sunny, southeast side of the ridges are found the noted Premiers Crus of Côte de Léchet, Vaillons (Les Lys), Beauroy and Montmains (sometimes rendered as Monts Mains). The shady sides of the ridges support only sparse vineyards interspersed with other crops and woods.

It is not clear when vines were first brought to Chablis, but in the 9th century monks from St. Martin of Tours cultivated vines in the area. During the 12th century the Cistercians of the abbey of Pontigny transacted for about 50 acres (20 ha) for winegrowing in the area. They also obtained a building known as Le Petit Pontigny which served as a press room, dormitory, and religious house. The little nondescript house still stands in Chablis. The ecclesiastical presence was important, but according to Professor Gérald Gilbank, by the end of the Middle Ages, the vineyards of Chablis were largely in the hands of local families, many of whose ownerships persist to this day.

TOPSOIL AND BEDROCK

Jonathan Swinchatt and David Howell's *The Winemaker's Dance* looks at everything from geology to winemaking style in examining how winemakers in the Napa Valley work with what Nature provides to tease out the expression of *terroir*. In the early, geologically-focused chapters, the authors emphasize the importance of bedrock, the underlying parent rock that pre-dates and prefigures younger, looser vineyard topsoils.

As the authors explain, in Napa the bedrock is often very close to the surface, close enough for vines to penetrate. More important, the topsoils of the Napa Valley were created largely from the parent rock of the mountains that define its boundaries. Where Wilson's Kimmeridgian chain is an example of one soil type forming the basis for hundreds of miles of fine vineyards, Swinchatt and Howell portray the Napa Valley as remarkably diverse, composed of endless variations that provide the potential basis for many distinctive wines. Many vineyards straddle multiple soil types, creating the need for variations in viticultural practices and cellar treatments.

FROM JONATHAN SWINCHATT AND DAVID HOWELL, *THE WINEMAKER'S DANCE*

We begin with bedrock, the solid rock that is the foundation for everything else. It is the underpinning that supports the mountains and extends beneath the flat valley floor and from which we read the history of the region. In much of the Napa Valley AVA (American viticultural area), on the mountain slopes and ridges, bedrock is at or near the surface, covered by perhaps a few inches of loose material or by accumulations of forest debris. In some places, it forms impressive cliffs that tower over the valley.

Tom Burgess remembers planting his vineyard on the steep slopes of Howell Mountain by cracking the bedrock with dynamite to make room for vine roots. Ric Forman used a thousand sticks of

TALKING DIRT

Most of us get by with a simple understanding of the taxonomy of soil categories: there are big rocks, little rocks, gravel, sand, and just plain dirt. Geologists, of course, have a multitude of ways of classifying the Earth's materials, and many of these show up in discussions of *terroir*.

Origins and formation: **Magma,** the molten material under the Earth's crust, solidifies first as **igneous** rock; it can be transformed by heat and pressure into **metamorphic** rock (the dominant form), or weathered and eroded into pieces, moved by water and wind, combined with the detritus of plants and animal skeletons to form **sedimentary** rock. Over time, given the right conditions, any one of these broad types can combine into other rock forms, or be converted into soil.

Size: Soil particles are classified, in descending order, as **boulders, cobble** (small rocks), **gravel** (of various sizes), **sand, silt** (dust-sized), and **clay.** Clay particles are sized and shaped such that they are only found combined or clumped together, not as single units.

Texture: Although a given piece of land may contain rocks and gravel, soil texture is generally described in terms of its composition of sand, silt, and clay. The best mixtures for vineyards contain some of each. Soil that is all sand cannot hold water, and is therefore not much use for agriculture (picture desert sand dunes); soil that is all clay can harden to the point that water cannot penetrate at all (like roof-tiles, and terra cotta flower pots.)

Layers: The thin surface layer of ground is known as **topsoil,** often containing plant debris, pebbles, insect droppings, worm castings and other decaying organic matter. The **soil** beneath (sometimes called **subsoil**), from a depth of a few inches to many feet, is loosely-textured, decomposed material (some of it organic) in which plants initially root and from which they obtain much of their water and nutrients. Beneath this is **bedrock** (sometimes called a **substratum**), hard rock with little or no organic content. Vine roots may be able to penetrate the bedrock through cracks and fissures in search of water and nutrients. Since portions of topsoil and subsoil are composed of weathered bedrock, it is often referred to as **parent rock.**

Horizons: Both soil and bedrock can contain multiple strata, reflecting a change in lithology, or physical characteristics. When mountains rise, when hillsides fold or warp, when rivers carve valleys, basins or canyons, layers of rock and soil end up being exposed in pockets, islands, strips, and deposits; how erosion and decomposition have carved the hillside or exposed the strata frequently determines the vine's environment. In some cases, in Burgundy for example, the exposed middle strata may prove to be the most fortuitous for grape growing.

Topography: Soil analysis invariably involves more than just the sum of its content; its texture, location, and aspect are all important features as well. The elevation of a vineyard, the angle of its exposure to the sun, the presence or absence of nearby features that channel wind, and other topographical factors can have an enormous impact on vine growth and grape quality. (Since these influences are climatic, they are addressed in chapter 4.)

dynamite in his upper vineyard to break and dislodge the volcanic rock that covered a layer of river gravel. When Jan Krupp developed the 650 acres of Stagecoach Vineyards in the region of Atlas Peak, he ripped an estimated half million tons of boulders, many twelve feet or more in diameter, from what he then considered very rocky soil. In other words, in the hills, bedrock is close to the surface and very much a factor in planting and growing grapes.

The loose material that lies atop the bedrock, usually just a thin cover in the hills but a thicker layer in the valleys, is created as weathering processes break down the bedrock. Water, made slightly acidic by contact with organic material, works its way stealthily along tiny cracks, fractures, and grain boundaries, slowly dissolving minerals and loosening fragments. Solid rock is transformed, slowly but continuously, into loose particles—boulders, gravel, sand, silt, and clay.

Water and gravity transport this loose stuff down stream courses into the valley below, where it mixes with other material carried, in this case, by the Napa River. Slowly, the valley fills with these sediments that are constantly reworked by the river as it meanders, eventually creating a flat flood plain. Each of the valleys—Napa, Chiles, Pope, Wooden—has been filled in this way, by erosion of weathered materials from the surrounding hills. The chemical and mineral composition of the sediments in the valleys thus reflects the bedrock in those hills.

Geologic maps record the distribution of different types of rock, together with descriptions of its character and orientation (horizontal, vertical, at an angle). Although the geologic map of the Napa region (Figure 13) (*not reproduced here—eds.*) looks as if someone had splotched color on the DEM (*digital elevation model—eds.*) with all the skill of a kindergartner using poster paint, a closer look reveals a more sophisticated story. Even though the information on this map has been simplified from the much larger original, it is nonetheless obvious that the bedrock of the region shows considerable diversity. One of the map's most distinctive features is the general northwest-southeast orientation of color patterns. This reflects the structural trend—the orientation of the various geologic elements that form the bedrock. "Geologic structure" is the formal jargon for what we will call the rock architecture, the geometric relationships between the various rock units—Franciscan formation, Great Valley sequence, Napa volcanics—that underlie the region.

To picture this architecture more clearly, geologists rely on cross sections, slices through the uppermost part of Earth's crust. Picture a slice of cake, with all its layers lying atop one another evenly. In an area that has been geologically stable for the past few hundred million years—the region around the Grand Canyon, for example—a cross section looks remarkably like such a piece of cake, with layers of rocks stacked in an orderly fashion. In areas of mountain building, a geologic cross section looks more like a tiramisu or a trifle, one that has been jiggled a bit or compressed, the layers broken apart, cake mixing with fruit and cream in disorder, though still with a sense of the original components.

Historically, geologic cross sections have been the most economically important element in the process of mapping the distribution of geologic units, providing a view, however subjective, of the regions beneath the surface that might contain mineral wealth. These diagrams come in three levels of detail and accuracy: cartoons or drawings constrained only slightly by the facts; diagrams that are tied more closely to rocks exposed at the surface; and balanced cross sections, a quantitative methodology based on rock properties and our knowledge of how rocks bend and break. Balanced cross sections are especially useful in areas where the rocks have been pushed around by the forces that move Earth's crustal plates and are consequently highly deformed.

In California, particularly in the Coast Ranges, the rocks have indeed been deformed, pushed together by movement along the San Andreas fault system, which runs northwest-southeast, parallel to the structural trend of the rocks. Wrinkles in the crust—the hills and valleys that reflect folded and faulted rocks beneath—generally form perpendicular to the forces that create them. If you push a rug from one end, folds will form parallel to the end you're pushing. If you push from the corner, the folds will cross the rug at a 45-degree angle. Compression associated with the San Andreas system was directed perpendicular to the fault trend, leading to folds that parallel the fault.

A cross section of the Napa region indicates that the Mayacamas Mountains are an elongate upwarp, or anticline, a fold in which the beds on opposite sides dip away from each other. This geometry puts the youngest beds on the outside of the fold. The Napa Valley is a downwarp, or syncline, in which the beds on opposite sides generally dip toward each other, the youngest beds lying in the middle. Other valleys in the region have a similar origin, as downwarps in the crust. The geometry on the northeast side of the Napa Valley is more complex. There, a series of rock slices, each hundreds of feet thick, bounded by faults and dipping to the northeast, are stacked like so many shingles.

Throughout their length, the Vaca Mountains consist of primarily one geologic formation, the Napa volcanics. These rocks also cover the Mayacamas Mountains of St. Helena. They occur as a thin band along the base of these mountains from south of St. Helena to the Yountville hills, and as a thicker band from there to north of the city of Napa. They reappear as a thin band at the foot of the mountains in the Carneros District. Most of the hills and knobs that dot the valley floor are made up of Napa volcanics, which are the dominant rock type in the area bounded by the Mayacamas and Vaca Mountains, the heart of the Napa Valley AVA.

Wineries that produce from grapes grown on volcanic soils often imply that these soils provide some special quality to the grapes and wine. The dominance of volcanic rocks, however, likely has no particular significance. Throughout the world, great wines are made from grapes grown on rocks as disparate as limestone and granite; volcanic rocks are just part of the mix. But winegrowers in Napa have commonly overlooked one aspect of the Napa volcanics that, in fact, is of considerable importance: the degree to which they vary in character.

These rocks run the gamut of volcanic lithologies, including tuffs, lava flows, mudflows, pyroclastic flows, and stream deposits made up of volcanic particles. Each has a different texture and structure, and each breaks down into somewhat different granular by-products. Airflow tuffs may weather in place to soft beds of clay, whereas the harder welded tuffs that resist erosion form cliffs and ledges. Stream deposits composed of boulders and pebbles from both volcanic and sedimentary sources differ from hard beds of frozen lava. And although the chemical composition of these rocks is grossly similar, local variation in the balance of chemical elements may well influence the character of grapes grown on diverse types of parent volcanic material. The general term "volcanics" risks masking the true nature of these complex rocks and their contribution to the most fundamental character of the Napa region: the diversity of ground and climate that creates local microenvironments.

While volcanic rocks may be the primary bedrock components of the Napa Valley region, Franciscan mélange and Great Valley sequence rocks have been equal contributors to Napa Valley terroir in position and prestige, if not in volume. Franciscan rocks cover the central Mayacamas Mountains from Oakville to St. Helena. Few wineries are located in the Mayacamas Mountains on Franciscan formation rocks (Cain and Long Meadow Ranch are the exceptions), but this stretch of bedrock feeds debris to a portion of the valley that is redolent with the aroma of great wines produced over many vintages, particularly on the Rutherford Bench, at the foot of the mountains west of the town of Rutherford. The vineyards of Georges de Latour, Inglenook (now part of Niebaum-Coppola), and BV, together with a variety of small but prestigious growers, have thrived in this area.

Rocks from the Franciscan formation also appear in a broad band north of Lake Hennessey, underlying the highlands that mark the southwestern border of Chiles Valley. Ribbon cherts are found along the road leading north from the east end of the lake. These rocks, composed mainly of the shells of radiolaria, accumulate as gel-like substances on the floors of deep ocean basins. Under pressure, these gels slowly compress. As they do, clay minerals segregate from the silica-rich gel and concentrate as thin layers that form the ribbon-like character seen in these outcrops.

You might wonder about the presence of Great Valley sequence rocks in the Mayacamas Mountains, far west of the Central Valley—the trough in which they accumulated. Originally, the Great Valley sequence was deposited as a fifty-thousand-foot thick pile of strata that extended continuously for several hundred miles. Subsequently, the forces exerted by the San Andreas fault bent, broke, and shifted the western extremities of this immense pile. Great Valley sequence rocks that now lie west of the San Joaquin and Sacramento Valleys exist as displaced masses, slices, and layers of rock, separate from one another, shuffled together with rocks of the Franciscan formation and the Napa volcanics like cards at a blackjack table.

From Oakville south, Great Valley sequence rocks form the surface of the Mayacamas Mountains, except for a fringe of Napa volcanics at their base. Mount Veeder supports a number of wineries (Hess Collection, Mount Veeder, Mayacamas) and growers, but the rocks here have also fed some historically rich wine ground in the valley, including the Oakville Bench, home to the great To Kalon Vineyard, now owned mostly by Mondavi; Martha's Vineyard, just south of To Kalon; and John Daniel's property, Napanook, now Dominus. Fine-grained sandstones of the Great Valley sequence are exposed in outcrops at the top of Oakville Grade and down the western side. The bedding here is diffuse—myriad fractures break the rock into a pebbly-looking surface. Individual beds of Great Valley sandstone, a foot or two thick, formed almost instantaneously as mud slurries (turbidity currents) flowed into the deep ocean basin that fronted the Sierra Nevada volcanic arc. Each of these beds represents, literally, a few hours of time.

Great Valley sequence rocks account for much of the bedrock surface in the northeastern half of the Napa Valley AVA. They are found along the northern boundary of Chiles Valley and beneath the hills that are the southern boundary of Pope Valley, where they also appear along the lower edge of the northern border. In addition, the Napa Valley AVA contains other small areas of Great Valley sequence rocks, including part of the hill occupied by the Pine Ridge and Silverado Wineries in the Stags Leap District, as well as the hill on the opposite side of the Silverado Trail. (The knob at Stag's Leap Wine Cellars, however, is a volcanic plug, part of the Napa volcanics.)

SOIL TYPES

Soil types are categorized in many ways, and any single type may well have half a dozen alternate names in different classification schemes or regions of the world. The following wide-ranging list covers most of the soil types encountered in discussions of vineyard *terroir.*

Alluvial: Soil composed of clay, silt, sand, and gravel, formed over time from mineral deposits left behind by the movement of water.

Basalt: Hard volcanic rock, lower in quartz than granite but rich in iron, thus often reddish-brown.

Calcareous: Soils rich in calcium carbonate, often from fossilized shells; prominent examples are *chalk* and *limestone;* many of the world's fine wine regions have limestone subsoils.

Clay: Refers to soil mixtures containing large amounts of clay; solid clay is not a viable soil type.

Granite: Hard, igneous rock, containing substantial amounts of quartz; may be broken down into sand and gravel.

Loam: A crumbly mixture of clay, sand, and silt; depending on the proportions, can be described as sandy loam, clay loam, silt loam, etc.

Marl: Mixture of clay with calcium carbonates and fossilized shells.

Porphyry: Grainy, decomposed granite.

Sand: Granular decomposition of rocks, stones, pebbles, etc., of varying textures and fineness. Its composition is highly variable; like clay, sand rarely makes up the entirety of a vineyard soil, though soils with a very high sand content and good bedrock can support vines.

Sandstone: Silica and sand compacted by pressure and time; easily fractured.

Schist: A class of metamorphic rock derived from different types of parent material. It is usually foliated, meaning it can flake or scale off, and is characterized by the visible evidence of its constituent minerals—"mica" schist, for example, is heavy in mica; graphite schist is heavy in graphite content.

Shale: Fine-grained, multi-layered sedimentary rock, also known as mudrock, which can form beds of fragments on the ground surface.

Slate: Fine-grained, homogeneous metamorphic rock created from shale deposits. Like its parent material, it's often "foliated," meaning it shows layers.

Tufa: A light, friable variety of limestone; not to be confused with *tuff,* a porous, volcanic soil.

The rock serpentinite—its ominous name foreshadowing its equally dark presence—also exerts an important influence on Napa Valley terroir. The mottled green sheen of the mineral serpentine resembles that of a serpent; serpentinite, a rock dominated by serpentine, has the bite of its namesake, for its high-nickel and high-magnesium chemistry is deadly to grapevines. Serpentinite forms when high-temperature water alters ocean crust rocks. The only major body of serpentinite that intrudes on premier grape-growing land in the Napa Valley region lies at the edge of the Mayacamas Mountains south of St. Helena, in the area of Zinfandel Lane. This region is drained by Bale Slough, long known as difficult ground on which to grow grapes. Within the Franciscan mélange of the Mayacamas Mountains, serpentinite does occur in small scattered patches. A wide band of serpentinite runs from south of Lake Berryessa to the northern boundary of the AVA, and a thin band extends along the southwestern boundary of Chiles Valley. The Lake Hennessey area contains three small patches of serpentinite, one along the southern margin of the lake, interbedded with ribbon chert.

BEDROCK AND GRAPEVINES

Geologist Kevin Pogue has made the *terroir* of Washington state's Columbia River Valley his special focus for several years. Pogue's work traces the geological history of the Columbia Basin—highlighted by cataclysmic volcanic activity and, more recently, by massive primeval floods that dispersed soil elements hundreds of miles from their origins. The result is one of the country's most distinct growing areas.

In this paper, presented to the 8th International Terroir Congress in Soave, Italy, in 2010, he focuses on the particular role of the basaltic bedrock underlying much of the region. Basalt, he argues, impacts not just soil nutrient chemistry, but other, less tangible factors, like grape cluster temperature, clearly tying the composition of the soil to fruit character.

FROM KEVIN POGUE, "INFLUENCE OF BASALT ON THE TERROIR OF THE COLUMBIA VALLEY AMERICAN VITICULTURAL AREA"

Worldwide, the percentage of vineyards planted in basalt-derived soils is relatively small. Notable viticultural areas with soils developed in weakly weathered to unweathered basalt include the Canary Islands, the Azores, and Sicily's Mt. Etna. Regions that host vineyards planted in older or more deeply weathered basaltic soils include western India, southern Australia, Oregon's Willamette Valley, south-central France, northern Italy, and Hungary's Badascony region. The world's largest government-designated viticultural region with basalt-dominated bedrock is the Columbia Valley AVA, which encompasses over 4,500,000 km² (11,000,000 acres) of the states of Washington and Oregon. The basalt was erupted during the Miocene Epoch from volcanoes associated with the hot spot that now lies beneath Yellowstone National Park in Wyoming. The Columbia Valley AVA presently contains over 2,700 hectares (*that number is now 16,000, or nearly 40,000 acres—eds.*) of vineyards that are located primarily on gentle slopes or on valley floors below 400 meters in elevation. Almost all of the vineyards are planted in loess derived from the deflation of sediments deposited by a series of catastrophic Pleistocene glacial outburst floods, known as the Missoula floods. At elevations below 330 meters, the loess commonly overlies sand, silt, and gravel deposited by the Missoula floods, while above this elevation the loess directly overlies basalt bedrock. Despite its age, the basalt is not deeply weathered due to the combined effects of the region's arid climate and the protective mantle of flood- and wind-deposited sediment. The most obvious effects of weathering are fracture networks filled with calcium carbonate and iron oxide-stained clays within the uppermost one meter of the basalt. The soils in most Columbia Valley AVA vineyards were derived by the glacial and fluvial erosion of regions dominated by granitic and metasedimentary bedrock that lie north and east of the Columbia Valley AVA. They are therefore rich in quartz, muscovite mica, and potassium feldspar, minerals not present in the underlying basalt bedrock. Since the thickness of these soils generally exceeds the rooting depth of the vines, basalt has had, until recently, almost no influence on terroir.

Over the last 10 years, viticulture in the Columbia Valley AVA has rapidly expanded. Vineyards have recently been planted at higher elevations, on steeper slopes, and in rocky, alluvial soils. The soils in these vineyards are commonly much thinner than those of the traditional valley floor sites, and therefore vine roots are able to directly interact with basalt bedrock or basalt-derived alluvium or colluvium. In preparation for planting, the thin soils are often mechanically ripped to a depth of 0.5 to 1 meter to increase water holding capacity and available rooting depth. The ripping process crushes the upper parts of the weakly weathered basalt bedrock and incorporates fractured basalt and basalt weathering products into the overlying sediments, significantly altering their mineralogy and chemistry. The introduction of basaltic minerals into soils derived from a granitic parent should increase the concentrations of elements that are present in higher concentrations in basalt, such as iron. Iron is an important nutrient for grapevines and unlike most elements, the concentration of iron in grapes and vineyard soils has been demonstrated to be directly related. Iron concentrations in musts, which vary according to soil iron content, have been shown to affect the stability, clarity, and color of wines.

The incorporation of fractured basalt by ripping also significantly alters both the texture and color of soils derived from fine-grained light-colored loess. In some recently planted Columbia Valley vineyards, ripping has produced soils with a very high ratio of rock to loess, and basalt is exposed over a significant percentage of the ground surface. The physical properties of these basalt-rich soils are very different from the loess-dominated soils that are typical of most Columbia Valley AVA vineyards. Unlike the highly erodible loess-dominated soils, the rocky coarse-textured basalt-rich soils require no cover crop. Relative to a vegetated ground surface, bare soil rich in basalt should absorb, store, and radiate more heat and conduct heat to deeper levels of the soil more effectively. Winegrowers have long recognized the thermal properties of basalt. In Germany's Forst region, it has even been imported to warm vineyard soils.

MATERIALS AND METHODS

To test the influence of basalt on the chemistry of Columbia Valley AVA soils, samples were collected from diverse sites that typify the range of its involvement. Soil samples were collected from: 1) ripped alluvial soils rich in basalt cobbles, 2) ripped alluvial soils with scattered basalt cobbles, 3) steep hillsides with thin loess and basalt colluvium (unripped), 4) steep hillsides with thin loess and basalt colluvium (ripped), and 5) gently sloping topography with thick loess. In addition, as a control, an artificial soil sample (sample #6) was created by crushing unweathered basalt. All samples were sieved to <1mm particle size. The availability of iron in all samples was determined by a commercial soil laboratory using diethylenetriaminepentaacetic acid (DTPA) as an extractant.

To measure the thermal effects of a basalt-covered ground surface, temperature data loggers were inserted into the interiors of grape clusters in two vineyards located 2.5 km apart. The surface

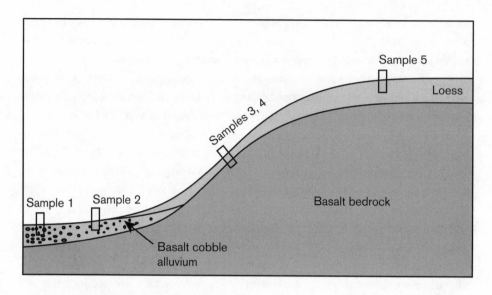

FIGURE 3.3
Diagrammatic cross section: Basalt bedrock and soil.

of one vineyard is covered almost entirely by basalt cobbles while the other by a combination of dry, grassy vegetation and brown, loess-based soil. Data loggers were placed in 4 clusters in each vineyard. Clusters were selected to be approximately 0.5 m above the ground surface and shaded by leaves from direct sunlight. Data loggers were also buried midway between two rows in each vineyard at depths of 5 cm and 25 cm. The ambient air temperature in each vineyard was recorded by a radiation-shielded temperature data logger positioned 1.5 m above the ground. Data were collected at various times during July and August of 2007.

RESULTS AND DISCUSSION

This study measured the concentration of available iron in each sample in parts per million. As expected, sample #5 from the thick loess soils showed the lowest concentration (9 ppm). Relative to the other soils, these soils contain virtually no basaltic component. The highest concentration of iron (30 ppm) was measured in the alluvial soil with scattered basalt cobbles (sample #2). Being farther from the main stream channel, this soil is older and more deeply weathered than its cobble-rich counterpart (sample #1), which had a 37 percent lower iron concentration (19 ppm). The sample of unripped thin loess-based soil from a steep hillside (sample #3) contained 16 ppm iron, reflecting minor colluvial input from the basalt bedrock. The 44% increase in available iron in sample #4 (23 ppm) relative to sample #3 is likely related to the incorporation of weathered basalt by mechanical ripping. Sample #6, the artificial soil created by crushing unweathered basalt, had an iron concentration of only 14 ppm, which emphasizes the critical role of weathering in the production of plant-available iron.

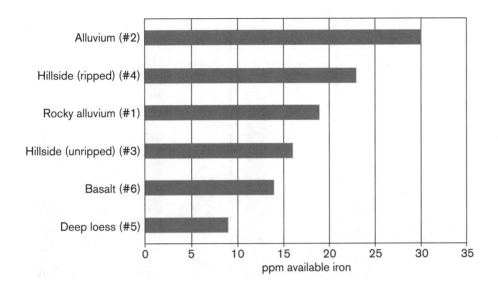

FIGURE 3.4

Chart: Available iron in each sample.

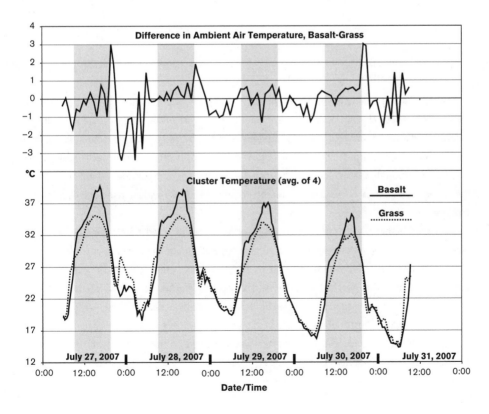

FIGURE 3.5

Grape cluster temperatures and difference in ambient air temperature in basalt- and grass-covered vineyards.

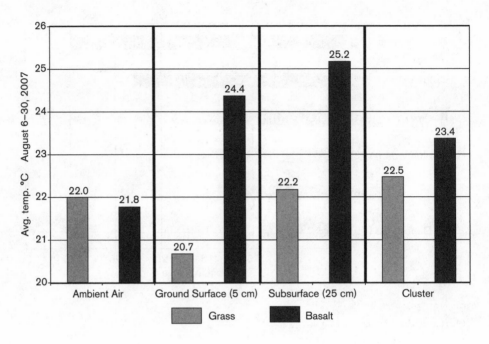

FIGURE 3.6

Chart: Comparison of average temperatures in basalt- and grass-covered vineyards.

CONCLUSIONS

The terroir of some Columbia Valley AVA vineyards is significantly influenced by the chemical and thermal properties of basalt. Vineyard soils in the Columbia Valley AVA that incorporate basalt bedrock or basalt alluvium show substantial increases in available iron relative to the more widely planted loess-based soils. The largest increases are observed in older alluvial soils and in mechanically ripped soils that incorporate weathered bedrock. Since the iron content of grapevines is directly related to the availability of iron in vineyard soil, increased iron should also be evident in the grapes and wines produced from basalt-rich soils.

Vineyards within the Columbia Valley AVA covered by fractured basalt bedrock or basalt-rich alluvium have higher average ground surface and subsurface temperatures than their grass-covered counterparts. From late morning to early evening, grape clusters in basalt-covered vineyards are heated to higher temperatures than clusters in grass-covered vineyards. The extra heat is derived from infrared radiation from the sun-warmed dark-colored basalt, not from conduction from heated air. Cluster temperatures within the basalt-covered and grass-covered vineyards rapidly equilibrate near sunset. No evidence was observed of the oft-cited ability of surface stones to store heat and release it after sunset, at least not to the above ground part of the grapevines. Due to advection, vineyard surface material appears to have little effect on ambient air temperature.

TERRA NOT SO FIRMA

Wild grapes can show up without any human intervention; vineyards cannot. Planting vines of the same grape variety in nice, straight rows isn't the only man-made alteration; for centuries, winegrowers have labored to change the soil itself, often in dramatic ways without which notable *terroirs* would likely not be so notable.

Although Bordeaux was already a winegrowing region in Roman times and already beloved by the British in the 12th century, most of its production came from areas regarded today as second-tier—Bergerac, Gaillac, Entre-Deux-Mers. Only with the assistance of Dutch engineers in the late 17th century were the swamps surrounding the Gironde River drained, revealing the beds of limestone and gravel that now comprise the Medoc. In the St-Émilion district of Bordeaux, underground pipes still provide soil drainage that prevents vine roots from being waterlogged. Without these immense feats of human engineering, the First Growths of Bordeaux would not exist.

In Champagne, the thin chalk soils have been enriched for centuries by additions of lignite ("black ash"), a low-grade form of coal that contributes minerals necessary for vine health. On the prized slopes of the Cote d'Or in Burgundy, erosion steadily sends topsoil to the bottom of the hills, and the *vignerons* dutifully haul it back up to the top. The steep slopes of the hillside vineyards of Germany's Mosel similarly see their loose slate soils slide ever downwards, only to be collected and carted, often by foot-power, back uphill.

Active human involvement in modifying vineyard soils, in other words, is a longstanding tradition of winegrowing, and an important contribution to the *terroir* of a place.

WHAT THE SOIL DOES, NOT WHAT IT IS

Perhaps the single most influential scientific examination of *terroir* came in a 1986 article in the journal *Experientia* by researcher Gerard Seguin of the University of Bordeaux, with the less-than-riveting title "'Terroirs' and Pedology of Wine Growing." Seguin adopts as his criterion for a great vineyard site its ability to deliver excellent wine from year to year, good weather and bad, drought years and monsoon years, under conditions that overwhelm lesser vineyards. Seguin is not attempting to explain a particular *terroir;* rather, he is looking for the unifying characteristics of high-quality sites. He then surveys and summarizes a broad body of research on the soil characteristics in fine wine regions, observing that vintage after vintage, excellent wine comes from a wide variety of soil types and chemistries, concluding that soil type must not be the sole explanation.

Searching for what might unite disparate soil *terroirs,* he argues that two factors stand out as most important: soil structure and water availability. In an examination of two prominent growing areas in Bordeaux with very different soils—the Medoc and St. Emilion—his research demonstrates that each solves the problems presented by climate variation in different ways—but with strikingly similar results, thanks to structure and water.

FROM GERARD SEGUIN, "'TERROIRS' AND PEDOLOGY OF WINE GROWING"

At the moment, not one single soil constituent or element may be said to be an absolutely decisive factor in wine quality. The only one which is generally associated with wine quality is active calcium carbonate; this is not surprising since many quality vineyards are situated on calcareous parent material, and the abundance of calcium is a well-known factor in the structural improvement of soils. It has a direct bearing on their physical properties, but its presence in soil is not indispensable since there are many quality vineyards which are situated on acidic soils, where there is absolutely no active calcium carbonate (Grands Crus Classés in the High-Medoc for example).

■   ■   ■

INFLUENCE OF GEOLOGICAL AND PEDOLOGICAL FACTORS

On a world scale, excellent quality wines are produced on the most diverse types of geological formations: schists (Porto, Moselle), chalk, limestone, marl or sandstone containing different amounts of active calcium carbonate (Champagne, Bourgueil, Chinon, Chablis, Saint-Emilion, Burgundy; Jerez, Rioja; Barolo, Barbaresco, Chianti, Marsala; Rheingau, etc.); clay (certain crus of Pomerol, Sauternes, etc.); sand (Nebbiolo d'Alba); schist granite and porphyry (Beaujolais).[1][2][3][4][5][6][7]

In the Bordelais, no geological formation (quaternary siliceous pebble-sand but also tertiary limestones and sandstones, clays . . .) can be said to be the best in quality. Moreover, on the same parent material, like the pebble-sand alluvials of the left bank of the Garonne, one can produce the excellent red wines of the Medoc and a part of the Graves, the dry white wines of the Graves and the sweet white wines of the Sauternais.

■   ■   ■

INFLUENCE OF SOIL TEXTURE AND STRUCTURE

A few examples chosen from the best Bordelais crus show that as in Burgundy[8] and in other wine producing areas, the quality of wine does not seem to be related to a definite textural type, since in these terroirs one may note considerable variations in the gravel and pebble content (from 0% to more than 50%) and in clay content, which may be almost negligible in certain soils, but may reach 60%, as in the best Pomerol cru.[9][10]

Soil structure seems to play a much more important role. In the majority of cases, the best terroirs are characterized by a high degree of macroporosity, permitting rapid water percolation, and thereby preventing stagnation at root level. Coarse soils (gravel-sand) are of course permeable; however, clayish soils are only porous when there is sufficient humus (at least 10% of clay content) and calcium in abundance to flocculate the clay-humus complex (one must remember that many quality terroirs are situated on parent material containing active calcium carbonate). Apart from the high degree of permeability and aeration which characterizes them these soils are easily penetrated by the roots of the vine and are very resistant to erosion, since the particles are bound by a clay-humus cement.

The macroporosity of the soil and of the parent material sometimes enables roots to attain a considerable depth; 5–7 meters for the pebble-sand soils of Medoc, Sauternes and Graves of the Bordelais,[11][12][13][14] and 4 meters through the diaclases of the 'albarizas' of Jerez.[15] In certain clayish soils, and especially in soils situated on compact non-karst limestones, root depth is sometimes limited (30–70 cm) in Burgundy[16] and Saint-Emilion.[17] The depth and the mode of colonization of the soil by the roots have repercussions of course on the mineral nourishment and the water supply to the vine.

■   ■   ■

Excellent wines are produced on acid, alkaline, or neutral soils, or again, on soils where the chemical constitution is balanced; others, however, are characterized by a deficiency in certain elements and important ionic disequilibrium.[18] Without doubt, the soils of the Premiers Grand Crus Classés of the Medoc are on average richer in organic matter (therefore in total nitrogen), in assimilable potassium and phosphorus but it is not because they are richer in these elements that they are Premier Grands Cru soils. On the contrary elements have been added to these soils because they belonged to very great crus where the owners have always taken the trouble to look after their soils. Indeed, these differences are evident on the surface but not at depth (beyond 60 cm) in those layers which are only slightly affected by chemical additives.[19][20] As our knowledge stands at the moment, it is impossible to establish any correlation between the quality of wine and the soil content of any nutritive element, be it potassium, phosphorus or any other oligoelement. If there were such a correlation it would be easy, with the appropriate chemical additives, to produce great wine anywhere.

WATER SUPPLY TO THE VINE

Factors that we have looked at so far (parent material, type of soil, texture, chemical properties, etc.) certainly have a bearing on the character and type of wines, but none of them seems to have a really deciding influence as regards quality, i.e. the pleasure one obtains on tasting or drinking wine.

Another method is to study the water supply to the vine; this does seem to have a bearing on the quantity and the quality of the harvest and the approach integrates numerous factors:

- climatic factors: rainfall, sunshine, temperature, relative air humidity, which together determine the evaporating power of the air, i.e. the potential evapotranspiration;
- edaphic factors: parent material, topography, water table, structure, and root arrangement, reserve of usable water, permeability, aeration and so on;
- biological factors: nature of cultivar and rootstock;
- human factors: techniques of soil upkeep, drainage, irrigation; training of the vine and particularly plantation density, trellising method, pruning, topping, etc., which result in a system in which water may evaporate.

■   ■   ■

THE WATER SUPPLY TO THE VINE IN THE GRAVEL-SAND SOILS OF THE HIGH-MEDOC

The study was carried out with a neutron moisture meter, which provided hydric profiles expressed in volumic humidity in relation to soil depth. The principle of the apparatus is the slowing down of neutrons by hydrogen in the water; the flux of thermic neutrons measured thereby is proportional to the water content of the soil.[21] As the measurements are made in a tube which remains permanently in the soil, it is possible directly to determine the humidity in the soil in situ, always at the same levels and at very great depths. We have therefore been able to study for the first time the mechanism of water supply in these soils where, in spite of the depth, the reserves of available water are relatively limited.

An interesting peculiarity of the majority of the soils of the Grands Crus Classés of the High-Medoc is the presence of a free water table situated within reach of the roots, and whose level goes down progressively from the spring to the beginning of the autumn. During the first part of the vegetative cycle, when the vine is in full growth, the radicles develop in the "drop-away" zone of the water table. They absorb the water which is stored in the micropores of the upper layers of the soil; this water is more or less replenished by rainfall. The radicles also absorb the water which is left in the soil by the water table as it drops away. Finally they have at their disposal water which is easily usable, which they absorb at the upper limit of the capillary fringe; the latter is a sufficiently aerated zone since the water content there is equal to the field-capacity (the micropores are full of water, while the macropores are full of air).

However, from August onwards, the vine stops growing and the dropping-away of the level of the free water table, which continues, is no longer compensated by a corresponding growth of the radicles. In these very coarse texture gravel-sand soils, where the height of capillary rise is limited to 30 or 40 cm, the role of the water table diminishes progressively. The water supply to the vine during grape maturation depends on rainfall, but also on the quantities of water stored in the soil at the moment of "veraison" (the point at which the berries turn from green to red but are still not ripe, i.e. the beginning of maturation).

In light of this, and particularly of the depth of soil exploited by the roots, it is not surprising that the old, deep-rooting vines resist drought well; this is true even during grape maturation when the water table no longer plays a role in the water supply to the vine. On the other hand, superficially rooted plants, particularly the young ones, suffer from a lack of water when the summer is a dry one.

However, when rainfall is particularly heavy, the vines are not very sensitive to excess humidity since the soils are characterized by a remarkably high degree of permeability. This is due to the coarse texture of the parent material which favors drainage, but is also due to the topographical location of the Grands Crus, on "croups" (small hills) next to the Gironde (Garonne estuary) and its small tributaries; these conditions facilitate the evacuation of drainage water. The hydric profiles recorded after heavy rainfall show that the rain water filters through very rapidly. The water content recorded 24 hours after heavy rainfall only rarely and temporarily exceeds those values corresponding to the field capacity, whereas in other types of soils which are rather impermeable and are badly drained, the roots would be soaked.

It must also be noted that in summer, water from rainfall remains localized in the upper layers of the soil; consequently plants with superficial roots are the most affected by rainfall, whereas, as we have just seen, it is these very plants which are the most sensitive to drought.

In the High-Médoc, but also in the majority of Grands Crus Classés situated in the north of the Graves region, soil permeability and root depth limit the well-known effects of excessive rainfall during grape maturation: a rush of water into the berries causing them even to burst and provoking the onset of common rot, the dilution of the sugars, and strong acidity. There is a much greater chance of harvesting mature and healthy grapes than in other wine-producing areas. In fact we have shown that the absorption of water (calculated by hydric profiles) after summer rain is clearly higher in soils with superficial roots, and that in consequence, the percentage of berries that burst and rot is much greater when the depth of the roots is shallow.[22]

Therefore in these soils where the old vines are deep-rooted, there are regulation phenomena which limit the effects both of extreme drought and heavy rainfall and which provide a hydric pattern not much different from one year to the next. The water content of such soils varies only slightly whatever the level of rainfall, unlike soils where the roots are shallow. Thus, the Grands Crus still produce quality wine when the months of August and September are marked by extreme climatic conditions (such as long drought or heavy rainfall). It is also clear why the best wines are produced by vines of a sufficient age (about 8–10 years in Medoc) which, having had the time to develop their root system in depth, thrive on these regulation mechanisms.[23] [24] [25]

WATER SUPPLY TO THE VINE IN OTHER TYPES OF TERROIRS

Given these conditions, the question may be asked whether similar mechanisms exist in other types of terroirs, particularly in the limestone soils of Saint-Emilion, where the roots are superficial and where drought could pose the vine serious problems.

Studies carried out since 1978 by Duteau et al.[26] have shown that the field capacity expressed in volumic humidity is about 25%, but that, in the light of the very shallow soil depth and the fact that the roots do not go beyond 70 cm, the reserve of usable water is low. However, during a dry summer the water supply is never too low since the parent rock, which is composed of very compacted limestone can, by capillary action, provide up to 35% (70% in 1985) of the water consumed by the vine between flowering and harvesting; this is quite considerable.

Similar questions may be asked about certain clayish soils of the Pomerol area; unlike the case with limestone soils, excessive water supply may be feared.

Such soils have a clay content as high as 60% and are poorly aerated. Thus the root system remains localized in the first meter of soil. Moreover, in these more or less asphyxiating conditions, many radicles die each year, as has been observed on cultural profiles. However, these shortcomings are apparent only in this particular case. The field capacity expressed in volumic humidity is situated between 25% and 45%, which provides good but not excessive water supply, since only a few roots absorb it, and this at a shallow depth. Moreover, the mobility of the water, which in clay is already low, diminishes progressively during the drying-out of the subsurface layers, with the result that water reaches the vine only with difficulty during grape maturation.

■   ■   ■

CONCLUSION

In-depth study of all the factors of the climate-soil-vine ecosystem is difficult; each has its own action but acts in synergia with or opposition to the others. The vine undergoes environmental constraints (climate, chemical, and hydric state of the soil) but, like most living things, adapts and reacts, sometimes even modifying its environment.

■ ■ ■

In the areas where the climate is fresh or temperate, the best crus are those in which the grape ripens completely but slowly; such crus regularly produce high quality wines in spite of climatic variations. In this respect, the soil intervenes by limiting the climatic and particularly the hydric extreme conditions; the best crus would seem to be those which, because of their topography, physical properties (structure, macroporosity, microporosity) and root system development, limit the effects of either heavy rainfall or drought. Then again, there may be natural factors which the hand of man cannot easily modify; this is doubtlessly why a Grand Cru cannot be situated just anywhere on any terroir.

LIVING SOIL

Just as wine cannot be reduced to its chemical profile, soil is a complex reality that resists simple characterization. Since 1989, French soil researcher Claude Bourguignon and his wife Lydia have operated the Laboratory of Soil Microbiological Analyses (LAMS), advising the owners of some of the most famed vineyards in France on how to improve their soils and let the underlying *terroir* shine through in their wines.

Although Claude Bourguignon has a solid academic background, he left the world of mainstream plant science to embrace a broader view of viticulture and geology, stressing the importance of living soil, the dangers of chemical farming, and supplementing a scientific approach with traditional ideas and practices. The Bourguignons' writings don't appear in standard peer-reviewed journals, and their books are hard to find in the French originals, let alone in English. While many of the Bourguignons' theories are considered controversial (Claude once compared the microbial life in the soils of Burgundy with that of the Sahara) their reputation among elite growers and within organic-biodynamic farming circles is formidable.

As the following excerpt from a 2010 article in the international wine journal *Tong* shows, the Bourguignons part ways with Gerard Seguin on whether soil *type* matters as much as *structure* and *drainage*. They argue that the mineral composition of a particular soil is reflected, if indirectly, in the aromas and flavors of the resulting wine (a topic we address more directly in chapter 7). The following also touches on two other concepts important to the Bourguignons: the necessity for all the life forms underground to have access to oxygen, and the importance of insects and microbes in truly healthy soil.

FROM CLAUDE AND LYDIA BOURGUIGNON, "SOIL SEARCHING"

Our laboratory has carried out a scientific study of the biology of soil that sheds new light on terroir and will help develop viticultural practices. It is soil and not vinification that prevents wine-makers

GOOD SOIL, BAD SOIL

The range of soils and subsoils that support high-quality grape growing around the world is tremendous, but there are a few criteria that nearly always have to be met. They are a lot like the criteria guiding Goldilocks in the home of the three bears: not too much, not too little, just right.

Water availability and drainage: Good vineyard soil provides enough water for the vines, but not too much, and particularly not too much during the final stages of grape ripening. Too little water can injure a vine severely and desiccate the fruit; excess water can rot roots and produce diluted fruit. Some sites come by this balance naturally, some through controlled irrigation.

Porosity and aeration: Vine roots need to be able to penetrate the soil and possibly the subsoil in search of water and nutrients; roots need a certain amount of air in the soil, as do the microbial and insect life. Sandy soils that are too porous can lose essential water and nutrients; soil that is too loose, with no cohesive strength, cannot support vines at all.

Fertility, nutrients, and organic matter: Soil fertility, provided by organic humus and mineral nutrients, is important for vine health. But excess fertility—including too much nitrogen and potassium—can result in excess vine vigor, huge canopies of foliage, and thinly-flavored fruit. Good tomato soil is not good vine soil.

The vast majority of the material in soil and bedrock is inorganic, mineral compounds of various kinds. But viable vineyard soil should include some organic matter—either from the original soil or from amendments with compost, etc.—that eventually breaks down into stable humus, an important source of vine nutrients. Critical inorganic compounds like mineral nutrients only make their way into plant roots through the action of living soil microbes, mycorrhizae, which serve as go-betweens linking root hairs and soils. Earthworms play an essential role in breaking up and aerating soil. (More on soil organics in chapter 5.) While discussions of soil often focus on minerals, they aren't the whole story.

The right combination of these elements—which can be achieved many ways—results in superior tilth, the ability of a soil to support agriculture in a healthy way.

from improving the quality of their wines. Despite what many experts believe, terroir is not a myth. The terroirs that produce great wines are few and far between, and they will only improve if their soil is treated with due respect.

Our planet has three types of rock: 90% metamorphic, 3% volcanic, and 7% limestone. The vine, originally found growing in Caucasian limestone, is a plant that thrives in lime-rich soils. All great red wines are produced on limestone soils and as these are relatively rare, not many places produce them. The underlying marl in the area of Bordeaux produces great wines, just as Jurassic limestone makes great Bourgognes and limestone marl produces excellent Barolos. As for the great whites, the wines of Alsace, Moselle and Anjou need metamorphic rock and Hungary's Tokays require volcanic rock. The United States has very little limestone, and its wine-makers would be well advised to concentrate on producing great white wines.

We've now determined that great wines need the combination of a temperate climate, limestone rock and well-oriented slopes, conditions that are seldom concurrent on the world's eight million hectares of cultivated vines. Wine-makers have to accept the unfair fact that terroir is undemocratic and that not everyone is lucky enough to have it. No amount of technical sleight-of-hand is going to make up for its absence. The planet is divided into zones for making table wines and zones for varietal wines, and then there are terroirs for making terroir wines.

The fourth dimension of terroir is the most complex. It's the contribution made by the soil, a subject of heated debate among experts who often are divided into those who take an ideological or a technical position, and those who look at it from a scientific point of view.

We chose to study the soils of Burgundy and in particular those of the Côtes de Beaune and Côtes de Nuits. The region offers several advantages: a homogeneous climate and orientation, and the fact that it has three Jurassic strata: Bajocian, Bathonian and Oxfordian. The region's grape varieties are homogeneous too. In AOC Burgundy, reds must be made from Pinot Noir and whites from Chardonnay. As a result, the taste of the wines cannot be attributed to blending. Lastly, and as early as 1936, Burgundy's wines were classified according to taste into *grands crus, premiers crus* and *villages,* and we planned to confirm or discard those findings.

We started by examining the aspect of the soil. Granulometry told us nothing, but after studying its clay content, we organised a soil classification index that covers 75% of the current (1936) classification. We found a distinct correlation between types of clay and a wine's colour, and by means of science and agronomy we were also able to confirm that the 1936 tastings leading to the Bourgogne crus classification had got it right. Scientific tests and subjective tastings are complementary tools that help us recognise that taste and agronomy are closely linked.

The chemical structure of a soil was the next thing we explored. According to its particular composition, every rock liberates elements which the vine roots then assimilate. Burgundy limestone and Morgon granite are rich in manganese, an element found in the region's wines. The fine-grained volcanic Lantinié granites are rich in barium and fluorine; the Brouilly's andesites are rich in zinc and copper. Each of these rocks releases its own taste of terroir.

But, our critics may object, a wine's aromatic molecules are carbonated and don't contain such oligo-elements as manganese, barium or zinc. But in reality these aromatic molecules are synthesised by enzymes. Enzymes are proteins with a metal co-factor, that is itself responsible for the enzymes' reaction to the ambient temperature. For instance, chlorophyll accomplishes photosynthesis with a magnesium cofactor, and most of the enzymes involved in breathing and the transport of oxygen (cytochrome, peroxydase, catalase, ferrodoxine, etc.) have an iron co-factor, while nitrogenase fixes nitrogen in the air with a cobalt co-factor, etc.

These oligo-elements that the vine extracts from rock can be incorporated into these enzymatic co-factors and thus participate in the synthesis of aromas. As we will discover, the answer to all these questions lies in vegetal physiology, but physiologists haven't yet listed all the enzymes and their co-factors involved in the synthesis of aromatic molecules.

Another interesting factor we examined is every soil's particular type of limestone. Our observations under the microscope showed that, depending on how the water circulates, some limestone is produced by rocks and others by microbial recarbonation. Red wines are planted on the former, whites on the latter. Soils rich in bacterial limestone contain very little free ferric iron and we know that this element is vital to the synthesis of anthocyanins.

Last of all, the biological dimension of soil involves three types of organisms: roots, fauna, and bacteria. In a terroir that spreads over a large area, a vine's roots must be able to sink deeply and they must twist. The more they twist, the longer it takes the sap to reach the grapes, and the richer will be the grapes' mineral content. But the roots need fauna if they are to dig into the ground, because the galleries built by these small creatures aerate the terrain.

The increasing use in viticulture of chemical weeding is killing off the creatures that used to eat the machine-cut grass of traditional cultivation. Now that they're gone, so is the soil's surface porosity and oxygen can no longer penetrate it. After 40 years of chemical weeding, the depth reached by vine roots around the world has dropped from an average of 3.50 metres to 50 cm.

A vast *terroir* needs three types of fauna: the creatures that live on top of the soil (bugs, ants, etc.); the worms that feed on compost made of wood cuttings from the terroir and on animal manure; and the creatures that clean the dead roots of the vine and allow the new ones to grow, and feed on intercalary winter cultures. This fauna needs plants with deep roots.

The last aspect of soil in the expression of *terroir* is microbiology. Bacteria feed the plants by turning the soil's nutritive elements into soluble matter. On dead soil, the wine-maker has to add chemical fertilizer. These fertilizers are the same across the world, and they tend to standardise a wine's taste. Vines use not only N [nitrogen], P [phosphorous], and K [potassium] but 28 other elements provided by the bacteria in the mother rock. These elements need oxygen, as we've already mentioned, if the bacteria are to form them, and oxygen can only penetrate the soil if there is plenty of fauna.

Wine is a perfect illustration of the notion of terroir. It helps us understand how the same plant, *Vitis vinifera,* in different terroirs uses fermentation to produce terroir wines. For all that, it's the wine produced by the fermentation of grape juice that produces the typical tastes of terroir: grape juice doesn't have these distinctive tastes. Between the transformation of grape juice into wine, various microbiological elements take place during fermentation that sharpen the differences between the wines of a same region. That's the magic of terroir.

ON THE OTHER HAND . . .

Randall Grahm's impassioned quest for that elusive great growth in the New World (see chapter 1) highlights the dilemma faced by those who crave *terroir* and don't have it. Save for winemakers born into families residing on one of those special spots for generations, the search for *terroir's* soil expression is uphill.

Or it may be altogether quixotic in the first place. In 2007, food researcher Harold McGee and chef Daniel Patterson got up on their soapbox in the pages of the New York Times to debunk claims of 'soil' or 'mineral' expression in the world's wines. They concluded that a

TRACE ELEMENT SIGNATURES

If the soil of a particular vineyard site really does leave its mark on a wine, it seems reasonable to assume that there is some way to analyze that fingerprint, to detect a chemical signature that can only originate from one place. Such was the intention when authorities attempted to prove the authenticity of a number of older bottles allegedly from Thomas Jefferson's personal collection, which came under suspicion in 2005, a story retold in riveting detail by Benjamin Wallace in his 2008 book, *The Billionaire's Vinegar*.

While it has proven difficult to discern a signature in single-plot and single bottles, researchers have been able to isolate regional differences with more success. Some of the best research in this area has been done in Canada, where geological researchers led by John Greenough unearthed statistically significant differences in trace mineral concentrations between major Canadian winegrowing regions. Greenough eschewed the usual vineyard suspects—potassium, phosphorous, nitrogen—because the levels of those chemicals partly depended on grape growing and winemaking practices. Instead he sought more esoteric chemical elements, rubidium, chromium, cesium, molybdenum and even uranium, present only in a very few parts per billion. The researchers' explanation for the regional pattern was climatic: warmer climates forced more evaporation of water through vine leaves, which stimulated more water uptake from the roots, leading to higher levels of the trace elements in the region's wines.

In another study, researchers were able to verify certain wine vintages by the level of nuclear fallout they contain. Dr. Graham Jones of the University of Adelaide in Australia and his colleagues studied the levels of Carbon-14 in the environment since the beginning of above-ground nuclear testing in the late 1940s, tracking their levels through the discontinuation of above-ground testing, until their eventual ban in 1963. Sure enough, wines from around the world in the affected decades could be precisely dated by the amount of Carbon-14 absorbed in each vintage.

While both trace element analysis and the "bomb pulse" calculations show promise for wine authentication, an important forensic tool in an era of increasing wine fraud, neither of these signatures has a direct link to any discernible sensory feature of the wines themselves.

---

1. John Greenough et al., "Regional Trace Element Fingerprinting of Canadian Wines," in *Fine Wine and Terroir: The Geoscience Perspective*, ed. R. W. MacQueen and L. D. Meinert (Geological Association of Canada, 2006), 127–35.

2. Graham Jones (presentation, American Chemical Society meetings, San Francisco, 2010).

sommelier's claim that a vine's mineral environment can be reliably detected is simply not true—or, at the very least, that isn't all there is to the story.

FROM HAROLD MCGEE AND DANIEL PATTERSON, "TALK DIRT TO ME"

It's hard to have a conversation about wine these days without hearing the French word terroir. Derived from a Latin root meaning "earth," terroir describes the relationship between a wine and the specific place that it comes from. For example, many will say the characteristic minerality of wines from Chablis comes from the limestone beds beneath the vineyards (although, when pressed, they generally admit that they've never actually tasted limestone). The idea that one can taste the earth in a wine is appealing, a welcome link to nature and place in a delocalized world; it has also become a rallying cry in an increasingly sharp debate over the direction of modern winemaking. The trouble is, it's not true.

When terroir was first associated with wine, in the 17th-century phrase goût de terroir (literally, "taste of the earth"), it was not intended as a compliment. Its meaning began to change in 1831, when Dr. Morelot, a wealthy landowner in Burgundy, observed in his "Statistique de la Vigne Dans le Département de la Côte-d'Or" that all of the wineries in Burgundy made wine essentially the same way, so the reason some tasted better than others must be due to the terroir—specifically, the substrata underneath the topsoil of a vineyard. Wine, he claimed, derived its flavor from the site's geology: in essence, from rocks.

In recent years, the concept that one can taste rocks and soil in a wine has become popular with wine writers, importers, and sommeliers. "Wines express their source with exquisite definition," asserts Matt Kramer in his book "Making Sense of Wine." "They allow us to eavesdrop on the murmurings of the earth." Of a California vineyard's highly regarded chardonnays, he writes, there is "a powerful flavor of the soil: the limestone speaks." The sommelier Paul Grieco, in his wine list at Hearth in New York, writes of rieslings that "the glory of the varietal is in its transparency, its ability to truly reflect the soil in which it is grown." In his February newsletter, Kermit Lynch, one of the most respected importers of French wine, returns repeatedly to the stony flavors in various white wines from a "terroirist" winemaker in Alsace: "When he speaks of a granitic soil, the wine in your glass tastes of it."

If you ask a hundred people about the meaning of terroir, they'll give you a hundred definitions, which can be as literal as tasting limestone or as metaphorical as a feeling. Terroir flavors are generally characterized as earthiness and minerality. On the other hand, wines with flavors of berries or tropical fruits and little or no minerality are therefore assumed not to have as clear a connection to the earth, which means they could have come from anywhere, and are thought to bear the mark of human intervention.

■ ■ ■

Since there's so little consensus among winemakers about how to foster the expression of place—what Matt Kramer calls "somewhereness"—in their wines, what are our wine experts tasting? How can a place or a soil express itself through wine? Does terroir really exist?

Yes, but the effects of a place on a wine are far more complex than simply tasting the earth beneath the vine. Great wines are produced on many different soil types, from limestone to granite to clay, in places where the vines get just enough water and nourishment from the soil to grow without deficiencies and where the climate allows the grapes to ripen slowly but fully. It's also true that different soils can elicit different flavors from the same grape. Researchers in Spain recently compared wines from the same clone of grenache grafted on the same rootstock, harvested and vinified in exactly the same way, but grown in two vineyards 1,600 feet apart, one with a soil significantly richer in potassium, calcium and nitrogen. The wines from the mineral-rich soil were higher in apparent density, alcohol and ripe-raisiny aromas; wines from the poorer soil were higher in acid, astringency and applelike aromas. The different soils produced different flavors, but they were flavors of fruit and of the yeast fermentation. What about the flavors of soil and granite and limestone that wine experts describe as minerality—a term oddly missing from most formal treatises on wine flavor? Do they really go straight from the earth to the wine to the discerning palate?

No.

Consider the grapevine growing in the earth. It takes in elemental, inert materials from the planet—air and water and minerals—and, using energy captured from sunlight, turns them into a living, growing organism. It doesn't just accumulate the earth's materials. It transforms them into the sugars, acids, aromas, tannins, pigments, and dozens of other molecules that make grapes and wine delicious.

"Plants don't really interact with rocks," explains Mark Matthews, a plant physiologist at the University of California, Davis who studies vines. "They interact with the soil, which is a mixture of broken-down rock and organic matter. And plant roots are selective. They don't absorb whatever's there in the soil and send it to the fruit. If they did, fruits would taste like dirt." He continues, "Any minerals from the solid rock that vine roots do absorb—sodium, potassium, calcium, magnesium, iron, a handful of others—have to be dissolved first in the soil moisture. Most of them are essential nutrients, and they mainly affect how well the plant as a whole grows."

Most of the earthy and mineral aromas and flavors that we detect in wine actually come from the interaction of the grape and yeast. Yeasts metabolize the grape sugars into alcohol, along the way freeing up and spinning off the dozens of aromatic chemicals that make wine more than just alcoholic grape juice. It's because of the yeasts that we can catch whiffs of tropical fruits, grilled meats, toasted bread and other things that have never been anywhere near the grapes or the wine. The list of evocative yeast products includes an organic sulfur molecule that can give sauvignon blancs a "flinty" aroma. And there are minor yeasts that create molecules called volatile phenols, whose earthy, smoky flavors have nothing to do with the soil but are suggestive of it, especially in wines from the southern Rhone.

■   ■   ■

Grape minerals and mineral flavors are also strongly influenced by the grower and winemaker. When a vineyard is planted, the vine type, spacing and orientation are just a few of many important decisions. Growers control the plant growth in myriad ways, such as pruning, canopy management or, most obviously, irrigating and replenishing the soil with manures or chemical fertilizers. The winemaker then makes hundreds of choices that affect wine flavor, beginning with the ripeness at which the grapes are harvested, and can change the mineral content by using metal equipment, concrete fermentation tanks or clarifying agents made from bentonite clay. Jamie Goode, a British plant biologist turned wine writer, describes in his superbly lucid book "Wine Science" how techniques that minimize the wine's contact with oxygen can increase the levels of sulfur compounds that may be mistaken for "mineral" character from the soil.

It's possible, then, that soil minerals may affect wine flavor indirectly, by reacting with other grape and yeast substances that produce flavor and tactile sensations, or by altering the production of flavor compounds as the grape matures on the vine.

The place where grapes are grown clearly affects the wine that is made from them, but it's not a straightforward matter of tasting the earth. If the earth "speaks" through wine, it's only after its murmurings have been translated into a very different language, the chemistry of the living grape and microbe. We don't taste a place in a wine. We taste a wine from a place—the special qualities that a place enables grapes and yeasts to express, aided and abetted by the grower and winemaker.

The answer lies in the complex relationship between tradition, culture, and taste. Those wine professionals have all spent vast amounts of time and energy learning what traditional European wines taste like, region by region, winery by winery, vineyard by vineyard. The version of terroir that many of them hold is that those wines taste the way they do because of the enduring natural setting, i.e., the rocks and soil. These wines taste the way they do because people have chosen to emphasize flavors that please them.

The pioneering French oenologist Émile Peynaud wrote nearly 25 years ago: "I cannot agree with the view that 'one accepts human intervention (in vinification) as long as it allows the natural characteristics to remain intact,' since it is precisely human intervention which has created and highlighted these so-called natural characteristics!" Modern European views of terroir recognize that typical local flavors are the creation of generations of growers and winemakers, shaping the vineyard and fine-tuning the fermentation to make what they feel are the best wines possible in their place. Typical flavors are expressions not of nature but of culture.

But culture, unlike nature, isn't static. It evolves in response to shifting tastes and technological advances. Over the past 30 years, the staid world of European winemaking has been roiled by an influx of American consumers, led by their apostle, the writer Robert Parker. In his reviews, Parker has brushed aside the traditional practice of judging wine according to historical context (that is, how it should taste), focusing instead on what's in the bottle. His preference for hugely concentrated, fruit-forward wines—the antithesis of distinctive, diverse terroir wines—has dramatically changed the economic landscape of the wine industry. Throughout the world, more and more winemakers are making wine in the style that Parker prefers, even in Europe, where this means abandoning distinctive local styles that had evolved over centuries. "Somewhereness" is being replaced by "anywhereness."

. . . Conjuring granite in Alsatian rieslings and limestone in Chablis puts that connection to the land right in the bottle, ours for the tasting.

If rocks were the key to the flavor of "somewhereness," then it would be simple to counterfeit terroir with a few mineral saltshakers. But the essence of wine is more elusive than that, and far richer. Scientists and historians continue to illuminate what Peynaud described as the "dual communion" represented by wine: "on the one hand with nature and the soil, through the mystery of plant growth and the miracle of fermentation, and on the other with man, who wanted wine and who was able to make it by means of knowledge, hard work, patience, care, and love." "Somewhereness" is given its meaning by "someoneness": in our time, by the terroirists who are working hard to discover and capture in a bottle the difference that place can make.

NOTES

1. J. Duteau, "Le vignoble de Pomerol et Saint-Emilion: climat, formations géologiques et grands types de sols; particularités du régime hydrique" (Colloque Franco-Roumain, Relations sol-vigne, INRA, Bordeaux, June 23–27, 1980).

2. J. Duteau, "Rôle des facteurs climatiques et géologiques en Bordelais," in *Actualités oenologiques et viticoles* (Paris: Dunod, 1981), 40–44.

3. M. Fregoni, "Effect of the soil and water on the quality of the harvest" (International Symposium on the Quality of the Vintage, Cape Town, 1977).

4. I. Garcia del Barrio Ambrosy, *La Tierra del Vino de Jerez* (Sexta S.A., 1979).

5. B. Pucheu-Plante, "Les sols viticoles du Sauternais. Etude physique, chimique et micro-biologique. Alimentation en eau de la vigne pendant la maturation et la surmaturation du raisin" (thèse 3ème cycle, Bordeaux, 1977).

6. J. Salette, R. Morlat, A. Puissant, C. Asselin, H. Leon, and M. Remoue, "Recherches de relations entre le milieu écogéopédologique et le type de vin: cas du vignoble rouge de la moyenne vallée de la Loire. Premiers résultats et principaux problèmes" (Colloque Franco-Roumain, Relations sol-vigne, INRA, Bordeaux, June 23–27, 1980).

7. G. Seguin, "Les sols gravelo-sableux du vignoble bordelais: propriétés physiques et chimiques; alimentation en eau de la vigne et conséquences sur la qualité des vendanges" (Colloque Franco-Roumain, Relations sol-vigne, INRA, Bordeaux, June 23–27, 1980).

8. S. Meriaux, J. Chretien, P. Vermi, and N. Leneuf, "La Côte viticole. Ses sols et ses crus," *Bull. Sci. Bourg.* 34 (1981): 17–40.

9. M. Guilloux, J. Duteau, and G. Seguin, "Les grands types de sols viticoles de Pomerol et Saint-Emilion," *Conn. Vigne Vin* 12/3 (1978): 141–165.

10. G. Seguin, "Les sols de vignobles du Haut-Médoc. Influence sur l'alimentation en eau de la vigne et sur la maturation du raisin" (thèse doctorat ès sciences naturelles, Bordeaux, 1970).

11. J. Compagnon, "Alimentation en eau de la vigne dans quelques sols des Graves. Influence sur le raisin" (thèse 3ème cycle, Bordeaux, 1970).

12. B. Pucheu-Plante, "Les sols viticoles du Sauternais. Etude physique, chimique et micro-biologique. Alimentation en eau de la vigne pendant la maturation et la surmaturation du raisin" (thèse 3ème cycle, Bordeaux, 1977).

13. G. Seguin, "Alimentation en eau de la vigne dans les sols du Haut- Médoc," *Conn. Vigne Vin* 3/2 (1969): 93–141.

14. G. Seguin, "Les sols de vignobles du Haut-Médoc. Influence sur l'alimentation en eau de la vigne et sur la maturation du raisin" (thèse doctorat ès sciences naturelles, Bordeaux, 1970).

15. I. Garcia del Barrio Ambrosy, *La tierra del vino de Jerez* (Sexta S.A., 1979).

16. S. Meriaux, J. Chretien, P. Vermi, and N. Leneuf, "La Côte viticole. Ses sols et ses crus," *Bull. Sci. Bourg.* 34 (1981): 17–40.

17. M. Guilloux, J. Duteau, and G. Seguin, "Les grands types de sols viticoles de Pomerol et Saint-Emilion," *Conn. Vigne Vin* 12/3 (1978): 141–165.

18. G. Seguin, "Influence des terroirs viticoles sur la constitution et la qualité des vendanges," *Bull. O. I. V.* 56 (1983): 3–18.

19. J. Duteau and G. Seguin, "Caractères analytiques des sols des grands crus du Médoc," *C.R. Acad. Agric.* (1973): 1084–1093.

20. G. Seguin, "Caractéristiques analytiques des sols de Grands Crus," in *Actualités oenologiques et viticoles* (Paris: Dunod, 1981), 44–47.

21. C. Andrieux, L. Buscarlet, J. Guitton, and B. Merite, "Mesure en profondeur de la teneur en eau des sols par ralentissement des neutrons rapides," *Industries Atomiques* 3/4 (1962): 63–70 and 5/6 (1962): 72–80.

22. G. Seguin, "Alimentation en eau de la vigne dans les sols du Haut-Médoc," *Conn. Vigne Vin* 3/2 (1969): 93–141; G. Seguin, "Les sols de vignobles du Haut-Médoc. Influence sur l'alimentation en eau de la vigne et sur la maturation du raisin" (thèse doctorat ès sciences naturelles, Bordeaux, 1970); G. Seguin and J. Compagnon, "Une cause du développement de la pourriture grise sur les sols gravelo-sableux du vignoble bordelais," *Conn. Vigne Vin* 4/2 (1970): 203–214.

23. G. Seguin, "Alimentation en eau de la vigne dans les sols du Haut-Médoc," *Conn. Vigne Vin* 3/2 (1969): 93–141.

24. G. Seguin, "Les sols de vignobles du Haut-Médoc. Influence sur l'alimentation en eau de la vigne et sur la maturation du raisin" (thèse doctorat ès sciences naturelles, Bordeaux, 1970).

25. G. Seguin, "Influence des facteurs naturels sur les caractères des vins," in *Sciences et Techniques de la Vigne,* vol. 1 (Paris: Dunod, 1971), 671–725; G. Seguin, "Caractères particuliers de l'alimentation en eau de la vigne, en 1973, dans un sol typique du Médoc. Conséquences sur la maturation du raisin," *C.R. Acad. Sci. Paris* 277 D (1973): 2493–2496; G. Seguin, "Alimentation en eau de la vigne et composition chimique des moûts dans les Grands Crus du Médoc. Phénomènes de régulations," *Conn. Vigne Vin* 9/1 (1975): 23–34.

26. J. Duteau, M. Guilloux, and G. Seguin, "Influence des facteurs naturels sur la maturation du raisin en 1979, à Pomerol et Saint-Emilion," *Conn. Vigne Vin* 15/1 (1981): 1–27.

4

CLIMATE
Limits and Variations

In traditional conceptions of *terroir,* climate plays a distinctly secondary role to the main actor, the soil. From this viewpoint, places with special *terroir* are first and foremost *places,* particular patches of earth that nurture superior grapes and wines. And yes, the terra-centric concede, there is some weather involved, too. The soil, by this logic, determines the essence of the wines produced from a place; the vagaries of weather produce the vintage variations.

At first glance, this all seems reasonable: after all, the vines are rooted in the soil, not in the air. The first time you visit the heart of Châteauneuf du Pape, you find yourself uncontrollably staring at the *galets roulés,* the quartzite cobbles, big as Idaho russets, that cover the vineyards, and immediately conclude that they must be the secret to those amazing, powerful wines. Only later do you notice how hot it is, or how strongly the *mistral* is blowing.

Because in many ways, and in many of the world's wine regions, climate trumps soil. Climate imposes its own set of absolute limits on grapes and wine. There are no great wines—not even any mediocre wines—grown in the Gobi Desert or the South Pole, and the reason is not the deficiencies of the soils. Grapes only have a chance of growing to proper ripeness in a small sliver of the world's land mass, basically the two temperate belts, one in each hemisphere. The range of climate configurations that support wine grapes is much narrower than the range of soil types amenable to vines. And while nutrients and other chemicals from the soil do support and even sometimes get inside the grapes, the most important single input comes from the weather. The sunlight triggers photosynthesis, the central process through which grapes ripen and develop their abundance of aromatic and flavor compounds and precursors.

Climate—the long-term trends in temperature, rainfall, and so on—and weather—short-term manifestations in a single growing season—wield their influence on both a very grand and a very intimate scale. Climate in broad strokes (macro-climate) is critical to the delineation of entire winegrowing regions; it makes the Napa Valley different from Sonoma, the Northern Rhône different from the Southern Rhône, the Mosel different from the Pfalz. Macro-climates can determine which grape varieties flourish in a particular region and which don't, and account for generalized differences in wine style—for example, why the Sauvignon Blanc from New Zealand is different from Sauvignon Blanc from the Loire Valley.

Within a macro-climate zone, specific vineyard meso-climates may depart significantly from the norm due to heavier rainfall, less frequent fog, stronger winds, cooler temperatures, unusual topography or other factors, and those differences can show up in the character of the wines. Then there is the micro-climate, the circumstances of weather with which the vine is most intimate, in the inches surrounding the clusters, under the leaf canopy, which may be several degrees cooler than outside in the direct sun, or much more protected from breezes. Climate is not a single thing, but makes its mark in many layers.

Our tour of climate and weather begins with a look at the broad limits within which grapevines flourish and then proceeds through the many disparate factors climate includes. Topography, the nexus of interaction between land and climate, gets a technical overview from viticulturist John Gladstones, and a more expressive elaboration by Robert Mayberry in his portrait of the Rhône Valley. Temperature gets its due, starting with the classifying scheme of degree days pioneered by Maynard Amerine and Albert Winkler in the 1940s, along with consideration of whether cooler climates set better conditions for the expression of *terroir*. Wind and humidity are examined by viticulturalist Markus Keller. Finally, climate is scrutinized from the point of view of the grape itself, in selections from Richard Smart and Rudolf Geiger.

Note: As the international wine community is acutely aware, changeability is no longer limited to yearly weather, but applies to longer-term climate trends as well. The implications of climate change for grapes, wines and *terroir* itself are examined in Chapter Nine, "The Future of *Terroir*."

LINES AND LIMITS

The vast majority of wine grapes grow in two geographical and climate bands, one in the Northern and one in the Southern Hemisphere. Hospitable climates are best found between 30° and 50° latitude, north and south, encompassing the warmer parts of the temperate zone. Because of elevation, marine influence, wind streams and other factors, some areas within the latitude limits have uncongenial climates, and some successful growing areas stretch the latitude boundaries.

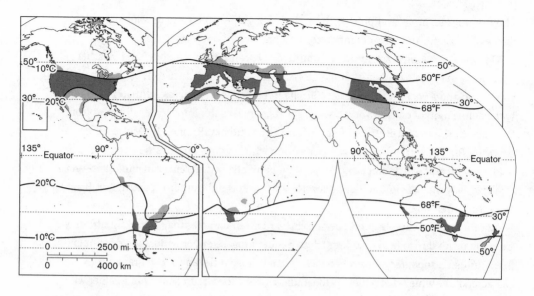

FIGURE 4.1
World Viticultural Areas.

These latitude boundaries more or less correspond with the contour lines (isotherms) marking mean annual temperatures between 10° and 20° Celsius (50°—68° Fahrenheit). The above diagram (figure 4.1), taken from Harm Jan de Blij's *Wine: A Geographical Appreciation*, overlays the isotherms on the latitude lines. The shaded areas are major grape growing regions (including table grapes). Note that these wine grape bands are defined by temperature; a given area may or may not have adequate rainfall, or the right soil types, for practical grape production.

Within these broad regions, temperature variation has a powerful influence on which grape varieties flourish and which produce a useable crop. The following chart (figure 4.2) was compiled and periodically updated by researcher Greg Jones, an expert on climate change and viticulture whose work appears later in Chapter Nine. The chart lays out the most favorable temperature ranges—expressed as averages for the growing season—for each grape, and also indicates the estimated length of the growing season required for each variety. If a given site is too hot or too cool, or if its available growing season is too short, a particular variety may not do well. Note that the optimum temperature ranges for different varieties differ by only a few degrees, which helps explain why within a growing region (a macro-climate) or even a single vineyard (a meso-climate), different varieties may be happier in different places.

This chart documents climate maturity for a number of varieties based on growing season temperatures. It represents the estimated span of varietal ripening potential that occurs within and across the groups.

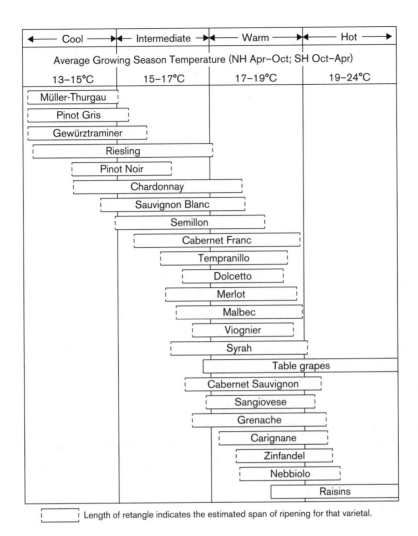

Length of retangle indicates the estimated span of ripening for that varietal.

FIGURE 4.2
Grapevine Climate/Maturity Groupings.

GEOLOGY, TOPOGRAPHY, AND CLIMATE

Climate and landscape are intricate, interdependent variables. Over time, climate factors (rain, wind, etc.) change the shape of the land, cracking solid rock, dredging canyons with rushing water, and eroding one shape into another, a process called, appropriately enough, weathering. At the same time topography, the surface shape and features of the land, creates a kind of obstacle course through which climate functions, tilting one patch of land toward the sun and another away, channeling wind through ravines, and stopping rainstorms dead in their tracks against mountain walls.

Climate interacts with topography on a broad, continental scale and in minute vineyard detail. In the world viticultural zones map reproduced earlier, notice how the optimal climate zone hugs the 30° southern latitude line for most of the map, except for a large, abrupt detour in the middle of Latin American. The reason: the towering Andes, making parts of Chile and Argentina hospitable for grapes and parts of neighboring Brazil inhospitable. On a smaller scale, little nips and tucks of topography can mean that the first five rows of a vineyard get morning fog and the next 20 rows do not.

The location, shape, size, relief, orientation, and altitude of each vineyard translate a given climate pattern and its annual weather variations into a real, concrete environment, not a statistical summary. In chilly Germany, the best slopes face south, taking advantage of every flicker of available sunlight, and generally command higher prices than neighboring vineyards with different exposure. Growing regions close to oceans (Bordeaux, Santa Barbara, Marlborough) have one set of maritime influences; regions further inland (Burgundy, Austria's Wachau, Alto Adige) have another set of continental influences. Similarly, large lakes and major rivers can buffer the effects of harsh climates, especially severe winters, and make grape growing possible. The Finger Lakes region of upstate New York would grow only apples without the moderating influence of Lake Ontario and the smaller Finger Lakes; the vines of Tokaji would wither without the subtle aura of Lake Balaton. Just as soil variation within a vineyard can affect the vines, so can topography-induced differences in microclimate.

Several key topographical factors are addressed in the following selection from John Gladstones's *Wine, Terroir and Climate Change*. The context for these excerpts is Gladstones's assertion of the primacy of temperature as a factor in vine and wine quality, and he goes into some detail on how topography can contribute to temperature variation.

FROM JOHN GLADSTONES, *WINE, TERROIR AND CLIMATE CHANGE*
4.1 Latitude

High and low-latitude vines have substantially different light and, to a lesser extent, temperature regimes. Summer days at high latitudes are long but sunlight intensities low, due to the oblique angle of the sun. Mostly there is also more cloud than at low latitudes, and because of the sun angle the clouds cast more extensive shadows. Light intensity through the grapegrowing season nevertheless remains much of the time at or close to that needed for maximum photosynthesis. Total seasonal assimilation can therefore be greater at high than at low latitudes with comparable temperatures, as Huglin and other authors have argued. Most of this difference accrues in mid-summer, when high-latitude vines not only enjoy very long days, but are also unlikely to be stressed by excessive heat, total or ultraviolet light intensities, or (as a rule) low humidities or water supply. Superior assimilate build-up then assures enough sugar supply to ripen moderate crops well into the autumn, despite fast-diminishing temperatures and sunlight; and usually, for a surplus to remain to support early growth in spring even if there is little further assimilation after harvest.

■   ■   ■

4.3 Topography, air drainage and frosts

The largest climatic effect of topography, apart from that simply of altitude on temperature, is on cold air drainage at night, and hence local susceptibility to autumn, winter and spring frosts. Other effects related to slope and aspect will be discussed in the following section.

Two types of frost are normally distinguished: advective frosts brought by the migration of cold (usually sub-polar) air masses to lower latitudes under certain synoptic conditions, and radiative frosts caused by local radiative heat loss from the ground or vegetative surfaces under clear night-time skies and still conditions.

■ ■ ■

Advective frosts are influenced chiefly by major topographic features that channel or obstruct air-mass movement. For instance the unimpeded flow of arctic air down the Mississippi/Missouri drainage basin in the USA can carry extreme winter cold as far south as the Gulf of Mexico; however, west coastal regions are protected by its containment behind the Rocky Mountains, while to a lesser degree the east coast shelters behind the Appalachian Range.

Cashman describes a typical advective frost event in south-eastern Australia. Here sub-polar air from the south may be further chilled by radiative cooling as it perches over the Australian Alps before spilling inland. Frost on the inland slopes and adjacent plains then depends on how medium-sized topographic features assist or obstruct its drainage. For instance the Whitlands High Plateaux region in Victoria is on a north-projecting ridge of the Alps, and partly escapes because the coldest air sinks and by-passes it down the valleys of the King River and Boggy Creek on either side. However, the risk beyond is increased in line with the efflux from the valleys.

■ ■ ■

4.4 Aspect and slope

The influence of aspect on effective temperature works jointly with those of latitude, steepness of slope, and time of day and year. Despite the complex trigonometry this entails, the principle for present purposes is simple. On flat ground the intensity of surface insolation at any time on a cloudless day is proportional to the sine of the sun angle above the horizon. (Greater loss due to a longer passage through the atmosphere at low angles somewhat intensifies this relationship.) Aspect determines whether a slope faces predominantly towards the midday sun (increased insolation) or away from it (reduced insolation). Steepness of slope in turn governs the relationship according to the sine of the sum of the sun angle above the horizon and that of a slope facing it, or of the difference in angles where the slope is facing away.

■ ■ ■

Both aspect and slope have their most critical effects at high latitudes, due to low sun angle and the fact that most temperatures are in the limiting range for vine phenological development. The effects diminish to negligible at the lowest viticultural latitudes where the growing-season midday sun is close to overhead and much of the season the temperatures are non-limiting. The adjustments (table 4.1)

TABLE 4.1   Guidelines to temperature adjustments, °C, for site factors other than altitude[a]

Adjustment factor	Min.	Max.	Mean	Diurnal Range
Slope and air drainage				
Strong cold air ponding	−1.6		−0.8	+1.6
Moderate cold air ponding	−0.8		−0.4	+0.8
Flat or free-draining valley	—	—	—	—
Moderate slope	+0.4		+0.2	−0.4
Moderate slope on isolated hill	+0.8		+0.4	−0.8
Steep slope	+0.8		+0.4	−0.8
Steep slope on isolated hill	+1.2		+0.6	−1.2
Inclination to midday sun[b]				
On moderate slope, <40° latitude	+0.2		+0.1	−0.2
On moderate slope, 40–48° latitude	+0.4		+0.2	−0.4
On moderate slope, >48° latitude	+0.6		+0.3	−0.6
On steep slope, <40° latitude	+0.4		+0.2	−0.4
On steep slope, 40–48° latitude	+0.8		+0.4	−0.8
On steep slope, >48° latitude	+1.2		+0.6	−1.2
Soil				
Deep, loamy, water-retentive	−0.4		−0.2	+0.4
Mod. stony or calcareous, free draining	+0.4		+0.2	−0.4
Very stony or calcareous, very free draining	+0.8		+0.4	−0.8
Proximity to/distance from water bodies				
Moderately closer to lake etc.	+ 0.4	−0.4		−0.8
Much closer to lake etc.	+0.8	−0.8		−1.6
Moderately closer to ocean	+0.8	−0.8		−1.6
Much closer to ocean	+ 1.2	−1.2		−2.4
Moderately further from lake etc.	−0.4	+0.4		+0.8
Much further from lake etc.	−0.8	+0.8		+ 1.6
Moderately further from ocean	−0.8	+0.8		+ 1.6
Much further from ocean	−1.2	+1.2		+2.4
Exposure to/isolation from prevailing cool winds[c]				
Moderate exposure		−0.8	−0.4	−0.8
Strong exposure		−1.6	−0.8	−1.6
Moderate isolation		+0.4	+0.2	+0.4
Strong isolation		+0.8	+0.4	+0.8

[a] Allow also 0.6°C per 100 metres altitude difference to both minimum and maximum. All adjustments fully additive. Mixed or intermediate ratings pro rata.

[b] Full allowance for directly or near-directly facing the midday sun. Opposite aspects negative, E and W aspects neutral.

[c] Full allowance for directly or near-directly facing the midday sun. Opposite aspects negative, E and W aspects neutral.

are understood as plus or minus: plus for slopes facing the midday sun, and minus for those facing away. They converge to nil for east and west aspects. Full positive adjustments can be made where slopes face approximately southeast to southwest in the Northern Hemisphere or northeast to northwest in the Southern Hemisphere.

Viticultural lore holds firmly that east-tending aspects are best for wine grapes. Good reasons exist to support this belief.

The first is that excessive rain and cloud come mostly with weather systems from the west. (East coasts are partial exceptions to this.) In European high-latitude climates these are primary limitations of viticulture. The sheltered eastern slopes of mountain ranges, with adiabatic warming of westerly winds as they descend, are the warmest and sunniest of their regions but without the often excessive heat and dryness of comparable topographies at low latitude. Alsace and Burgundy in France, the Rhine Valley in Germany, the Willamette Valley in Oregon, and all viticultural regions of the South Island of New Zealand are good examples.

Second, an easterly aspect has earliest exposure to the morning sun, warming both vine and soil at the time of day when temperatures are lowest and most limiting. Photosynthesis can start and accelerate early, while fruit can gain all the benefits of sun exposure without over-heating. By contrast a westerly exposure maximizes heating when air temperature is already at its highest and most likely to cause injury. An easterly aspect promotes effective temperature equability, a westerly aspect detracts from it. This difference may, however, be moderated in west-coastal areas where a westerly aspect catches afternoon sea breezes.

4.5 Soil and above-ground microclimate

The daytime accession of sunlight to the soil surface, and its reflection or absorption and later reradiation as long-wave heat, is an aspect of terroir that recent research has too much neglected. Aerial measurement and mapping of night-time long-wave radiation has indeed long been used in Germany and elsewhere, to evaluate sites for frost risk and overall warmth, but good reasons exist to think that the implications are wider. It is easy to appreciate that heat absorbed by sun-facing, rocky slate soils of the Mosel Valley warms the vines and fruit at night and allows Riesling to ripen in a climate that otherwise would be too cold. But why do very rocky soils that do much the same epitomize the best of Châteauneuf du Pape, in France's warm southern Rhône Valley; or likewise the sun-facing rocky terraces of Portugal's hot Douro Valley?

An apparently universal superiority of stony or rocky soils for quality viticulture, regardless of climate, has been acknowledged for centuries. More must be involved than merely extra heat to enable ripening.

One factor is undoubtedly that a stony or rocky surface helps to protect against erosion, making viticulture possible on mesoclimatically desirable slopes that would otherwise be too steep. Yet the same quality advantage of stony soils extends to many flat locations as well.

▪   ▪   ▪

Two major factors influence the energy balance between soil surface and the air and vines above it.

The first, and more problematic, is soil surface colour. A light-coloured surface reflects a material part of incoming light energy, with a correspondingly reduced proportion absorbed and converted to soil heat. Net consequences for the vine and fruit are mixed. Extra light to the shaded parts of the vine is an advantage for photosynthesis and probably for the ripening berries. On the other hand in warm and hot climates it can mean a harmfully enhanced energy load through the hottest times of the day. Dark surfaces absorb more of the energy (depending also on their thermal conductivity and heat storage capacity, as discussed below) and re-radiate most of it later when temperatures are cooler. The daytime difference may not be critical for white grapes, which themselves reflect away a fair proportion of incident light, but can be so for red grapes. Their red pigments absorb much of the light and convert it to heat, adding to their heat load. Nor do berries cool by transpiration as can leaves, as they have very few stomata. Perhaps these are reasons for the widely believed association, at least in warm to hot areas, of white soils for white grapes and red soils for red grapes. Perhaps also the spectral quality of the reflected light plays a role.

More certainly dark-surfaced soils that reflect little light and convert nearly all to heat, to be re-radiated later under cloud or in the evening, are regarded as a general factor for successful viticulture in cool climates where night temperature most clearly limits ripening. They warm the evenings and reduce temperature variability. Examples here include the dark slates of the Mosel, and the "bituminous" soils of Württemberg and of the Meuse Valley in southern Belgium.

The complementary second factor, which is critical in the examples just cited, combines a soil's heat conductivity and its volume capacity for storing heat. A low value of each means that the surface heats shallowly to high temperature and re-radiates its heat intensely and more or less immediately. This is clearly undesirable. A conductive soil with good heat storage capacity, on the other hand, absorbs to depth and retains most heat with least rise in surface temperature. The stored heat is re-radiated more steadily and over a longer period. This both benefits night-time ripening processes and reduces the risk from morning frosts.

Solid rock conducts heat many times faster than still water, although its volume heat storage capacity is only about half. Still water in turn conducts many times faster than still air, which is a very effective insulator. So is organic matter, especially if loose and dry. A stony or rocky soil with relatively little organic matter absorbs, conducts and stores most heat during sunshine and returns most at other times. Moderate dampness of the underlying fine soil (maintained in part because surface stones and rocks block evaporation) helps in soil heat storage through increasing both conductivity and volume heat storage capacity. Moderate soil compaction helps further through a reduction of insulating air spaces.

All these factors, singly or in combination, improve vine temperature relations above ground. The resulting greater temperature equability benefits the vine and fruit quality across *all* viticultural climates.

■　■　■

4.6 Proximity to water bodies

Sea and lake breezes have already been partly covered in Section 3.4. The following are supplementary remarks on the immediately local effects of inland lakes and rivers. These gain significance because they can part-offset the limitations of some inland environments: for instance the sustained heat and atmospheric dryness of inland Australia, or winter-spring cold in the cool continental climates of North America and central and eastern Europe.

The role of medium-sized lakes in North America is well established: notably the Finger Lakes in New York State, USA, and Lake Okanagan in British Columbia, Canada. In both cases successful viticulture is confined to the often steep slopes immediately overlooking the lakes. Lake influence in these cases grows by the fact of their elongated shapes, allowing cumulative effects on winds or breezes blowing along them. Temperature inertia of the deep lakes, as of larger water bodies, delays budburst in spring until the worst threat of frost is over, and extends ripening later into autumn. The main effect during ripening is to raise night and minimum temperatures, which according to present argument should benefit grape and wine quality.

European opinion also attributes benefits to sunlight reflected from rivers, such as from the Mosel and Rhine, onto their steep northern banks. Such effects would be especially strong during ripening at or after the equinox, when the sun angle is low. It is a logical supposition for cool, high-latitude environments.

The benefit of proximity to lakes and rivers in hot, summer-arid environments is for the opposite reasons of afternoon cooling and humidifying, although night warming and reduction in spring frost risk also play a part. Even modest-sized water catchment dams can ameliorate such climates slightly for vines sloping down to them. However the biggest benefits come downwind of larger permanent lakes and rivers. As with elongated lakes, breezes blowing along rivers cool and humidify cumulatively. The east bank of Lake Mokoan in north-east Victoria, Australia, is an example of the first type, while in the same region Rutherglen and Wahgunyah get cooling afternoon breezes that blow up the Murray River from the elongated Lake Mulwala. They are marked enough to be known locally as "The Wahgunyah Doctor." Similar breezes are thought to be significant for the Nagambie Lakes viticultural sub-region of Victoria's Goulburn Valley, and may also be so upstream and to the east of Lake Eppalock in the Heathcote region.

Benefits such as these are relatively greatest in regions that are well inland as well as being warm and dry. Not only do high evaporation rates maximize air cooling and humidifying within the local convections created, which are themselves strong because of the wide temperature contrasts developed daily between land and water; the convections are also less likely to be overridden by strong regional winds as experienced nearer the coast.

THE ROMANCE OF TERRAIN

Describing the interaction of climate and landscape can be more clinical than captivating. Not so this excerpt from the introduction to Robert Mayberry's *Wines of the Rhône Valley,* in which he surveys both the geography and the wine styles of a maze of regions and sub-regions by way of their underlying topography. Mayberry's exposition of the interactions between soil, elevation, the network of rivers, and the driving force of the *mistral*— the strong, cold, dry, north wind that blows through the Rhône Valley like a wind tunnel— should make any wine lover thirsty.

FROM ROBERT MAYBERRY, *WINES OF THE RHÔNE VALLEY*
LOCATION OF LES CÔTES DU RHÔNES MÉRIDONALLES

At Donzère (Drôme), on its course south from Lyon, the Rhône suddenly emerges from a gorge between cliffs and flows out into a spacious valley, really a basin. Its floor spans plains and ascends plateaus and slopes along tributary rivers to either side of the Rhône all the way from the southern pre-Alps on the east to the foothills of the Cévennes mountains on the west. From here south to about Beaucaire (Gard), the vineyards of *les Côtes du Rhône méridionales* stretch about as far from east to west as they do from north to south.

If one has followed *route nationale* 86 down the river from St.-Péray, one has already left granite behind for limestone country. South of Donzère one enters a zone where the gray-green of olive trees, wild thyme, rosemary, and lavender and the windowless north sides of tile-roofed farmhouses speak clearly of a Mediterranean climate. On the plateau of Donzère, in the area of the adjacent appellation Coteaux du Tricastin, and again near Orange (Vaucluse), the mistral wind achieves record levels of velocity. Not only is the mistral the prime reason the air of the south is bright and the climate dry, it is also a factor in making the differences of *terroir* among southern Rhône wines occur both horizontally and vertically. Horizontal differences reflect origin on the east or west bank of the Rhône river. These will be spoken of later. Vertical differences are those that reflect the terrain. In broad outline, two fundamental associations with the terrain provide the underlying logic of *appellation d'origine contrôlée* in the southern Rhône. One is between topography and soil. The other connects topography with the hierarchy of specificity and quality in the AOC system.

■   ■   ■

DIFFERENCES BETWEEN THE TWO BANKS
OF THE RHÔNE

Vertical differences among the southern Rhône *terroirs* reflect features of the landscape that occur throughout the Côtes du Rhône *méridionales.* Horizontal differences reflect a variation in climate between the two banks of the Rhône. This variation is largely a story, again, of that river's tributaries.

The climate on the left bank (Vaucluse and southern Drôme) is drier, partly because, like the Rhône River itself, its left bank tributaries, the Lez, Aygues, and Ouvèze, originate in the Alps. This means that their rise and fall in water-level tends to parallel that of the Rhône. They run high with alpine melt-water when the Rhône is running high, and are almost dry when the Rhône is running low.

An additional reason for less ambient humidity on the left bank is the orientation of the tributaries there. They flow from northeast to southwest, the lower Ouvèze even north-northeast to south-southwest. So the valleys of these rivers, especially that of the Ouvèze, become like the Rhône, corridors for the mistral. The concentrating effect is particularly felt on the north-facing slopes and on the open *terrasses* (the two coincide at Gigondas and on the north side of Châteauneuf-du-Pape). The broad expanse of wide-open spaces jutted into by higher plateaus, hills, and the jagged peaks of the Dentelles de Montmirail, give the area an atmosphere something like the American Southwest, most noticeable perhaps at sunset when the whole landscape shades from crimson to lavender.

The right bank tributaries, Ardèche, Cèze, and Tave, originate in the Cévennes. Their rise and fall in water-level is therefore less parallel to that of the Rhône than is the rise and fall of the left bank tributaries. The consequence is a little more and more constant ambient humidity in the Gard, a tendency that is reinforced by the geographical orientation of these tributaries *vis-à-vis* the mistral. The right bank tributaries flow nearly west to east, athwart the mistral, which enhances the sheltering effect for south- and east-facing slopes. This and a more general irregular hilliness in the Gard, compared to the definite contrast of broad *terrasses* with slopes in the Vaucluse, may be the prime reason why red wines of the Gard are often both firmer and more delicate, leaner and more Bordeaux-like, than the sometimes more robust, sometimes huge and less sharply defined reds from the Vaucluse. Other differences, both natural and cultural, seem to have accumulated to reinforce this primary distinction.

On the east bank, many exposures face the westward quadrant, and hence the heat of afternoon sun. This is to the advantage of alcohol and to the detriment of acidity. On the west bank, many exposures face the eastward quadrant, providing morning light at the same time the air is richest with carbon dioxide. This is to the advantage of aroma. (The eastern zone of Châteauneuf-du-Pape also sees eastward exposure.) Finesse is also contributed by a greater prevalence of sandy soil in the Gard, which makes erosion a cultivation problem in some areas.

TEMPERATURE AND ITS SUPPORTING CAST

Of all the elements of climate that can affect a vineyard, temperature gets top billing, since it determines whether grapes can be grown at all and, to a large degree, which varieties work best where. The first system for characterizing grapevine temperature zones was proposed in 1944 by two University of California researchers, Maynard Amerine and Albert Winkler, who were seeking a way to advise growers on the best places for planting different varieties in the state. Their results were published in a seminal study, *Composition and Quality of Musts and Wines of California Grapes*. In this publication, they summarized the temperatures during the seven-month growing season in a new calculation they called Degree Days (or Growing Degree Days, often abbreviated DD). Each day's heat contribution, from April 1 to October 31, was measured as the mean temperature, minus 50° Fahrenheit. They called this figure a Heat Summation Unit (HSU). So for a day with a high of 83° and a low of 57°, the mean temperature would be 70°, and after subtracting the minimum 50°, the Degree Day contribution would be 20 HSU.

Summing up weather data for the entire growing season, Amerine and Winkler came up with five temperature-based growing regions for California—and easily applicable to the rest of the world. Based on tastings from across the state, they were able to formulate suggestions about which varieties and wine styles had the best chance of succeeding in which areas. (One of their more prescient recommendations was for Chardonnay, whose acreage in 1944 was negligible. The authors felt that this low-profile grape would be a propitious choice in three of the five climate regions.)

A number of alternate approaches for characterizing climate regions for wine grapes have been developed, all of which have their advantages and limitations. The index developed by Pierre Huglin, for example, includes an adjustment factor for latitude, which affects available hours of sunlight and the angle of the sun. New Zealand viticulturalist David Jackson, whose thoughts about the advantages of cooler climates are excerpted below, advocates a categorization based on the mean temperature for the month before normal grape ripening in a particular place.

Each region's range is broad and general, and delimited by a single degree, all of which renders the Degree Day system a rather blunt instrument. And yet it has proven to be extremely useful in shaping the direction of the modern California wine industry, where it continues to be a starting point for planting decisions today. The following discussion of the concept and its applications comes from Winkler et al.'s *General Viticulture*, originally published in 1962.

FROM ALBERT WINKLER, JAMES COOK, W. MARK KLIEWER, AND LLOYD LIDER, *GENERAL VITICULTURE*
DEGREE DAYS

Broadly, climate limits grape production to the Temperate Zone and further limits the highest development of individual varieties to localized areas within this zone. For example, the table-grape variety Tokay (Flame Tokay) attains its best development in the warm, dry area within a radius of five miles surrounding the city of Lodi, California. Similarly, the tablegrape variety Emperor is restricted to a hot, dry district in eastern Fresno, Tulare, and Kern counties. Comparable examples for given varieties, though seldom so restricted, are to be found in most countries that produce table grapes.

The centuries of experience and research of European growers and enologists have definitely established the effect of climate on wine grapes. Climate influences the rates of change in the constituents of the fruit during development and the composition at maturity. Moderately cool weather, under which ripening proceeds slowly, is favorable for the production of dry table wines of quality. Cool weather fosters a high degree of acidity, a low pH, and a good color, and in most table wine varieties it brings to the mature fruit optimum development of the aroma and flavoring constituents and the precursors of the bouquet and flavoring substances of the wines. The combination of specific environmental conditions with the qualities of a given variety has made possible the Rieslings of Germany, the Clarets and Burgundies of France, the Chiantis of Italy, the Constantias of South Africa, and other renowned wines. If, however, varieties lack special character, even the most favorable climate will not endow the resulting wines with good quality. Table wines from such varieties will be improved—because of a better balance of the sugar, acid, tannin, and flavor of the grapes at maturity—but they will still lack such special qualities as the bouquet and freshness that are characteristic of premium-quality wines.

In warm climates the aromatic qualities of the grapes lose delicacy and richness, and the other constituents of the fruit are less well balanced; hence the resulting table wines, even from the best grape varieties, cannot compare with the best wines of cooler regions. In very hot regions, where growth and ripening changes proceed with great rapidity, the taste of most dry wines is harsh and coarse, and the other components are so poorly balanced that usually only common dry table wines can be made.

The abundance of heat in some regions, which makes them poorly suited for dry wines, makes them ideal for such dessert wines as port, muscatel, and sherry. With abundant heat the varieties especially suited to the production of such wines attain their most nearly perfect development. Large summations of heat, especially just before and during ripening, favor a high ratio of sugar to acid in the fruit, and the ill effect of the heat on aroma and flavor is less than in the case of table wines.

CLIMATIC REGIONS

In the investigations of factors affecting wine quality, begun in 1935, it was found that the geographic regions of California that had served well for locating table and raisin grapes did not delineate the effects of climate sharply enough to serve similarly for wine grapes (Amerine and Winkler, 1944). In some instances the geographic regions showed widely varying climatic factors, especially in heat summation, and in other instances there were no differences, in certain factors, between one region and another. For example, the range of heat summation in the Sacramento Valley is almost identical with that in the San Joaquin Valley, whereas there is a wide diversity of heat summations within the North Coast region.

Therefore, available climatological data of United States Weather Bureau stations and the stations of the Department of Viticulture and Enology, University of California, were summarized for the areas where grapes are produced successfully for wine making. Then the principal climatic factors of the areas were correlated with the analytical data and quality scores of the matured wines of the areas. The only factor of climate that proved to be of predominant importance was temperature.

Other factors, such as rainfall, fog, humidity, and duration of sunshine, may have effects, but these are much more limited than the effect of heat summation. Amount and time of rainfall definitely restrict the production of natural raisins to certain areas. Rainfall, fog, and humidity also influence the development of organisms that may have a marked effect on production costs and, as in the case of *Botrytis cineria* (noble rot), may, in certain varieties, be beneficial. Further, these factors of climate influence temperature, but no data show that they have a direct effect on the balance of the composition of the fruit at maturity, except when the noble rot intervenes. This finding, together with the very marked effect of temperature when expressed as heat summation above 50°F from April through October, led Professors Amerine and Winkler (1944) to use heat summation as a basis for segregating the grape-producing areas of the state into five climatic regions.

Heat summation, as used here, means the sum of the mean monthly temperature above 50°F for the period concerned. The base line is set at 50°F, because there is almost no shoot growth below this temperature. The summation is expressed as degree-days. For example, if the mean for a day is 70°F, the summation is 20 degree-days, and, if the mean for June is 65°F, the summation is 450 degree-days (15 degrees times 30 days). The importance of heat summation above 50°F (10°C) as a factor in grape quality has also been indicated by Koblet and Zwicky (1965). They found that degree Brix was more closely correlated with heat summation above 50°F than with total heat summation or hours of sunshine, under the cool condition of Wädenswil.

The heat summations for the climatic regions are: I, less than 2,500 degree-days; II, 2,501 to 3,000 degree-days; III, 3,001 to 3,500 degree-days; IV, 3,501 to 4,000 degree-days; and V, 4,001 or more degree-days. Some characteristics of the climatic regions in California and their adaptation to important wine-producing localities follow.

Region I.   This region contains restricted areas of fertile soils. As a rule, hillside slopes and valley areas of moderate productivity are available for vines. The early maturing premium-quality dry table wine varieties attain their best development here. Heavy-bearing varieties should not be planted, since their production cannot compete with that of warmer districts with more fertile soils.

Region II.   An area of great importance. The valleys can produce most of the premium-quality and good standard white and red table wines of California. The less productive slopes and hillsides vineyards cannot compete in growing grapes for standard wines, because of lower yield, but, nevertheless, can produce favorable yields of fine wines. Irrigation is beneficial in the areas of low rainfall.

Region III.   Another important region. The lands are generally level or slightly sloping and fertile, except for some that are gravelly or stony and of only moderate depth. The moderately warm climate favors the production of grapes of favorable sugar content—sometimes with low acid, as may occur in warm years. Excellent red wines of later maturing premium quality varieties are the rule here.

TABLE 4.2   Heat summation as degree-days above 50°F

For the period April 1 to October 31 at various county locations
in California and a few foreign locations

Station and county or country	Heat summation	Station and county or country	Heat summation
Climate Region I Locations			
Trier, Germany	1700	Coonawarra, Australia	2170
Geisenheim, Germany	1790	Beaune, France	2300
Reims, France	1820	Santa Cruz, Santa Cruz	2320
Lompoc, Santa Barbara County	1970	Bordeaux, France	2390
Salem, Oregon	2030	Geneva, New York	2400
Climate Region II Locations			
Auckland, New Zealand	2540	Napa, Napa	2880
Budapest, Hungary	2570	Hollister, San Benito	2890
Yakima, Washington	2600	Asti, Italy	2930
San Luis Obispo, San Luis Obispo	2620	Kelseyville, Lake	2930
Petaluma, Sonoma	2740	Sonoma, Sonoma	2950
Climate Region III Locations			
Oakville, Napa	3100	Healdsburg, Sonoma	3190
Hopland, Mendocino	3100	Milan, Italy	3310
Paso Robles, San Luis Obispo	3100	Livermore, Alameda	3400
Calistoga, Napa	3150	Cloverdale, Sonoma County	3430
St. Helena, Napa	3170	Ramona, San Diego	3470
Climate Region IV Locations			
Florence, Italy	3530	Cape Town, South Africa	3720
Sao Paulo, Brazil	3540	Sydney, Australia	3780
Mendoza, Argentina	3640	Clarksburg, Yolo	3860
Lodi, San Joaquin	3720	Auburn, Placer	3990
Climate Region V Locations			
Ojai, Ventura	4010	Shiraz, Iran	4390
Modesto, Stanislaus	4010	Fresno, Fresno	4680
Perth, Australia	4020	Algiers, Algeria	5200

NOTE: *The entries above are a representative selection of those contained in the original table in* General Viticulture, *and are edited for clarity.—eds.*

White wines of fine quality may be produced in limited areas on the lighter soils and on slopes with exacting vineyard management. Excellent natural sweet wines and good white and red wines can be produced on the more fertile soils. The best of the port varieties of moderate productivity will produce excellent port-type wines. Irrigation is beneficial in the areas of low rainfall.

Region IV. The soils in region IV are generally fertile. Most vineyards are irrigated and are capable of producing large crops. Some of the soils on the slopes of the east side of the Intermediate Central Valley region and in San Diego County are less fertile and less productive, but their grapes may be of better quality. Natural sweet wines are possible here, but in warm years the fruit of the most acceptable varieties tends to be low in acid. The white and red dessert wines produced here are of good quality. The white and red table wines are satisfactory if produced from the better, high-acid varieties.

Region V. This region embraces the Sacramento Valley from Sacramento to Redding and the San Joaquin Valley from Merced to Arvin. The soils are the most uniformly fertile in California. Except for a few vineyards near Redding and some in the lower foothills, the region is entirely in the highly productive irrigated interior valleys. Of the regions that can produce wine grapes, this has the hottest climate. Standard red and white table wines of varying qualities can be made from the better high-acid varieties. The white and red dessert wines produced can be very good.

The climatic regions into which the grape-producing areas of California have been divided is a basic advance in the development of variety-climatic relationships. The climatic conditions of the present regions merge from one into the other, so the boundaries are not definite. Neither are conditions uniform within a given region. Nevertheless, the figures for heat summation in regions I, II and III provide a valuable indicator of the quality level of the wines of the premium varieties produced in the coastal areas of California. Similarly, heat summation provides a basis for differentiating the interior valleys into regions; some limited areas there have the same heat summation as the warmer coastal regions, but quality levels of the interior valley wines of the same varieties and same climatic regions are not comparable. The wines of the interior areas tend to be flat, unbalanced, lacking in varietal aroma and finish. These deficiencies are likely owing to factors of climate, such as the heat summation above a given temperature, say 70°F, differences in heat summation during the ripening period, and lower relative humidity. For instance, three region III locations, each with a heat summation of 3,400 degree-days above 50°F, have the following degree-days above 70°F: Livermore (coastal area) 15, Jamestown and Mokelume Hill (both interior areas) 295 and 353, respectively. The average yearly degree-days above 70°F for four well known locations—namely, Bakersfield with 1,168, Fresno with 913, Lodi with 311, and St. Helena with 72—when related to the scores of table wines from those areas further indicate the injurious effects of higher and higher temperatures on wine quality. Working with Tokay over a period of 13 years, Winkler and Williams (1939) found that the heat summation during the ripening period had a marked influence on table grape quality. In years with near 700 degree-days during ripening the fruit was very good, while in years when the heat summation above 50°F was near 600 degree-days the fruit scored very poor.

■   ■   ■

Seasonal conditions.   In addition to the general effect of climate there is a seasonal influence that is more or less marked according to the position of the producing region in the Temperate Zone. Wines of best quality are usually produced in the hot years of the coolest regions, whereas in the warm regions the cool years produce the higher-quality wines. Deviations from optimum conditions for maturing of the fruit are greater and most frequent in the coolest regions. Because California's present grape acreage is primarily in the warm part of the zone adapted for grape production, the belief is prevalent that every year is a "vintage year" in California. If this term simply designates years in which the grapes attain full maturity, such a belief is correct. By general as well as historical usage, however, the term "vintage year" properly designates a year of outstanding quality. To say that California wines of all years are of outstanding or superior quality, or even to say that the wines of all years are of equal quality, is not in keeping with the facts.

IS COOLER BETTER?

A lot of experts think so. Historically, most prized viticultural regions have been on the cool side; most recently, a lot of the "heated" criticism directed against high alcohol wines from warmer climates—particularly in California—comes from critics who think that alcohol mars the signature of *terroir*. There are some obvious exceptions, like the sun-baked wines of Spain's Priorat, but the preference for cooler climes remains a majority sentiment among devotees of *terroir* expression.

In their 1994 handbook, *The Production of Grapes and Wine in Cool Climates,* David Jackson and Danny Schuster made the case rather directly:

FROM DAVID JACKSON AND DANNY SCHUSTER, *THE PRODUCTION OF GRAPES AND WINE IN COOL CLIMATES*

It is not yet fully understood why cool climates generally produce the best quality table wines, but the evidence suggests that it is the lower temperatures in the autumn which are of special significance. In warm climates ripening of grapes occurs early, when the weather is still warm or even hot. These hot conditions cause rapid development of sugars, rapid loss of acids, and high pHs. The consequences of these developments will be discussed later, but at this stage it is sufficient to note that the juice is often unbalanced with respect to sugar, acid, and pH, and the grape appears to have had insufficient time to accumulate those many chemical compounds which add distinction to the wine. A cool autumn—often with considerable diurnal temperature variation—slows down development; better balances can be achieved, and more aroma and flavour constituents are accumulated. ■

■   ■   ■

FOR THIS position, Jackson was roundly criticized, in particular by those who advocated for the quality and virtues of warm climate wines. In *Climate,* his subsequent work on the subject, he was more circumspect, but his preferences aren't hard to tease out.

WHAT IS A COOL CLIMATE?

In no other fruit crop does climate appear to play a more important role than it does for grapes. Apples, for example, are a temperate crop, although they mature better in areas where summers are not too cold and the climate not too wet. Being temperate fruits they do not perform well in subtropical or tropical climates. Different temperate climates produce higher or lower apple yields and quality may also be related to climate factors. Growers may be concerned with weather events such as late spring frost, hail or sunburn. Most of these aspects also affect the grape grower but, in addition, he/she needs to consider that heavy rain near harvest will cause fruit splitting followed by disease, early autumn frost may cause leaf fall and even berry death, and short cool seasons will severely delay ripening.

Our special concern with climate and particularly heat accumulation is because the growing season for grapes is relatively short. For example, remaining with the apple comparison, apples flower in spring and the fruit has until autumn to grow and ripen. Thus they will grow in areas of Canada, England, Holland and Germany which are at a relatively high latitude and have a relatively short growing season.

Grapes in the same areas often have to struggle to ripen, because, although grape buds burst at the same time as apples, grape flowers do not appear until mid-summer when considerable vegetative growth has occurred. Thus the grape must grow and mature in only half the growing season that is available to apples.

An explanation is needed as to why the topic of cool climates is special and worthy of a monograph series. Here are some features to consider:

1. Definition. A grape is growing in a cool climate when the mean temperature in the month before harvest is 15°C or below.

2. Grapes in cool climates have larger variation between vintages than those in warm climates.

3. In cool climates, poor vintages are mostly due to cooler-than-normal conditions resulting in lowered ripeness of the grapes. In warm climates cool seasons may result in later ripeness but not necessarily poorer quality

4. Higher yields are normally achieved in warmer climates because buds are more fruitful. The vine in such climates has a greater capacity to ripen a higher crop. Yields in cooler climates are usually lower and if high yields are obtained the vine may not be able to ripen them to a satisfactory level.

5. Warm climates produce a higher level of sugar in the berries, which increases the alcohol level in wines, resulting in a wine with more body. Grapes in warm climates often ripen with low acid levels which may need to be supplemented in the winery. Sometimes high pH levels in the wine cause wine-making problems.

6. Cool climates usually produce grapes with lower sugar levels than warm climates, even if the grapes attain full ripeness. In cool vintages it may be necessary to add sugar

(a process known as chaptalisation). Acid levels may be high and in cool seasons they may have to be adjusted in the winery. The pH levels are seldom too high, which is usually considered to be a bonus.

7. Cool-climate wines although of lower body are often said to be more delicate and elegant. Higher acids give a sense of freshness not found in warm-climate wines.

8. Of the dozen or so world-recognised quality grapes there is a tendency to find more whites that are successful in cooler climates and more reds that are successful in warmer climates. Pinot noir is a notable exception—producing excellent "Méthode Traditionelle" wines in cool climates such as Champagne, and fine Burgundy-style wines in slightly warmer but still mild climates. If a district is found to produce consistently fine wines from Cabernet Sauvignon or Shiraz (Petite Syrah) it is almost certainly a warm climate.

9. In cool climates it is common for viticulturists, in their selection of clones of specific varieties, to select an earliness characteristic. Rootstocks may be chosen for their ability to ripen the grapes before the norm.

10. Warm climates are more forgiving than cool climates and more skill is required by the cool-climate viticulturist to produce quality grapes.

11. Fortified wines are seldom produced in cool climates.

BETTER WINES, OR DIFFERENT?

In *The Production of Grapes in Cool Climates,* Danny Schuster and I were criticised for saying: 'a cool climate is one which will have the possibility to produce table wines of distinction'. As such it is not an incorrect statement except that it implies that grapes in warm climates cannot do this. They can and they do. Wines should be compared like with like. Those who prefer cool-climate wines, simply prefer cool-climate wines, they are expressing no superior judgement.

Nevertheless, the impression remains that many of the best table wines are produced in cool climates. Becker (1985) says:

> Under cool conditions white wines tend to be fresher, more acidic and of finer bouquet and aroma. In warmer zones the aroma forfeits its freshness and the wines have more alcohol and often lack balance. High quality white wines generally come from cooler zones whereas the optimum for red wines is somewhat warmer. In very warm regions wines are high in alcohol and short on taste and aroma; such zones are suitable for dessert wines.

Coombe (1987) suggests that this is too simplistic. According to him all we can accurately say is that . . . 'in hot regions the increase in sugar and decrease in malate are more rapid, especially the latter. The evidence that hot regions give grapes that lack compounds contributing to fine flavour is indirect.' Rapid ripening means it is easier to miss the correct picking day for grapes, high temperatures at crushing and during fermentation may increase liability to oxidation faults. It could be argued that, traditionally, because more care has been lavished on vine management in cool climate districts the production of quality grapes is more likely to result.

Graham Due (1995) has argued that 'there is no causal relationship between climate and quality . . . what is certain is that climate does affect the style of the wine: warm climates produce wines higher in alcohol and extract.' In principle the author agrees with this.

We shall not consider the debate any further except to repeat that there will be differences in style between cool and warm regions. While it is pointless to argue which is better, we can be sure that management of vines and wines is different in the two situations and demands different approaches to viticulture and oenology. ∎

IN THE same article about living soil excerpted in Chapter Three, soil scientist and *terroir* campaigner Claude Bourguignon is far less restrained.

FROM CLAUDE AND LYDIA BOURGUIGNON, "SOIL SEARCHING".

Temperate climates are the only ones suited to the production of fine wines. Over the centuries, wine-makers have created grape varieties suited to local weather variations; Pinot Noir and Chardonnay in the northern temperate zones, Cabernet and Sauvignon in central areas, and Grenache and Marsanne near the Mediterranean.

If the weather is too hot, wines tend to be heavy and to contain higher alcohol levels. Fine wines require moderate heat with sunny summers and, starting in September in the north and March in the south, sunny days and fresh nights. Sugars mature early in the summer and phenolics mature in late summer, something New World wine-makers often ignore, planting varieties without paying

---

FOOLING MOTHER NATURE

While the debate over warmer vs. cooler climates continues, nearly everyone agrees that you can't grow grapevines in the tropics. Except, of course, for those who grow them there anyway, including the burgeoning winegrape industry in thirsty India.

In the Nashik region of Maharashtra state, not far from Mumbai, a traditional center of table grape growing, a near-Mediterranean climate prevails—for half of the year, from September through March, between the monsoons and the truly hot season. Though India lies in the Northern Hemisphere tropics, the grape growing calendar matches that of the Southern Hemisphere. Nashik is also in the rain shadow of a mountain range on the western end of the 2000-foot plateau, meaning that monsoon rainfall is limited and mostly over by the start of the growing season. The vines require careful pruning at the beginning and end of the growth cycle, since they have no true dormant period, and some varieties seem not to work at all. No one is claiming that Nashik is the next Priorat, but the success of a wine industry where before none could be imagined testifies to the ability of dedicated growers to take advantage of climate niches in the most unlikely places.[1]

---

1. Tim Patterson, "In Search of Grapevines and Terroir," *Wines and Vines*, May 2014.

heed to these basic facts. Cabernet, for instance, is not obviously suited to the Napa Valley. The climate there resembles that of Languedoc, more suited to Grenache and Mourvèdre.

BEYOND TEMPERATURE: WIND AND HUMIDITY

Temperature during the growing season is, of course, not the only critical variable. At least two other elemental forces, wind and humidity, can make or break a vineyard site and impact an entire vintage.

Each can be responsible for a number of effects, positive and negative. The Salinas Valley in California's seaside Monterey County gets more than its share of overnight fog, but turns into a wind tunnel every morning around eleven o'clock as ocean winds move inland, drying the vines and making them less hospitable for the growth of microbial diseases. In Australia's Barossa Valley, on the other hand, windbreaks are often necessary to shelter vines and improve grape quality. The generally dry growing season in California is a decided advantage over the constant high humidity in many East Coast vineyards and parts of Europe, but constant dryness combined with high heat makes vines transpire moisture out into the atmosphere and can induce water stress.

In this selection from *The Science of Grapevines: Anatomy and Physiology*, Washington State University viticulturalist Markus Keller summarizes the roles these two aspects of climate can play in the vineyard, for better or for worse. Wind does a lot more than cause leaves to flutter in the breeze; humidity, too, has a much more complex relationship with the vine than it would first appear.

FROM MARKUS KELLER, *THE SCIENCE OF GRAPEVINES: ANATOMY AND PHYSIOLOGY*
5.2.3. WIND

Strong wind (>6 m s^{-1}) can induce physical damage in addition to a reduction of shoot length, leaf size, and stomatal density. However, even if it is not strong enough to induce visible damage, wind mechanically disturbs shoot growth, leading to shorter but thicker shoots. This so-called thigmomorphogenetic response intensifies with increasing height above the ground. Wind during the day seems to inhibit shoot growth more than at night. The reduction in shoot growth is also more severe on the windward (into the wind) side of the canopy than on the leeward (away from the wind) side and intensifies with increasing number of wind perturbations rather than with increasing wind speed in each single event. Moreover, the shoots are often displaced away from the wind so that the canopy becomes lopsided, which has consequences for the intensity and duration of fruit exposure to sunlight. Similar thigmomorphogenetic responses may also be triggered by other physical influences, such as repeated bending or touching by passing vineyard workers, animals, or machinery.

As discussed in Chapter 3.3, wind decreases a leaf's boundary layer resistance, which increases transpiration and enables evaporative cooling. This is an advantage under warm conditions because leaves tend to heat up rapidly when the wind speed drops below approximately 0.5 m s^{-1}. Under otherwise similar conditions, the temperature of sun-exposed leaves tends to vary inversely with wind speed, whereas shaded leaves track the air temperature. To avoid excessive water loss and dehydration in stronger wind, however, grapevines respond to wind speeds greater than 2.5 m s^{-1} by partly closing their stomata. Although this strategy may conserve water, it also reduces photosynthetic CO_2 assimilation and increases leaf temperature. This may at least partially explain why vines growing in areas with frequent strong winds often produce fewer and smaller clusters and lower fruit-soluble solids. Conversely, reduced wind speed and air mixing in sheltered vineyards could also decrease photosynthesis because leaves may deplete the CO_2 in the air surrounding the foliage. Moreover, wind speeds less than 0.5 m s^{-1} result in humid canopies, which favors disease development. For instance, the powdery mildew fungus E. necator requires only 40% relative humidity for germination; this threshold is easily exceeded within the leaf boundary layer, where the fungus resides.

Wind moving down the rows in a vineyard creates less turbulence and movement of leaves than wind moving across rows. This may decrease water loss, especially under dry conditions, and reduce the negative effect of stomatal closure on photosynthesis. In addition, wind speed at the center of a canopy is often less than 20% of the speed at the exterior. Although this may not be important in terms of stomatal effects on CO_2 assimilation (which is light limited), it has implications for drying of leaves and fruits after rain; interior surfaces dry more slowly than exposed surfaces. Nevertheless, even mild winds can move leaves inside a canopy sufficiently to briefly increase their light exposure. The resulting sunflecks on these leaves can account for a temporary rise in photosynthesis, improving the overall carbon balance of the canopy.

5.2.4. HUMIDITY

Intuitively, one would expect transpiration by leaves and berries to increase the humidity inside the canopy, with subsequent implications for the development of fungal diseases. This is a subject that has been very little studied. Increases in humidity of less than 10% have been recorded, and the significance of these increases is not well understood. However, a decrease in relative humidity from 95 to 50% increases the vapor pressure deficit between the leaf and the surrounding atmosphere more than 10-fold. Moreover, humidity strongly depends on air temperature because increasing temperature also sharply increases the vapor pressure deficit. Therefore, as the vapor pressure deficit increases (e.g., on west-facing leaves in the afternoon), stomatal conductance tends to decline in an attempt to control excessive water loss by transpiration, which will limit photosynthesis. This means that an increase in relative humidity effectively reduces the transpiration rate and increases CO_2 assimilation. Due to the modulating effect of leaf layers, this benefits exterior leaves more than interior leaves.

Humidity also affects leaf growth. A high vapor pressure deficit reduces the growth rate of leaves by decreasing the rate of cell division and cell expansion, even when there is no soil water deficit. Therefore, leaves growing in low humidity remain smaller than leaves grown in high humidity. Vines growing in dry climates tend to have more open canopies than vines growing in more humid climates, even when they are equally well watered.

In addition to air humidity, even small changes in vine water status can have an effect on canopy microclimate. Water-stressed vines often have higher canopy temperatures than fully irrigated vines because there is less transpirational cooling of waterstressed leaves and because such vines have a sparser, more open canopy.

THE GRAPE'S-EYE VIEW OF CLIMATE

Broad generalizations about climate can help growers decide what to plant, or clarify stylistic differences between a warmer and a cooler region. For individual vines, however, climate comes down to the air surrounding its clusters and the dirt supporting its roots. What's going on ten rows away, let alone at the nearest weather station, is largely beside the point. The leaves of a particular vine may roast in the sun at midday; in the shade of the canopy the grapes could be ten degrees cooler; and in a heat-trapping soil, the plant's roots might experience yet a third climatic variation—a micro-micro-climate.

A slim volume of research reports, *Sunlight Into Wine,* published in 1991, drove home the importance of micro-climate—the conditions immediately surrounding the grape-vines—and became one of the most important works on viticulture in the last fifty years. Most of this handbook, written by Australian-born, Cornell University-trained Richard Smart and vineyard engineer Mike Robinson, is devoted to the training and management of grapevine trellises and canopies to get just the right amount of sunlight at just the right time. Here we reprint the first few introductory pages, which establish that the immediate microclimate is a critical environment, both for the quantity and quality of sunlight.

FROM RICHARD SMART AND MIKE ROBINSON, *SUNLIGHT INTO WINE*

"A day without wine is like a day without sunshine" . . . so says the old proverb. This handbook discusses specifically the relationship between sunlight and wine.

Wine is a product of sunlight. Grapevine leaves use energy from sunlight to change carbon dioxide (CO_2), an atmospheric gas, into sugars. This process is called photosynthesis. From the leaves the sugars move to the fruit. At a desired ripeness, grapes are harvested and crushed at the winery to produce juice as a first step in winemaking. Yeast cells convert sugars in the juice into alcohol during the process of fermentation and the juice is transformed into wine. And so, the close association between sunlight and wine can be seen. Warm and sunny climates cause grape juice to have a high sugar content and so produce wines with higher alcohol.

This handbook is, however, concerned with indirect relationships between sunlight and wine—that is, the effect of sun exposure of grape clusters (bunches) and leaves on wine quality. To explain this relationship, we should first think of some Old World ideas about factors that affect wine quality. From these regions we often hear phrases like "a struggling vine makes the best wine," or "'low grape yield gives high wine quality." A common feature of both "struggling" and "low-yielding" vines (the two are often synonymous) is that they grow few leaves. The leaves that do grow are smaller than for non-struggling or high-yielding vines. These vine canopies are then less shaded, and most leaves and fruit are well exposed to the sun. This observation leads to the important question—does a vine need to be struggling or low-yielding to make high quality wine, or, perhaps, is it necessary for the leaves and clusters to be well exposed to the sun?

This handbook is based on a yes answer to the second part of the question. Research in New Zealand and, indeed, in many countries has shown that dense, shaded canopies reduce wine quality. Changing the canopy so that clusters and leaves are better exposed to the sun has been shown to improve wine quality and yield.

This result contradicts the firmly held traditional beliefs listed earlier. We make no apologies for this. There is clear scientific and commercial experience to support our viewpoint.

In the pages to follow, we expand on these ideas. First, we discuss how grapevines function, and respond to their climate. Then the concept of canopy microclimate is discussed, to illustrate the factors that are important in leaf and cluster exposure.

Following this we show how the winegrape grower can assess his or her own vineyards to answer the question "Do I have a canopy problem?"

Then finally, some practical canopy management solutions to the problem are presented. Generally the answer involves a change in trellis method. Details of the construction of the most important trellises are provided, along with tips to carry out the task in a cost efficient manner.

Readers will be able to use ideas presented here to increase yield and to make improved quality wine by a process we call "winemaking in the vineyard."

GRAPEVINE CANOPY MICROCLIMATE
Three levels of climate

Early observers of winegrapes noted a strong effect of climate on both yield and quality. To fully understand these effects we need to distinguish between three levels of climate based on the ideas of Geiger, the German climatologist.

Macroclimate (or regional climate)　This is the climate of a region, and the general pattern is described by a central recording station. The macroclimate definition can extend over large or small areas, but typically the scale is tens of kilometres depending on topography and other geographic factors, e.g., distance from lakes or ocean.

Mesoclimate (or topo or site-climate)　The mesoclimate of a particular vineyard will vary from the macroclimate of the region because of differences in elevation, slope, aspect or distance from moderating factors like lakes or oceans. Mesoclimate effects can be very important for the success of the vineyard, especially where climatic conditions are limiting. For example, in the Mosel Valley (West Germany), a region marginal to ripen grapes, vineyards are planted on hillsides facing south to promote sunlight absorption. Another example is in New York State where the vineyards are located on lake shores to reduce winter freeze injury. Vineyards planted at high elevations can lead to cooler conditions for improved table wine quality, compared to the hotter valley floor. Examples of this effect are found in the Barossa Ranges of South Australia and Napa and Sonoma Valleys in California. Differences in mesoclimate can occur over ten to hundreds of metres, or by up to several kilometres.

Microclimate (or canopy climate)   Microclimate is the climate within and immediately surrounding a plant canopy. Measurements of climate show differences between the within canopy values, and those immediately above it (the so called "ambient" values). A simple example is sunlight. Sunlight can be measured at the centre of a dense canopy as one percent or less of the values measured above the canopy. Microclimate differences can occur over a few centimetres.

There is confusion in the popular literature (and on wine bottle back labels!) about the term microclimate. Often the word microclimate is used incorrectly when in fact mesoclimate would be more accurate. Thus, it is better to say "this vineyard has a special *mesoclimate* due to its aspect," rather than "a special *microclimate*." When we consider the climate of a vineyard we are concerned with the macroclimate or mesoclimate. When we consider the climate of or within an individual vine, or part of a vine like a grape cluster, microclimate is appropriate.

Why do canopy microclimates differ?

The canopy microclimate is essentially dependent on how dense (or crowded) is the canopy. Let us now consider each climate element in turn, to show how they are affected by canopy density.

Sunlight quantity   The amount of sunlight falling on a vineyard varies with latitude, season, time of day, and cloud cover. Sunlight intensity is commonly measured in units that correspond to the ability of plants to use sunlight in photosynthesis. Consequently, the intensity is often termed "photosynthetically active radiation" (or PAR). The units are amounts of energy per unit area per unit time, i.e., micro Einsteins per square metre per second, μE $m^{-2}s^{-1}$. A bright sunny day might give readings over 2000 μE $m^{-2}s^{-1}$, and overcast conditions can reduce this value to less than 300 μE $m^{-2}s^{-1}$.

Values of sunlight intensity measured in the centre of dense canopies can be less than 10 μE $m^{-2}s^{-1}$ although above canopy (ambient) values are over 2000 μE $m^{-2}s^{-1}$. The reason for this large reduction is that grapevine leaves strongly absorb sunlight. Measurements show that a leaf in bright sunlight (say 2000 μE $m^{-2}s^{-1}$) will only transmit 6%, so 120 μE $m^{-2}s^{-1}$ pass through to the next leaf layer in the canopy. A third leaf in line would receive only 7 μE $m^{-2}s^{-1}$ and would be in deep shade. This simple example ignores reflection of light between leaf layers.

Sunlight quality   Not only is the amount of sunlight altered in the canopy, but so is the colour spectrum that makes up sunlight. Plant leaves absorb only a part of the sunlight—especially in the so called "visible range" (400–700 nm wavelength). This is the part of the sunlight spectrum we can see. So, as sunlight passes through the canopy, there is less sunlight in the visible range relative to the remaining wavelengths of the spectrum. An important consequence is that the ratio of red light (660 nm) to far red light (730 nm) declines in the canopy. Plants like grapevines respond to the red:far red ratio via their phytochrome system, and this is important in affecting, for example, fruit colour development. Although shoots, stems, petioles and fruit also absorb sunlight, by far the greatest reason for shade in grapevine canopies is the sunlight absorbed by leaves.

Temperature   Temperatures of grapevine parts are generally at or near air temperature. This applies unless they are warmed by absorbed sunlight, or cooled by evaporation of water, as takes place with transpiration from leaves.

Elevated tissue temperatures due to warming by sunlight are most obvious on sunny and calm days. For example, grape berries exposed to bright sunlight on calm days can be warmed up to 15°C (27°F) above the air temperature. Wind cools because it removes some of the absorbed heat from the leaf.

In contrast, grapevine leaves do not warm as much as berries because leaves are cooled by transpiration, whereas berries have little transpiration to show this effect. Transpiration is a process whereby water is vaporized within the leaf surface and escapes through stomata (pores) on the bottom of the leaf. The vaporization or change from liquid to gas requires energy (heat) which cools the leaf in the same way that an evaporative air conditioner works. As long as the vines are well supplied with water, leaves exposed to full sunlight are typically less than 5°C (9°F) above air temperature. If the vine is water-stressed, the temperatures can be higher. It is interesting to note that leaves on the shaded side of canopies can even be below air temperature. These leaves still transpire but do not have the heat absorbed from the sun.

At night, the outside parts of the canopy can lose heat to the atmosphere, especially on clear calm nights. This is known as longwave cooling. Under these conditions, exterior leaves and berries can be cooled to 1–3°C (2–5°F) below air temperatures.

Humidity   Transpiration by leaves can lead to a slight buildup of humidity inside dense canopies. If the canopy is open, then the ventilation effect of even a slight breeze can reduce the humidity difference from the canopy exterior to interior. However, even small differences in humidity can be important for the establishment of fungal diseases like *Botrytis*.

Wind speed   Wind patterns around a vineyard are complex, and there is an interaction between wind direction and row orientation. Similar to sunlight, wind speed is very low in the centre of dense canopies. This occurs because leaves slow down air flow. Measurements in New Zealand show that wind speed in the centre of dense Semillon canopies can be less than 10% of the above canopy value.

VINES AS CLIMATE CREATORS

To understand climate on a very small scale, you must consult Rudolf Geiger's classic text, *The Climate Near the Ground,* still regarded as the foundation of microclimatology. Written in 1927 and periodically updated until a final edition in 1960, Geiger traces in intricate detail the reciprocal influences between soil and the adjacent air layer, the effects of humidity on temperature, and the constant interplay between sun, soil, air and vegetation, both above and below ground. One can get a glimpse of this climatic complexity in

Geiger's chapter on "The Microclimate of Gardens, Potato Fields, and Vineyards," wherein he explains how the grapevines influence their own climate surroundings.

FROM RUDOLF GEIGER, *THE CLIMATE NEAR THE GROUND*

From the macroclimatologic point of view, German vineyards lie near the northern limit of the area in which vines can be cultivated. They are therefore by compulsion dependent on situations which have a microclimate that is especially sunny, warm, and free of frost. An introduction to living conditions on a vineyard terrace has been given by O. Linck in a very readable, well-illustrated book which should be of interest to microclimatologists as well as botanists.

The microclimate of a "wine mountain," as it is called in German, is made up of many individual factors. To begin with, the term "mountain" indicates that the sunny slopes of hills are selected for the cultivation of grapes. The climate of terrace vineyard is therefore that of a slope. This type of climate is altered artificially by terracing. These flatter surfaces, on which the vines are planted, are easier to work and are bounded on the side toward the mountain by a stone terrace step. In many places the vineyards are divided by stone walls. These have been built through centuries by the laborers in the vineyard, collecting loose stones and building them into walls at the sides. They run down into the valley and form a shelter against the wind, hence providing warm spaces, which however do not impede the flow of colder air down into the valley. If the holes between the stones have not been blocked by fine accumulated dust, these stone walls have low thermal conductivity. They therefore become strongly heated during the day and act as sources of heat; they are correspondingly cool and moist deeper down, and in addition to promoting the growth of xerophytic surface flora they also support deep-rooted bushes and even trees by means of which their protective influence is increased.

The type of country in which the hill is situated is of equal influence. If its foot lies near a river or a lake shore, extra warmth is obtained through specular reflection. If the hill is topped by a cold plateau, there will be an increased risk of night frosts through cold air flowing down at night. As a protection against this, the upper boundaries are often shut off by thick hedges or woods. According to R. Weise, the locations safest from frost in Franconia are surrounded by a crescent of slopes from SE through E round to W, and are open only toward the SW.

But the vines themselves, and the manner in which they are trained, also play a decisive role in creating the microclimate of the wine mountain, or, as it is called in the wine-producing country, the Wingerts or Wengerts. This aspect of vineyard climate, determined as it is by plants, is clearly the easiest to investigate because this part of the hill is level. This is usually the case in the research reports that follow, but there are also exceptions to this rule.

The first instrumental measurements were made in 1928 by R. Kirchner in the Palatinate wine district. They became known at a much later date, when K. Sonntag began his research there. He recognized that a basic distinction had to be made between the climate of the rows of vines and that of the open lanes between them. The sun is able to penetrate to the ground in the lane, which runs N-S, producing high surface temperatures and a large temperature gradient close to it, similar to the conditions in an open-planted vegetable field. In the rows of vines the highest temperatures

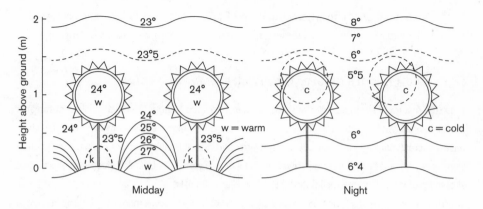

FIGURE 4.3
Radial heat depiction at midday and at night.

are found below the outer active surface where the foliage gives protection from the wind, but these highest values are naturally much lower than in the lanes. At night, the lowest temperatures are found at the level of the radiating surfaces of the leaves (not at the ground in the lanes, which are shielded), which means that the dew formed is collected to the advantage of the plants. "Even outside the vineyard," K. Sonntag wrote, "an iron pole was dry from the ground up to the level of the stems, but was covered with water droplets above the level of the leaves." This double influence of the outer active surface of the vines and of the solid ground can be seen also in the temperature profiles measured by Y. Tsuboi, Y. Nakagawa, and I. Honda in Japanese vineyards.

R. Weise, O. Jancke, and H. Burckhardt, in particular, advanced our knowledge of vineyard climate considerably. The influence of the height of the vine, with the distance between the rows constant, was investigated by Burckhardt in a level vineyard at Mussbach in the Weinstrasse district of Germany. The top of the foliage of the higher-trained Sylvaner grape reached 150 cm, while that of the lower was 90 cm above the ground and the distance between neighboring rows was 120 cm in both groups. Simultaneous readings with six psychrometers, including adjacent open ground, in August 1952 gave the following picture: the temperature decrease above the ground was found to exist in the N-S lanes between 10:00 and 14:00 only. In the higher growth, the temperatures were lower both by day and at night in comparison with the adjacent bare land, because of extensive shading. In the low growth there was hardly any difference between the lane temperatures and the bare land. The humidity gradient showed a lapse with height both by day and at night, greater in the low vines than over bare land, and greater in the high vines than in the low.

The crown area of the vines was subject to radiative heating and evaporation loss by day. In the higher vines the first of these was greater because of its greater leaf area, while in the lower vines the second factor was the larger. The temperature is therefore higher in the area of the grapes, in the taller vines, compared with the temperature at the same level in the lanes between them, and also higher than that of the shorter vines. However, since the grapes in the latter are closer to the ground, this balances out. Grapes of both taller and shorter vines enjoy the same kind of temperature in the warm hours at midday. This explains why the fear of the vintner that the yield of grapes will be reduced if they are trained too far from the heat-dispensing ground is not justified. Water-vapor pressure is, however, always higher over the vines that are trained higher because the mass of transpiring leaves is greater.

This result agrees with the observations of R. Weise that in the Würtzburg wine country the higher form of training (Frankish stem training) does not show a loss of heat during the day, in comparison with the lower form (Frankish head training). The first form, however, allows the cold night air to flow away more easily in the comparatively foliage-free space in the lower layer of air, and therefore runs less risk of frost damage than the second form. This has been proved recently by R. Weise in the published results of measurements he made inside the vine shoots.

Then in August 1957 H. Burckhardt investigated the difference between an open form of planting, with a distance of 3 m between the rows of 2.1-m-high vines, which is desirable because of the advantages it offers in ease of working, and a normally trained vineyard with 1.2 m between rows of vines 1.1 m high. The microclimate of the lanes in the widely planted vines was more balanced, and the air more settled, than in the normal style, and gave the cold night air a better chance to flow away. However, the temperature was rather lower during the day in the area where the grapes were growing than in the normal vineyard, which is confirmed by the greater acidity that can be observed in grape juice from open vineyards. The relative humidity was also somewhat higher, and this increases the risk of infection by fungus diseases.

N. Weger demonstrated the practical consequences of the considerable differences in the microclimate of two vineyards only 3 meters apart in the Geisenheim area. The difference was apparent in the quality of the grapes and in the incidence of pests. Practical viticulture may derive great advantages from such measurements, provided they are made with the necessary instruments, and the requisite amount of time is devoted to them.

ON THE OTHER HAND . . .

This chapter makes it obvious that climate is critical to the success of grape-growing, and that the endless combinations and permutations of climate make ideal regions rare indeed; it also illustrates how maddening it can be for those not fortunate enough to plant their vines in the best spots.

An inhospitable climate has been a source of great frustration for British winegrowers, world style-setters in all things vinous for the last two or three centuries. Despite being great consumers of wine, the inhabitants of the UK have not fared well as producers (except in recent years, thanks to climate change; see chapter 9.)

Ever resourceful, the British have been in the vanguard of greenhouse grape cultivation, through which the grower can create whatever climate he or she chooses, shielded from the outside world. The Royal Horticultural Society offers lots of information on indoor cultivation on its web pages and sponsors an annual competition. Helpful videos abound on the web. Nurseries advertise varieties especially well suited for indoor cultivation, which usually means either Black Hamburg or Muscat of Alexandria. Needless to say, *terroir* rarely comes up in these discussions.

5

GRAPEVINES
Bringing Terroir *to Life*

Most invocations of *terroir* make reference to special features of the soil or the climate in a certain place, and how they can (almost) literally be tasted in the glass. But the standard accounts are conspicuously silent about the contribution of the grapevines themselves. Everyone knows grapes are necessary for making wine, but once vines are planted and bear some fruit, their involvement is simply taken for granted, while the evangelists of *terroir* sift through the dirt and ponder the influence of sea breezes.

Such a view is palpably ridiculous, not to mention patently unscientific. Naturalists in antiquity, Aristotle among them, thought of plants as extensions of the soil, drawing in not just water and nutrients, but the plant's matter itself, its stems and leaves and flowers and fruit, all of it somehow manifestations of the ground below as it found its way into the glass. But starting in the middle of the nineteenth century, scientists began to develop a very different understanding of plant growth and development, centered on the action of photosynthesis. Instead of sucking their essences out of the earth, plants, in the presence of light, transform water and carbon dioxide into sugar, oxygen, chemical energy, and fruits and flowers. Show us a rock or a wind current that can do that, and we'll eat this book.

In the vineyard, grapevines do all the work, taking what they need—or what they can get—from the soil and the weather. By way of a complex genetic code, vines roll through a cycle of flowering and fruiting and dormancy, year after year, creating in the process hundreds of aromatic compounds and precursors from scratch. For any wine, whether it expresses *terroir* or not, virtually all of the flavors in the bottle are the product of the vines.

Grapevines are major players in the exposition of *terroir*, not merely passive transmitters of signals from the physical environment. Vinegrowing constitutes one of the most complex agricultural systems in the world, the intricate matrix of interactions linking climate, land, and human intercession, all of it mediated through the plant's physiology. The distinctive flavors of New Zealand Sauvignon Blanc may largely be due to climate factors, but the only reason anyone can taste them at all is because some grape berries served them up. To leave the labors of the grapevines out of the *terroir* equation is like critiquing a film by focusing on the sets and soundtrack, and ignoring the actors and the action.

Our survey of grapevines and *terroir* starts with an acknowledgement that those who research the science of the grapevine, viticulturalists, tend to be skeptical of the entire concept. The tenor of that doubt is expressed here in a piece by grapevine researcher Mark Matthews. Next, because most wine lovers have scant knowledge of how grapevines work, we've reprinted a not-too-technical overview of vine physiology by viticulturalist Ed Hellman. *Terroirists* often embrace older assumptions that may not survive modern scrutiny of how vines work: advantages of deep roots and dry-farming; a necessity for ancient soils and older vineyards; beliefs that low yields are necessarily better and that vines must "struggle." Two of the most important modern approaches to viticulture are summarized in selections on balanced vine growth by Oregon State professor Patty Skinkis and an overview of deficit irrigation from the viticulture website of the Australian state of Victoria.

Next, Cornell researcher Gavin Sacks takes up the question of how much grape variety matters for *terroir* expression. Napa winemaker John Williams muses on the virtues of sustainable farming and how it facilitates *terroir* expression; following this, we report on the results of a years-long field trial that measured the possible contributions of organic and biodynamic farming practices to conveying place. The concluding selection, by wine archaeologist Patrick McGovern, reminds us that above all, grapevines became a vector for *terroir* because they motivated growers, providing a prodigious source of sugar, and, of course, alcohol, a profoundly effective social and cultural lubricant.

THE VITICULTURAL CRINGE

In all the books and articles depicting soil and climate factors in *terroir*-expressive wines, the literature explicitly linking viticulture and *terroir* is scant. And since traditional and historical discussions of *terroir* have been perpetuated without the benefit of viticultural science, or have repeated dubious assertions about vinegrowing that don't hold up to scrutiny, many viticulturalists view the whole topic with indifference and disdain. Patty Skinkis, viticultural researcher at Oregon State University, captured this skepticism well in an email: "I think the concept arose in a time when we knew much less about grapevines and plant physiology in general. For this reason, those of us who are trained viticulturists/plant physiologists cringe at the word *terroir*."

Similarly, Washington State University viticulturalist Markus Keller chimed in with this: "I believe the concept of *terroir* is mostly meaningless, because something that explains everything (in one word!) ends up explaining nothing. It certainly isn't a useful approach to trying to figure out why some wines are better than others—and how to make those 'others' better as well."

One of the most vocal, pointed critics of the entire intellectual edifice of *terroir* is Mark Matthews, a professor of viticulture at the University of California at Davis who shared his unique perspective at a January 2006 conference on *terroir* sponsored by the Robert Mondavi Institute at UC Davis, and who wrote the 2016 book *Terroir and Other Myths of Winegrowing*, published by University of California Press. Unlike experts who seek to fine-tune our understanding of the concept of *terroir*, Matthews considers it unscientific and suggests dumping it altogether. He generously prepared the passage below in response to our request.

FROM MARK MATTHEWS, "A BRIEF HISTORY OF PLANT BIOLOGY IN RELATION TO TERROIR"

Humans make things up to explain to themselves what they experience and "the nature of things." Some vineyards seem to produce better wines. We want and need explanations. Science is one way of doing this, and it is by any account a remarkably successful way. However, grape and wine production is much older than Science. We necessarily have considerable received "knowledge" of winegrowing. Wine is a traditional product and terroir is a traditional explanation for those wines. Tradition is important in the wine trade and, for many, in wine appreciation. Tradition has an appeal derived from familiarity and/or nostalgia, but tradition is repetition—an inherited pattern of thought or action. It forestalls progress, and is not necessarily related to a truth about the natural world.

Aristotle said that the roots of plants were as the mouths of animals, and this was the paradigm—plants grew by taking in soil as food—for most of history. Our understanding of the natural world has progressed enormously since Aristotle: his ideas on the human body and the solar system took major hits from Vesalius and Copernicus in the same year, marking the outset of the Scientific Revolution. Plant biology, especially in the context of winegrowing, has been a laggard. Did plants grow from magically transmutated soil or water? Did plants use the soil mineral fraction or humus (decaying organic matter) as food? It was not until 1850 that the most advanced "natural philosophers" assembled the elements of photosynthesis and came to understand how the plant body was formed from water, air, and light energy.

By that time, the reputations of the top (first growth) wines of Europe were well established. How could those distinctions be understood? Only as soils that were transmutated to give different characters to the fruit and wines. Hence, "terroir." The concept originated in the faulty belief that the plant and its fruit and their wines literally come from the soil.

It is a fact seldom acknowledged by the Keepers of the Terroir that from its earliest use in conjunction with wine flavor as *goût de terroir* (taste of terroir), terroir was invoked as a pejorative. One example from the Antoine Furetière's *Dictionnaire Universal* 1690:

Terroir = land considered by its qualities. The plants, the trees grow well only if it is their own terroir . . . Vines need a dry terroir, with stone and rocks . . . We say that the wine has a taste of terroir when it has some unpleasant quality, which comes from the kind of soil where vines were planted.

A terroir wine was a bad wine, with the implied or explicit sense that the off flavors are directly transferred from soil to berry to wine. Sometimes the problem was attributed to the particular manure used.

The transformation from a pejorative to superlative use of terroir did not emerge from new understanding of how plants interact with soil or other aspects of the environment to give us good wine. Apparently, the re-branding was the result of regional (and laudable) efforts to better or protect the livelihood of producers, first for peasant producers in post-phylloxera reconstruction, and later in the face of New World competition for fine wine production. The regional promotion led to the eventual establishment of appellation systems with new terms like Grand Cru. The social history of this phenomenon is well-researched, including for example, by Marion Demossier on Burgundy, Robert Ulin on Bordeaux, and Kolleen Guy on Champagne. Demossier describes the promotional program as an example of "producing tradition," producing a favorable impression of regional wines once dismissed as "terroir." By the mid 20th century, the use of the term terroir was in transition, as shown by this reference in Renouil's 1962 *Dictionnaire du vin:*

Terroir (taste of), terroité. This taste corresponds to the characteristic taste given to the wine by the soil. Sometimes the taste of terroir which characterizes a wine is a key element which makes it particularly appreciated. But often also, it brings to the wine an unpleasant taste which can have been communicated by too much manure, by a particular manure, ...

The second development was the (in)famous "Judgment of Paris" in 1976 in which wines that were not top French wines were mistaken for top French wines by top French wine experts. This exercise has been repeated several times, with the consistent result being the difficulty in resolving the different origins among top wines. The earth shook and the status quo was challenged. Use of terroir in conjunction with Old World wines took off after the Judgment in Paris, as those threatened by New World competition claimed patrimonial places in the wine hierarchy based on the physical environment—i.e., soils.

The Keepers of the Terroir harken back to its long use in winegrowing, particularly to "the taste of terroir." New World wines couldn't have it (yet), but old is not evidence of truth. It's been a vague, "certain kind of soil" for 350 years. The idea that flavors come from soils has no basis in empirical observations. Indeed, there is little evidence of flavors even moving to the berry from other places in the plant. Almost all flavor and aroma are synthesized in the berry itself.

Recent decades have shown acquiescence to a larger source of wine flavor than soil—to recognize that aspects of the leaf and fruit environments play important roles in fruit growth and ripening. This terroir definition is a synonym for environment. A fundamental concept in biology for a century has been that the phenotype, the fruit composition, is the result of the variety (genotype) and its interaction with the environment. Some catching up is good, but investment in sorting out the details has been limited. The work by many viticulturists to define spaces, "terroirs," viticulture zones, etc., carries the implicit assumption that there are such definable spaces in terms of the plant environment and that they occur in economically viable units. This work is done without the necessary grounding research that shows how the environmental parameters affect winegrape growth and composition and the ultimate wine sensory attributes—the proverbial cart before the horse.

The question of the physical basis for the terroir/appellation system issue has been hotly contested. The results are in: the consensus of a wide range of scholars, from anthropologists, historians, geographers, economists, in addition to crop scientists, is that the attribution of distinct wines of appellations to the physical environment and to soils in particular is not tenable on the basis of present scientific evidence. The prominent French historian Roger Dion wondered why the French preferred to agree that the qualities of their wines were the effect of natural privilege, "of a particular grace accorded to the soil of France, as though our country would derive greater honor by receiving from Heaven, rather than from the painstaking labors of man, this fame for the wines of France."

Rather than get more specific about what it is in soil-plant interactions that are important as in the normal progression of science, the use of terroir in the world of wine has broadened—perhaps to the point of obscurity. The modern terroir concept and its appellation system are about much more than environment. Terroir in the broader and perhaps contemporary sense is a social geography term. Traditional products associated by consumers and producers with geographical areas have sociological and economic value. And it turns out that the wine laws of Europe have been step by step acknowledging that. Dev Gangjee of the London School of Economics reported that European countries have spent decades gradually distancing themselves from a system with physical geography having the greatest importance. He says that French wine law has gone further:

"A historical perspective therefore allows us to appreciate the French AO regime's eventual rejection of 'conditions particulières de climat et de terroir' as the sole or sufficient basis for protection, while also exposing an alternative basis founded on intergenerational, collectively generated savoir faire."

Thus, present arguments from the EU in World Trade Organization negotiations are for continued "terroir" or Geographical Indication protections on the basis of special *cultural* products, and it is the culture that needs protection.

It is indeed lovely that truly fine wines are different. Terroir is a vague, anachronistic, and inadequate explanation of those differences.

HOW GRAPEVINES WORK

Perhaps the starting point for bringing grapes and vines into the *terroir* conversation is to appreciate the complexity, resourcefulness and adaptability of these remarkable organisms. While most wine lovers and *terroir* aficionados have at least a general understanding of winemaking practices, few have a working knowledge of vine physiology, the collection of processes that generate the raw material for wine. Consequently, we decided that this chapter had to include a primer on how grapevines work, which will, we hope, provide a baseline for the more specialized pieces that follow.

The following selection comes from the first chapter of *Oregon Viticulture*, written by Edward Hellman, formerly on the faculty at Oregon State University and now at Texas A&M/Texas Tech. The introductory chapter on "Grapevine Structure and Function" is an excellent overview of the subject, introducing essential vocabulary and providing a concise overview of a very intricate topic in minimally technical terms. (The book gets quite a bit more dense after this chapter.) This excerpt covers only a small part of the anatomy and physiology of the vine, mainly focusing on flowers and fruit. But given that this will be uncharted territory for most readers, it is still one of the longest passages in this book.

FROM EDWARD HELLMAN, "GRAPEVINE STRUCTURE AND FUNCTION"

Flowers and Fruit. A *fruitful shoot* usually produces from one to three *flower clusters* (inflorescences) depending on variety, but typically two under Oregon conditions. Flower clusters develop opposite the leaves, typically at the third to sixth nodes from the base of the shoot, depending on the variety. If three flower clusters develop, two develop on adjacent nodes, the next node has none, and the following node has the third flower cluster. The number of flower clusters on a shoot is dependent upon the grape variety and the conditions of the previous season under which the dormant bud (that produced the primary shoot) developed. A cluster may contain several to many hundreds of individual flowers, depending on variety.

The grape flower does not have conspicuous petals; instead, the petals are fused into a green structure termed the *calyptra* but commonly referred to as the *cap*. The cap encloses the reproductive organs and other tissues within the flower. A flower consists of a single *pistil* (female organ) and five *stamens,* each tipped with an *anther* (male organ). The pistil is roughly conical in shape, with the base disproportionately larger than the top and the tip (the *stigma*) slightly flared. The broad base of the pistil is the *ovary,* which consists of two internal compartments, each having two ovules containing an embryo sac with a single egg. The anthers produce many yellow *pollen* grains, which contain the sperm.

The time during which flowers are open (the calyptra has fallen) is called *bloom* (also flowering or anthesis) and can last from one to three weeks depending on weather. Viticulturists variously refer to *full bloom* as the stage at which either roughly one half or two-thirds of the caps have loosened or fallen from the flowers. Bloom typically occurs between 50 and 80 days after budburst in Oregon.

When the flower opens, the cap separates from the base of the flower, becomes dislodged, and usually falls off, exposing the pistil and anthers. The anthers may release their pollen either before or after capfall. Pollen grains randomly land upon the stigma of the pistil. This event is termed *pollination*. Multiple pollen grains can germinate, each growing a pollen tube down the pistil to the ovary and entering an ovule, where a sperm unites with an egg to form an embryo. The successful union is termed *fertilization,* and the subsequent growth of berries is called *fruit set*. The berry develops from the tissues of the pistil, primarily the ovary. The ovule together with its enclosed embryo develops into the seed.

MAJOR PHYSIOLOGICAL PROCESSES

Photosynthesis

Grapevines, like other green plants, have the capacity to manufacture their own food by capturing the energy within sunlight and converting it to chemical energy (food). This multi-stage process is called *photosynthesis.* In simple terms, sunlight energy is used to split water molecules (H_2O), releasing molecular oxygen (O_2) as a byproduct. The hydrogen (H) atoms donate electrons to a series of chemical reactions that ultimately provide the energy to convert carbon dioxide (CO_2) into carbohydrates (CH_2O).

■　■　■

Photosynthesis occurs in chloroplasts, highly specialized organelles containing molecules called *chlorophyll,* which are abundant in leaf cells. The structure of a leaf is well adapted to carry out its function as the primary site of photosynthesis. Leaves provide a large sunlight receptor surface, an abundance of specialized cells containing many chloroplasts, numerous stomata to enable uptake of atmospheric carbon dioxide, and a vascular system to transport water and nutrients into the leaf and export food out.

The products of photosynthesis are generally referred to as *photosynthates* (or assimilates), which include sugar (mostly sucrose) and other carbohydrates. Sucrose is easily transported throughout the plant and can be used directly as an energy source or converted into other carbohydrates, proteins, fats, and other compounds. The synthesis of other compounds often requires the combination of carbon (C) based products with mineral nutrients such as nitrogen, phosphorus, sulfur, iron, and others that are taken up by the roots. Starch, a carbohydrate, is the principal form of food energy that the vine stores in reserve for later use. The carbohydrates cellulose and hemicellulose are the principal structural materials used to build plant cells. Organic acids (malic, tartaric, citric) are another early product of photosynthesis and are used directly or converted into amino acids by the addition of nitrogen. Amino acids can be stored or combined to form proteins.

■　■　■

Sunlight. The process of photosynthesis is obviously dependent upon sunlight, and it is generally assumed that between one-third and two-thirds full sunlight is needed to maximize the rate of photosynthesis. The optimization of sunlight captured by the vine is an important component of canopy management that not only affects the rate of photosynthesis but also directly influences fruit quality. Sunlight exposure on a vine is highly dependent upon the training system and the shoot density and can be influenced by the orientation of the rows and row spacing. The term *canopy management* encompasses many vineyard practices designed to optimize the sunlight exposure of the grapevine.

Other Environmental Influences. The rate of photosynthesis in grapevines is also influenced by leaf temperature; the apparently broad optimum range of 25–35ºC (77–95ºF) may be attributable to differences in grape variety, growing conditions, or seasonal variation. Leaf temperature can be highly dependent upon vine water status but otherwise cannot be influenced to the same extent as sunlight exposure in the canopy, so it is of less concern to vineyard management.

Water status of grapevines can have a strong impact on photosynthetic rate through its control over the closing of leaf stomata, the sites of gas exchange critical for photosynthesis. A *water deficit* exists when the plant loses more water (via transpiration) than it takes up from the soil. One consequence of water deficits is the closure of stomata, which reduces water loss but also reduces the uptake of CO_2 necessary for photosynthesis. The extent of stomatal closure, and therefore the impact on photosynthetic rate, is related to the severity of water deficit. Vines are considered to be under *water stress* when the deficit is extreme enough to reduce plant functions significantly. The major impact of water deficits on vine photosynthesis is the reduction of leaf area.

Inadequate supply of certain nutrients (nitrogen and phosphorus) may also limit photosynthesis directly, or indirectly by reduced availability of elements (iron and magnesium) for the synthesis of chlorophyll.

Absorption of Water and Nutrients

Water. The suction force of transpirational water loss is transmitted throughout the unbroken column of water in the xylem all the way to the roots, providing the major mechanism by which water is taken up from the soil. Water is pulled into the root from the soil. Young roots absorb the majority of water, primarily through root hairs and other epidermal (outer layer) cells. But older suberized ("woody") roots uptake water at a lower, but constant, rate. Water then moves through the cells of the inner tissues of the root and into the xylem ducts, where it continues its movement upward, reaching all parts of the vine, and is eventually lost via the stomata.

The effect of transpiration on the rate and quantity of water uptake is obvious, but new root growth is also necessary because roots eventually deplete the available water in their immediate area and soil water movement is slow at best. Therefore, conditions that influence root growth affect the rate of water uptake.

Nutrients. Mineral nutrients must be dissolved in water for uptake by roots. Nutrient uptake often occurs against a concentration gradient; that is, the concentration of a mineral nutrient in the soil solution is usually much lower than its concentration in root cells. Thus an active process, consuming energy, is required to move nutrients against the concentration gradient. Active transport is a selective method of nutrient uptake, and some nutrients can be taken up in much greater quantity than others. Nitrates and potassium are absorbed several times as rapidly as calcium, magnesium, or sulfate. There are also interactions between nutrient ions that influence their absorption. For example, potassium uptake is affected by the presence of calcium and magnesium. In rapidly transpiring vines, nutrient uptake also occurs by mass flow (a passive process) with water from the soil solution.

MAJOR DEVELOPMENTAL PROCESSES

Fruit Growth

Berry development commences after successful pollination and fertilization of ovules within a flower. Flowers with unfertilized ovules soon shrivel and die, while those remaining begin growth into berries. Many of these tiny berries *abscise* (drop off) within the first two to three weeks. Following this drop period (called *shatter*), the retained berries generally continue to develop to maturity. Commonly, only 20–30% of flowers on a cluster develop into mature berries, but this is adequate to produce a full cluster of fruit.

The berry develops from the tissues of the pistil, primarily the ovary. Although pollination and fertilization initiate fruit growth, seed development seems to provide the greatest growth stimulus, as evidenced by the relationship of fruit size to the number of seeds within the berry. The maximum number of seeds is four, but lack of ovule fertilization or ovule abortion reduces the number of developing seeds, generally resulting in smaller berry size.

Berry growth occurs in three general stages—rapid initial growth, followed by a shorter period of slow growth, and finishing with another period of rapid growth. A graph of grape berry growth thus appears as a double sigmoid pattern. Berry growth during the first stage is due to a rapid increase in cell numbers during the first three to four weeks, followed by two to three weeks of rapid cell enlargement. During this stage the berries are firm, dark green in color, and rapidly accumulating acid. Seeds have attained their full size by the end of the first growth stage.

The middle stage, called the *lag phase,* is a time of slow growth. The embryo is rapidly developing within each seed, and the seed coat becomes hardened. Berries reach their highest level of acid content and begin to accumulate sugar slowly. Toward the end of lag phase, berries undergo a reduction in chlorophyll content, causing their color to change to a lighter green.

The final stage of berry growth coincides with the beginning of fruit maturation *(ripening)*. The beginning of ripening, referred to by the French term *veraison,* is discernable by the start of color development and softening of the berry. The color change is most easily visible on dark-colored varieties, but "white" varieties continue to become lighter green, and some varieties turn a yellowish or whitish-green color by harvest. Softening of the berry and rapid sugar accumulation occur abruptly and simultaneously. Berry growth, occurring by cell enlargement, becomes rapid again in this final stage of ripening. It is thought that most of the water entering the berry after veraison comes from phloem sap, since xylem at the junction of the berry and its *pedicel* (stem) appears to become blocked at this time.

During ripening, acid content declines and sugar content increases. It is widely believed that flavors develop in the later stages of ripening. Berries begin to accumulate sugar rapidly at the start of the ripening period, and the rate tends to remain steady until accumulation slows as the end of the maturation period is approached. Sugar is translocated as sucrose to the fruit, where it is quickly converted into glucose and fructose. Both sugars and acids primarily accumulate in cells constituting the *pulp* (flesh) of the berry, although a small amount of sugar accumulates in the skin.

The skin (epidermis) and the thin tissue layer immediately below it contain most of the color, aroma and flavor constituents, and tannins contained in the berry. Thus, all things being equal, small berries have greater color, tannins, and flavor constituents than large berries because the skin constitutes a larger percentage of the total mass of small berries. Seeds also contain tannins that can contribute to the overall astringency of wine.

The chemical composition of grape berries is complex, consisting of hundreds of compounds, many in tiny quantities, which may contribute to fruit quality attributes. The single largest component is water, followed by the sugars fructose and glucose, then the acids tartaric and malic. Other important classes of chemical compounds within grape berries include amino acids, proteins, phenolics, anthocyanins, and flavonols. The reader is referred to a review of the biochemistry of grape ripening by Kanellis and Roubelakis-Angelakis (1993) for a thorough discussion of this topic.

Berries are considered to be fully *ripe* when they achieve the desired degree of development for their intended purpose, and they are generally harvested at this time. Ripeness factors of the fruit that are typically considered when scheduling harvest are sugar content, acid content, pH, color, and flavor. The combination of these factors determines the *fruit quality* of the harvest. Ripening processes in the fruit cease upon harvest, but while fruit is on the vine ripening is a continuous process. So there is usually a short time, influenced by weather, during which the fruit remains within the desired ripeness parameters. Berries can become *overripe* if harvest is delayed until the fruit has developed beyond the desired range of ripeness. Consider also that ripeness parameters can vary considerably depending on the intended use. For example, Pinot noir grapes for sparkling wine production are harvested much earlier, at lower sugar and higher acid content, than Pinot noir for non-sparkling red wine. Thus, the terms "fruit ripeness" and "fruit quality" do not have absolute values but are defined subjectively.

Fruit ripening can be delayed, and the attainment of desired ripeness parameters inhibited, by an excessive *crop load* (amount of fruit per vine). A vine that is allowed to produce more fruit than it can develop to the desired level of ripeness is considered to be *overcropped.* Severe overcropping can negatively impact vine health as well as fruit quality by precluding the vine from allocating adequate photosynthates to weaker sinks: shoots, roots, and storage reserves. Viticulturists generally seek to attain *vine balance,* the condition of having a canopy of adequate, but not excessive, leaf area to support the intended crop load to the desired level of fruit ripeness.

Climatic factors, particularly temperature, have long been recognized to have a major influence on the fruit quality of grapes and subsequent wine quality. The principal effect is on the rates of change in the constituents of the fruit during development and the composition at maturity. Hot climates favor higher sugar content and lower acidity; cool climates tend to slow sugar accumulation and retain more acidity. Grape varieties tend to ripen their fruit with a desirable combination of quality components most consistently in specific climates. Thus, some varieties, such as Pinot noir and Gewürztraminer, are considered to be "cool climate varieties," whereas others such as Carignane and Souzão are considered to be "warm" or "hot climate varieties." A few varieties, most notably Chardonnay, are capable of producing high-quality wines in different climates by adjusting the wine style for the varying expression of fruit characteristics in each climate. The relationship of climate, and in particular temperature, to fruit ripening and wine quality has been incorporated into methods of matching grape varieties to climate; Winkler's heat summation (degree days) system for California is one such system, and there are other more elaborate methods. *Phenology* is the study of the relationship between climatic factors and the progression of plant growth stages and developmental events that recur seasonally.

Thus, the first step in the production of high-quality winegrapes is the selection of a site with appropriate climatic characteristics for fruit ripening of the varieties to be grown. Vineyard practices, including training systems and canopy management, are utilized to optimize the sunlight and temperature characteristics of the canopy for fruit ripening. In cool climates, canopy management practices that provide good exposure of leaves and fruit to sunlight have generally improved grape and wine composition. Vines in which the canopy interiors are well exposed to sunlight usually produce fruit with higher rates of sugar accumulation, greater concentrations of anthocyanins and total phenols, lower pH, and decreased levels of malic acid and potassium compared to vines with little interior canopy exposure. Improved fruit quality under such circumstances may be due to higher temperatures in addition to better sunlight exposure, but it is extremely difficult to separate these factors.

VITICULTURE AND THE *TERROIR* MINDSET

Traditional *terroir* discourse rarely focuses on viticultural practice. One may speak of a wine's charms in terms of its soil character, or the patterns of a site's vine-cooling sea breezes; rarely, however, will you hear a devoted *terroiriste* sing the praises of canopy management, cover crop, irrigation or pest control—but such factors may lie at the heart of a wine's compelling flavors.

But the traditional *terroir* mindset does embrace a handful of generalizations about what grapevines need in transmitting *terroir* to the glass. Most often, these shibboleths are tossed off in a knowing sentence or two—"Vines must have deep roots in order to fully express a place," or "Low yields are essential for *terroir* expression"—before the conversation moves on. These convictions are deeply held in some quarters, constituting a kind of *terroir* belief system. Often the best way to shed light on age-old viticultural floklore is to see if it holds up when compared with current research and practices.

MODERN SCIENTIFIC PRACTICES

Most of the traditional assertions about *terroir* don't hold up to viticultural scrutiny. Conceptually, they arose as beliefs in a pre-scientific era, when grapegrowers knew little about plant physiology and nothing about photosynthesis. So, what *does* account for high quality and strong *terroir* expression in the vineyard?

Modern viticultural research and experimentation has provided a multitude of answers and resulted in an astonishing improvement in the quality of grapegrowing in recent decades. The fact that the world is awash in perfectly drinkable mass-market wine owes as much to improved grapegrowing as to improved winemaking. And at the premium level, the segment of the wine industry where distinctive wines speak of place, our collective understanding has advanced markedly.

In most thoughtful discussions of how the vine expresses terroir, two key ideas get articulated: "vine balance," and "water management," by means of timely, moderate water stress. Both of these ideas have gone from the experimental fringes of the industry to mainstream practice in a few short decades.

BALANCED VINES

Vine balance is vastly more sophisticated than the simple premise that great wines need low yields; most modern winemakers put the focus on the output of individual vines, not on that of sections of land. Winemaker Doug Fletcher of Chimney Rock in the Napa Valley explains it this way.

"High end customers would want to know how many tons per acre we got off our vineyards. They thought they were asking a quality question, and had some number in mind for high quality wines; anything over that number meant lesser quality. I told them that ratio, tons per acre, didn't tell them what they wanted to know. Of course, I'd get this look of disbelief."

Instead, Fletcher measures balance in terms of what he calls "a yield-to-pruning-weight ratio." For every gram of berry weight, he explains, he ensures that the plant has produced about 12 square centimeters of leaf area. "We get about 1.25 to 1.5 pounds of fruit per foot of cordon length," he explains, thereby avoiding any confusion that may arise over vine spacing. Of course, this was an unsatisfying answer to most people (sommeliers especially). So to prove his point, he developed the following story, which he's been telling for 15 or 20 years.

"In the early grape growing days in Napa, everyone spaced their Cabernet Sauvignon vineyards with 12' between rows & maybe 6' between plants," he says. "The average yield per acre was about 3.5 tons per acre. One day a sales manager approached the vineyard manager about yields. To get high scores, he told him, you had to lower yields. 'All the wine critics say you can't get more than 2 tons per acre if you want to make great wine,' he told him. So the following winter, the vineyard manager pulled out every other row. Sure enough, the tons per acre dropped to 1.75 tons per acre, just what the critics wanted to see."

The concept of balanced vines, outlined here by Oregon State University professor/researcher Patty Skinkis posits that in order to ripen fruit properly, vines need to reach an equilibrium between fruit growth and vegetative growth. That relationship, she maintains, can be calculated and measured. Vines that go out of range—including some that are intentionally unbalanced in pursuit of winemaking goals—are at war with their environment, and it shows.

FROM PATTY SKINKIS, "BASIC CONCEPT OF VINE BALANCE"
THE BASICS

Vine balance is the central concept in the study of viticulture where vine physiology and vineyard production merge. This concept was introduced long ago and has been further defined by viticulture researchers around the world. Examples of vine balance are central to vineyard production and have arisen from research during the past century. Nelson Shaulis developed the balanced pruning method which evaluates pruning decisions based on bringing a vine into balance. Numerous researchers have further defined this concept of balance by trying to understand the impact of production practices on sustained vine growth and development, including impacts of canopy management, effect on flowering and fruit set, influence on dormancy and winter hardiness, yield, and fruit quality. Studies continue today to further understand production and physiology as related to vine balance.

The concept of vine balance is often easier to understand in theory than it is to perfect in practice. Trained viticulturists learn that vine balance is defined as the state at which vegetative and reproductive growth lead to the most "balanced" vine. Unlike wine balance, which is a more subjective concept, vine balance measures vine growth capacity through fruit yields and vine size (leaf area or dormant pruning weight), and has some generalized guidelines to understand whether a vine is in good shape (i.e., vigorous or weak). In the most straightforward sense, *vine balance has been defined and calculated as the ratio between vine yield and vine size,* representing reproductive and vegetative production of the vine, respectively. This relationship is known as crop load and is calculated by taking a vine's yield and dividing it by the dormant pruning weight. There are two different equations for crop load, the Ravaz Index and the Growth-Yield Relationship. The distinction between the two equations is as follows:

- **Ravaz index =** Yield/Pruning Weight, where the yield from the current harvest is used against the pruning weight in the following dormant season. Note: This is the most commonly used crop load calculation.
- **Growth-Yield Relationship =** Yield/Pruning Weight, where the pruning weight from the dormant season is compared against the yield of the following season.

The resulting number is a ratio, and, therefore, is without units. Research to date indicates that vine balance is obtained within the range of five to ten for *Vitis vinifera.* This range is rather large but is required, as there are different levels of vine balance based on the vineyard site and production goals. Crop load ratios that are lower than the optimum range are under-cropped (low yields and larger vine), while numbers at the high end of the spectrum are over-cropped (more fruit and smaller vine). Ultimately, being at either end of the spectrum can lead to unsustainable vine growth and production.

Vine leaf area is another metric that has been used for understanding vine balance, comparing yields to leaf area. Many trials have been conducted over the years to define leaf area required to ripen fruit adequately. Kliewer and Dokoozlian (2005) found that 0.5–1.2 m^2 leaf area was required to ripen 1 kg of fruit, depending on the training system. While this metric is important to understanding vine size, it is less practical to monitor, as these data are far more tedious and time consuming to gather than pruning weights. Also, leaf areas are often highly manipulated in the vineyard through canopy management and can be harder to interpret in vineyard management records.

Vine balance is complex and dictated by the soil, environment, and overall production capacity of the vineyard. Since there are so many factors that play a role in vine balance, there are no clear guidelines as to how to create a balanced vine across all sites; rather, the methods for achieving vine balance depend on the following factors:

- Environment
- Soils
- Water availability
- Cultivar
- Rootstock
- Vineyard design (spacing, trellis system)
- Vineyard floor management
- Nutrition
- Diseases and pests
- Management practices
- Production goals

Vine-Dictated Balance

The grapevine, like other organisms, will reflect its environment. Management methods play a role in vine balance but are secondary to the inherent growth potential of a vine based on the vineyard site (soils, precipitation, climate) and the plant material (cultivar and rootstock). Vineyards on deep, fertile soils with good water holding capacity will be able to grow larger vines and produce higher yields than vineyards on a site with shallow or coarse soils with poor water holding capacity. That is, those vines with relatively unlimited resources in soil moisture, mineral nutrition, and climate (sun and heat units) will be able to produce the maximum amount of carbon for fruit production, vegetative growth, and reserves for the next season's growth. Conversely, those vines with limited resources will be weaker and have less carbon available for the current season's growth and less in reserves for the following seasons. In these divergent scenarios, vine spacing, pruning, trellis, and yields will need to be determined to allow the vine to reach a balance of canopy and fruit growth on that site. The vine-dictated balance will ultimately determine the vineyard design and management decisions made.

Viticulturist-Dictated Balance

The viticulturist-dictated balance of the vineyard is determined by the production goals, vineyard management methods, and economics of the business. Those production practices that do not adhere to a vine's production capacity will inevitably result in unsustainable vineyard production practices, as there will be reduced vine health and reduction in yields and/or fruit quality. For example, severely limiting vine yields through crop thinning for high quality winegrape production may be under-cropping vines to a point where they may become overly vigorous and may lead to problems such as poor fruitfulness/fruit set, shaded canopies, increased disease incidence and canopy management costs. In vineyards where there is competition for water resources by poor vineyard floor management and/or improper irrigation and high yields, there can be problems such as reduced fruit quality (lack of ripening), weak wood, poor bud fruitfulness, and poor bud break. Therefore, the vineyard balance really depends on the viticulturist reading the vine's capacity and maximizing that production capacity and/or quality with reasonable and economic inputs.

Unbalanced Vines

Vines that are not balanced have significant changes in various factors, all of which can affect vine size and yields. Vines that are overly vigorous often have poor bud fruitfulness, reduced fruit set, and lower yields. Unless adequately managed, these vines can spiral into an overly vegetative state. In addition, these vines can have poor winter hardiness and low bud survival in areas with more severe winters. Simply cutting off excess canopy of vigorous vines through hedging and leaf removal can help modify the microclimate, but these methods do not bring a vine back into balance. Care must be taken to address the cause of excessive vine vigor, which may be fertility, soil moisture, and/ or cropping level.

Weak vines can experience similar maladies as overly vigorous vines. There can be reduced bud fruitfulness, leading to reduced yields. Weak vines may also have reduced bud break in spring and poor shoot development due to limited reserves for adequate shoot growth in spring and limited carbon availability during the growing season. In very weak vines, shoot growth can be stunted in early spring and into the growing season. As a result, shoots may not grow enough to support the fruit and will require that the crop be thinned to allow for the reserves to be replenished with carbon.

DEFICIT IRRIGATION

The single most significant interaction between soil and vines is the uptake of water. Vines aren't particular about where their water comes from, but they are remarkably sensitive to how much water is available and when it presents itself during the growth cycle. Based on research originally conducted on fruit trees, viticultural researchers and commercial growers have become more and more sophisticated in their understanding of the intricacies of water relations. At certain points in the annual cycle, abundant water is essential; at other points, too much water is detrimental. Timing is everything, not just for the survival of the plants, but for berry development and fruit quality.

Many of the great *terroirs* of Europe had the good fortune to possess a combination of soil structure and climate patterns that made the right amount of water available at the right time more or less naturally. (This is the central finding of Gerard Seguin's work on how different soil types in Bordeaux solve the water problem in different ways—see selection in chapter 3.) If one year is drier than the last, *voila!* you have vintage variation. In drier places, like many New World vineyard regions, water availability can be controlled through irrigation management, often delivered with more precision than natural events.

The process has become known as regulated deficit irrigation (RDI), which utilizes soil and plant moisture measurements to calculate watering schedules. Mild water stress is part of the regimen—where more water evaporates through the leaves than is coming in through the roots. A related approach, partial rootzone drying (PRD), alternates irrigation between the two sides of a vine row, moistening one and drying the other, sending mixed signals to vine growth hormones and promoting the same outcome as RDI.

Here is a broad discussion of the place of RDI in viticulture, taken from the website of the Department of Environment and Primary Industries of the Australian state of Victoria. Its author is Ian Goodwin.

FROM IAN GOODWIN, "MANAGING WATER STRESS IN GRAPE VINES IN GREATER VICTORIA"

The wine growing regions in Greater Victoria normally have relatively high winter and spring rainfall, and drier conditions from mid-summer to autumn. In vineyards that rely on dams filled from small, localised catchments, irrigation water can be a limited resource in seasons where rainfall is below average. In other vineyards on reticulated irrigation schemes, water entitlements can be reduced below 100% in drought years. Combined with low soil moisture reserves from lack of winter and spring rainfall, water stress may be unavoidable. Irrigations must be rationed to minimise or avoid water stress at critical times, so that the best yield and quality is produced from the amount of water available for irrigation.

Where vineyards have access to a permanent and unlimited water source, irrigations can be managed so that water stress is controlled to increase fruit quality. This type of irrigation management is often referred to as regulated deficit irrigation (RDI). Using RDI, water stress is avoided during critical growth stages but then controlled at other times by irrigating at less than the full water requirement of the vines.

Good irrigation scheduling techniques are essential for both drought management and RDI. Water stress must be monitored to determine if irrigation is necessary during a drought or if stress levels are appropriate for RDI. This is best achieved by measuring soil moisture. Over-irrigation should be avoided because it wastes water and is detrimental to RDI management. Above all, accurate records must be kept for future reference and fine-tuning of an irrigation scheduling system.

Water stress

Water stress is a physiological reaction of a vine to insufficient water supply. Some of the physiological responses of grapevines include reduced cell division, loss of cell expansion, closing of leaf stomata, reduced photosynthesis and, in the worst case, cell desiccation and death. Most of these responses are dynamic (as the level of water stress increases so does the response). For example, leaf stomata (which affect photosynthesis and hence potential sugars in the fruit) do not completely close at the first signs of water stress, but slowly close as water stress increases.

The physiological reaction of a vine to water stress will affect the growth and development of the shoots, leaves, and fruit depending on the timing and level of water stress during the season. Generally, water stress during the season will affect the most active growth processes occurring in the vine at that time. For example, berry cell division is most active immediately after flowering, so that water stress at this stage could significantly reduce berry size at harvest, since berry size is to some extent dependent on cell number.

Water stress may also have less obvious or indirect effects on fruit yield and quality. For example, reducing berry size increases the skin to juice ratio, which may increase the concentration of anthocyanins and phenolics in the must and wine of red grapes. Water stress may affect the chemical breakdown or formation of important berry acids and flavours. Indirectly, water stress may reduce the shading of fruit. Shading has been shown to decrease fruit colour and the concentration of tartrate and soluble solids, and increase pH and the concentration of malate and potassium. The incidence of disease may also be reduced through opening up the canopy and keeping the bunches loose because berries are small.

Ideally water stress should be measured by monitoring one or more physiological responses of a vine such as leaf water potential, stomatal conductance or cell expansion. However, the techniques for these are difficult, expensive and time consuming. With current technology it is more straightforward to measure soil moisture, rather than plant water stress directly. Theoretically, this should be a good indicator of plant water stress, however, it is recommended that vine performance should also be observed to adjust soil moisture levels for irrigation scheduling.

■   ■   ■

Avoiding excessive yield loss

Yield is most affected by high levels of water stress during flowering and fruit set. During flowering, bunches can be completely desiccated by high levels of water stress, resulting in complete yield loss. During fruit set, high levels of water stress can reduce yield by up to 50% if berries drop or do not develop properly.

From fruit set to veraison, water stress can cause yield losses of up to 40%, mainly due to a reduction in berry size. Both cell division and cell enlargement can be affected. Some results indicate that yield can also be reduced in the following season because of lower bud fruitfulness.

For four to six weeks after veraison, yield losses of similar magnitude (i.e. 40%) have been measured in response to water stress and are attributed to a reduction in berry size. Berries are growing solely from cell enlargement at that stage. From six weeks after veraison to harvest, yield is least susceptible to high levels of water stress.

Maintaining soluble solids

Accumulation of soluble solids in the fruit mainly occurs after veraison over a period of four to six weeks. Considerable experimental evidence shows that high levels of water stress during this period will significantly reduce the concentration of soluble solids (Brix) at harvest.

The concentration of soluble solids at harvest has been shown to increase significantly if high levels of water stress occur during berry set. The effect is most likely due to a reduced crop load. Using water stress during fruit set to increase soluble solids at harvest is not recommended because the quality benefits are outweighed by the potential total crop loss if water stress occurs during flowering.

DOES VARIETY MATTER?

The most enthusiastic proponents of traditional *terroir* sometimes imply—or say out-right—that the grape variety planted in special places isn't particularly important, that great *terroir* can conceivably produce a great wine no matter what is planted there (barring obvious limitations: no one would claim Alsace is a great spot to grow, say, Nero d'Avola.)

But the most prestigious traditional *terroirs* are generally identified with one grape, or perhaps one red and one white: Pinot Noir and Chardonnay in Burgundy, Nebbiolo in Barolo, Riesling in Germany, Sangiovese in Tuscany, Tempranillo in Rioja, Cabernet Sauvignon in Bordeaux. The same goes for most of the newer *terroir* claimants: Sauvignon Blanc in Marlborough, Grenache in Priorat, Cabernet Sauvignon in Napa, Shiraz in the Barossa. At the same time, some of the world's most widely planted grapes are not honored for their *terroir* expression much of anywhere: Spain's Airen, Italy's Trebbiano, and Germany's Müller-Thurgau immediately come to mind, along with the many useful but low-profile *teinturier* (red-fleshed) grapes used for blending—Alicante, Souzão, Dornfelder, Rubired. And then there's Thompson Seedless, a *Vitis vinifera* table grape that still makes a double-digit percentage contribution to California wine production, but has never been terribly adept at revealing *terroir*.

Variety clearly does matter, in at least two ways. First, the climate in a given vineyard site poses limits on what will grow there at all, let alone grow well (for a reminder, you may wish to revisit the Gregory Jones chart in chapter 4). Second, some grapes are simply more sensitive to their environments than others. The stylistic range of excellent, distinctive wines made under the umbrella of Pinot Noir is broad indeed; Pinot Grigio, not so much.

One of the varieties long regarded as exquisitely responsive to the place where it is grown is Riesling, and work done by Gavin Sacks and his colleagues at Cornell University helps give that reputation an objective basis in grape chemistry. When Riesling, the flagship grape of the Finger Lakes, is compared to other aromatic varieties, it shows a much higher number of aromatic compounds and compound precursors at or near sensory threshold levels. The implication is that, depending on subtle variations in climate, soil, and viticulture, Riesling can take on many more guises than its simpler relatives. The following selection comes from a presentation Sacks made at the Australian Wine Industry Technical Conference in July, 2013.

FROM GAVIN SACKS, ET AL., "'TELL ME ABOUT YOUR CHILDHOOD'—THE ROLE OF THE VINEYARD IN DETERMINING WINE FLAVOUR CHEMISTRY"

Abstract: Assuming that wines produced from the same grape variety but produced under different growing conditions have consistent sensory differences, the growing conditions must have affected grape chemistry in some consistent manner. In many varieties, these differences in chemistry are not readily detectable by sensory evaluation because compounds important to wine aroma are derived from non-volatile grape precursors. Understanding the link between environment, flavor precursors, wine chemistry, and eventual wine flavor is of importance to winemakers interested in either mitigating or enhancing differences among grapes. Riesling wines make for a particularly interesting case study because i) Riesling often has several compound classes at or near sensory threshold, and thus sensory perception should be susceptible to differences in growing conditions, and ii) many grape-derived odorants potentially important to Riesling including monoterpenes, volatile polyfunctional thiols, and C13-norisoprenoids, are derived from non-volatile precursors. For some cultural practices and aroma compounds, like leaf removal around veraison and the concentration of 1,1,6-trimethyldihydronaphthalene (TDN, "petrol" aroma) in wine, there is a solid body of evidence to correlate vineyard practices to grape and wine chemistry. For other compound classes like monoterpene glycosides or S-conjugate precursors of thiols, the correlation of growing conditions and wine chemistry is more ambiguous, either because the effects of cultural practices seem to be relatively small, or else precursor concentrations are poorly correlated with wine concentrations. Studies are further complicated by the fact that changes in cultural practices do not generally affect single flavor compounds.

■   ■   ■

Introduction: Depending on the reference cited and the criteria used for "identification," approximately 700 volatile compounds have been characterized in wine, of which a subset of about 50 explain the major features of most wine aromas. It is doubtful that any of these volatile compounds is truly unique to a particular varietal wine (or, for that matter, to wine as opposed to other foods). Rather, concentrations of volatiles in wines differ from each other quantitatively instead of qualitatively, since at least a few molecules are likely to be detectable following sufficient analytical struggles.

As an example of this phenomenon, important odorants representing the major compound classes found in three varietal aromatic white wines are listed in Table 1, along with sensory characteristics and typical ranges of odour activity values (OAV), compiled from several sources. An OAV is calculated as the ratio of compound's concentration to its odour detection threshold. OAVs are a crude measurement of the contribution of a compound to the aroma of a wine (or any other foodstuff), since they ignore masking and additive effects. Allowing for this caveat, three observations can be made from Table 1.

First, while a compound or compound class may be higher on average in one varietal wine vs. another (TDN in Riesling, 3-MH in Sauvignon Blanc, linalool in Muscat), there are no unique compounds.

Second, Riesling has a number of grape-derived aroma compounds at or just above their sensory thresholds (OAV between 0.5–10), at least in comparison to other aromatic white wines. For example, Muscat wines typically have 1 compound class (monoterpenes) well in excess of odor threshold. The flavour of Riesling is thought to be more dependent on site or region than many other varietals (Robinson 2006), and speculatively (that is, very speculatively), this balance of peri-threshold odorants may explain this phenomenon.

TABLE 5.1 Grape-derived aroma compounds in three aromatic white varieties

Representative odorant(s)	Compound class	Aroma	Sensory threshold in model wine	Odor activity value (OAV) range in young wines			Precursor in grape?
				Riesling	Muscat	Sauv. blanc	
TDN (1,1,6-trimethyl-dihydronaphthalene)	C$_{13}$-norisoprenoids	Petrol, kerosene	2 µg/L	0.5–10	low	low	Glycosides
Linalool	Monoterpenes	floral	50 µg/L	0.1–5	20–40	<1	Glycosides or monoterpene polyols
Geraniol			130 µg/L				
3-mercaptohexanol	Polyfunctional thiols	Citrus, passionfruit	60 ng/L	2–15	2–15	15–300	S-conjugates
IBMP (3-isobutyl-2-methoxypyrazine)	Methoxypyrazines	Bell pepper	2 ng/L	low	low	1–28	None (primary)
4-vinylguaiacol	Volatile phenols	Medicinal, smoky	180 µg/L	low	<0.5–10		Hydroxycinnamic acids or glycosides*
4-vinylphenol			130 µg/L				
cis-3-hexenol	C$_6$ alcohols	Grassy	400 µg/L	0.5–1	n/a	0.5–2	Unsaturated fatty acids

NOTE: Other grape-produced odorants that could contribute to aromas of these wines include:

Sugar degradation products, e.g., Furaneol and homofuraneol ("cooked sugar" odour), particularly in botrytised wines

o-aminoacetophenone ("UTA" note) from indole acetic acid precursors

1,8-cineole ("eucalyptus taint")

Fungal metabolites or degradation products of pesticide residues

Dimethylsulfide ("canned corn") from S-methylmethionine

*Concentrations in smoke-tainted grapes could be much higher.

Third, the majority of important grape-derived odour compounds in Riesling are absent from the fruit, but are instead released from non-odorous precursors during fermentation and storage. Of particular importance are:

- Glycosides, in which the precursor includes a sugar molecule
- S-conjugates, in which the precursor contains the amino acid cysteine, the tripeptide glutathione, or related cysteine containing dipeptides

■   ■   ■

SACKS GOES on to examine what researchers do and do not know about how specific viticultural practices affect the concentrations of various aromatic compounds and precursors. Here is his discussion of TDN, the compound that sometimes gives Riesling a "petrol" character.

TDN is well known to contribute to the characteristic aroma of bottle-aged Riesling, where it can reach concentrations over 50 ng/mL. Our group has recently re-evaluated the odour threshold of TDN in a white wine and determined it to be 2 ng/mL—a factor of 10 below the previous reported threshold. We also determined that TDN was in excess of this threshold in 31 of 32 young Riesling wines, while it was below threshold in most non-Riesling red and white wines. The median TDN concentration was nearly 5-fold higher in Riesling as compared to other wines (5.7 vs. 1.2 ng/mL). Thus, TDN may contribute to varietal character of young Riesling wines, along with other compound classes shown in Table 1. The recognition threshold of TDN (the point where wine takes on a "petrol" aroma) is still not determined.

Beyond cultivar selection, several viticultural factors have been related to higher concentrations of TDN precursors ("potential TDN") in grapes or higher TDN in finished wines.

- Greater exposure of clusters to sunlight, e.g., through leaf removal or artificial shading
- Warmer climate
- Less nitrogen fertilization
- Less irrigation and lower water potential

Of these cultural factors, the best established is the effect of cluster light exposure, which typically results in a 2-fold increase in potential TDN. Greater cluster exposure should result in higher berry temperatures, but light exclusion treatments that avoid temperature changes still increase TDN in wine, further indicating that light exposure has a direct effect. The effects of the other factors could potentially be confounded with light exposure by reducing canopy growth, but could be mediated through other mechanisms. For example, lower N availability can directly affect carotenoid accumulation and inter-conversion, which may have a role in C13-norisoprenoid production.

ORGANIC VITICULTURE AND *TERROIR*

In the past couple of decades, organic and biodynamic farming practices have earned tremendous credibility in winegrowing around the world. Organic viticulture basically calls for eliminating synthetic fertilizers, herbicides and pesticides, opting for more "natural" approaches like cover crops, beneficial insects (that eat the bad guys), and composted organic matter. Biodynamics goes a step further, adding to the basic organic approach several low-dosage preparations alleged to improve soil and plant vitality and rules for timing vineyard and winemaking operations according to astral and lunar calendars. Biodynamics is the more controversial of the two approaches, but still counts among its adherents a number of important producers of highly respected (and high-priced) wines.

There's plenty of controversy, plenty to debate in these practices. Our concern here is not whether they're better for the planet, or whether some of the odder biodynamic practices make any sense at all, or whether it matters that Rudolf Steiner, the father of biodynamics, didn't want people to farm wine grapes at all, since he practiced and preached abstinence. The question for our purposes is whether these approaches to grapegrowing help in the quest to express the *terroir* of a place.

Plenty of anecdotal testimony to this effect can be found, in the form of winemakers insisting that since they converted to organics/biodynamics, their wines taste better. And proponents argue more broadly that getting all those conventional chemicals out of the way can only strengthen the relationship between vine and earth, letting the *terroir* speak.

At a 2003 symposium titled "The Science of Sustainable Viticulture," held as part of the annual meetings of the American Society for Enology and Viticulture, panelist John Williams, founder and first winemaker at Frog's Leap Winery in the Napa Valley and an early convert to organic farming, offered a mix of science and philosophy that remains fresh and engaging more than a decade later.

JOHN WILLIAMS, "WINE QUALITY AS INFLUENCED BY SUSTAINABLE PRACTICES"

People become interested in the sustainable agricultural movement for a variety of reasons. Many share a concern for the environment and envision greater stewardship of the land. Others may be attracted to issues of farm worker safety or perhaps community concerns over pesticide use. Some may be attracted to the lower costs and higher productivity of sustainable farming. Then there is the attraction of positive public relations that come with ecologically sound practices. Frog's Leap Winery has been farming sustainably for over fifteen years now and we dry-farm over two hundred acres of certified organic vineyards. Through our grower program, we have influenced the organic certification of a dozen different vineyards covering over eight hundred acres. I can say that we have benefited in a very positive way from all the reasons for farming sustainably mentioned above and more. It remains, however, that the best reason to farm sustainably is the improvement of wine quality.

Simply put: the principles of organic farming and sustainable practices are the single most important tools you can employ to improve wine quality. Most, if not all, challenges you face in winemaking can be positively influenced in a significant manner by a closer relationship with the natural eco-systems responsible for growing your grapes.

Now, I am fully aware that I just used the word none of us ever thought we would ever hear at an ASEV convention and I know that many are more comfortable with the word sustainable than the word organic. In truth, however, the principles of organic farming are the fundamental and inseparable bases for sustainable agricultural systems. It is the deep respect and care for the miraculously complex soil organism that cannot be separated or ignored in this discussion. A healthy soil enriched by the organic matter of compost and cover crops and left alive by avoiding toxic herbicides, fumigants, nemadocides, and the like, is the building block of great wine.

Here is the major point: A healthy soil produces a healthy vine; a healthy vine produces healthy fruit; healthy fruit produces healthy wines: deep in color, deep in flavor, and deep in their natural character.

Pick nearly any problem in winemaking today and you will find with a minimum of research a deep connection to farming practice.

Having a problem with stuck fermentations? You are not the only one. Linda Bisson and Christian Butske's paper on "Stuck and Sluggish Fermentations," (AJEV Vol. 51) lists sixty-four reference papers on this subject. Their conclusions, among other reasons: "Poor yeast nutrition"; "The harvesting of grapes later and at a higher sugar content" to produce more flavor; Toxic substances in the must and low amounts of assimilable nitrogen. It may not be immediately apparent to everyone, but to the vintner using organically grown grapes, you would give a knowing nod. We used to have those problems until we built natural balanced soil fertility, stopped using sterol inhibitors, stopped spraying with dimethoate and started getting flavors at 23 and a half Brix instead of 28 and a half Brix.

Got sulfides? Park, Bolton and Nobles' most recent paper on the subject (AJEV Vol. 51 Issue 2) points at the causes. Three of the five most frequently associated causes of hydrogen sulfide production during fermentations are related to nutritional stress and other adverse must conditions. These conditions can be directly related to nutrition status of the grapevines themselves. As humans, we know that if you only eat beans you're going to stink up the place. The winemaker's solution? Add nutrients to the must. The organic growing solution? Develop the incredibly complex microbial world in your soil that degrades the organic matter in a balanced and measured way, promoting low vigor and high nutritive value. Think you can get enough nitrogen in your must by adding more N fertilizer to your vineyard? Spade, Nagel, and Edwards (AJEV Vol. 46) conclude that yes, you can, and suggest a minimum rate of 56 kg N/ha/year. They do point out though that this rate might have some adverse effects since it would surely lead to excessive vine growth, delayed fruit maturation, increased juice and wine pH, herbaceousness, poor color development and a lack of flavor. Maybe if you add the fertilizer to the irrigation water . . . there's a good idea . . . fertigating grapevines instead of promoting balanced soil nutrition is like giving a coke and a candy bar to your child for dinner. The immediate result might be impressive, but the eventual output is not.

There have been a minimum of ten papers in the AJEV in the last ten years on the effect of nitrogen fertilization in grapevines. Most of the papers deal with the downside—excessive vigor, poor color development, increased pH, problems with rot, high arginine correlated with ethyl carbonate production. But some papers dealt with the upside—increased yield—great, lots of bad fruit. Jackson and Lombard in their review "Environmental and Management Practices Affecting Grape Composition and Wine Quality" (AJEV Vol. 44 Issue 4) modestly conclude that, "while the evidence does not allow us to pinpoint the exact nature of the response to nitrogen fertilizer, it is such that growers can be cautioned against its excessive use." They go on to sum up the effects of soil on quality in the same paper in four ways: (1) the effect on moisture availability to the plant due to its moisture-retaining capacity; (2) effect of nutrient availability; (3) effect on microclimate due to heat-retaining and light-reflecting capacity and (4) the effect on root growth due to penetrability. Any organic grower would tell you that all four of these effects can be positively improved using sustainable farming techniques.

Truthfully, we could go on and that's even before we get into racking by the moon and bottling using tidal charts. Pretty soon, I'll be trying to convince you that organic farming will cure hair loss and improve your love life.

So far, we've only been talking about making better wine. Now, let's get to the really good part—increasing wine quality.

Now, I have to admit that when assigned to do this talk, I had to look up the definitions of a couple of words. The first word was "science"—since this session is "The Science of Sustainable Viticulture" and since it's been a few years since I was behind the test tubes in Professor Ough's laboratory, my Webster's definition of "science" heartened me. Mr. Webster defines "science" as "knowledge, especially that gained through experience." Any organic grower will tell you about learning the hard way, and I can tell you they are not very impressed with the help they have obtained until lately from the university, from Ag. Extension, from their county ag. people or from ag. suppli-ers. Until recently, these entities have been openly dismissive of organic farming.

The second definition I had to look up was "quality." Here's what Webster's has to say about that: "Quality: (1) the essential character or nature of (2) an inherent or distinguishing attribute."

Now we're getting somewhere, for, as important as it is to promote making wine better, the exciting part of this talk is about bringing out the inherent qualities of a wine and the place it is grown. If you believe, as I do, that the essence of winemaking, the Holy Grail as it were, is to make wines that deeply reflect the soil and climate from which they emanate, it seems self-evident that you would want every molecule, every enzyme, every ester, every flavonoid, every protein, every essence, to be derived from the soil in which the grapevine is grown. And if you achieve that, the product of that vine will imbue the essential character of its place. Real quality wine.

Without soil-based flavors, we, as winemakers, are stuck with trying to manufacture those flavors on our own, Thus, ridiculously excessive overripe grapes, spinning cones, esterifying yeasts, reverse osmosis, super malo-lactic cultures, micro-oxygenization, mega-purple, flying winemakers, and two hundred percent new oak.

Just because some of our hapless wine critics think that this is what wine is all about, I, for one, am not sure people who truly love wine are buying this, and these days I mean that literally. I person-ally think that these wines taste all the same, that they deliver little pleasure and that they develop little, if any, nuance in the bottle.

If you want real pleasure in a wine, if you believe that a wine has a soul, if you believe in the natural quality of wine and its dependence on place, you will be left with the inescapable conclusion, as have I, that we need to grow our vines deeply in their soil. We need to retain adequate amounts of organic matter back to our vineyards. We need to, at all costs, promote the vastly complex biological life in our soil, and that means the complete banishment of toxic herbicides and, yes, I include glyphosates *[known commercially as Roundup—eds.]*. We must promote soil structure and moisture retention capacity to significantly reduce, or, better yet, eliminate irrigation. We must cultivate our vines to promote deep root growth. We must banish chemical fertilizers and other growth stimulants. We must return natural soil fertility to complete balance for, in doing so, we will eliminate excessive vigor and all of its related problems. We must promote as much bio-diversity in our farming system as is possible, as this will be our best defense from a host of problems from gopher damage to Pierce's Disease. We must support our farm workers through liveable wages, healthy work environments, adequate healthcare, and respect for their contribution. We must be conservative in our use of resources in every way. And, we must be respectful of our larger community, to our nature and to our God. Then, at the end of *that* day, we must raise our glass and toast ourselves—for those who have the privilege of growing grapes and making wine sustainably are the truly blessed.

MEASURING AGRICULTURAL ALTERNATIVES

There's very little science to support the case that organic and biodynamic viticulture bolster *terroir* expression. Most of the testimony, in fact, is purely anecdotal, and the reason for this is simple: Controlled experiments are devilishly hard to pull off. Plots of land, after all, are contiguous, but the agriculture employed on those plots is rarely the same. If a conventionally farmed plot sits next to an organic plot, what keeps the pesticides used in conventional farming from drifting over and contaminating the organic grapes? If the biodynamically farmed block lies downhill from the conventionally farmed block, the differences in the wine might be the result of differences in soil drainage, and not of differences in farming practices. And how many growers are willing to put up with the extra effort of running two systems at once, at least one of which they may not believe in?

Some rigorous research has been conducted, however—with mixed results. An extensive study on 12 acres of Merlot vines in Mendocino County was conducted from 1996 through 2003. Half of the vineyard was farmed organically, half was farmed biodynamically, all according to a randomized block design. Farming practices and winemaking protocols were identical, the only difference being that the biodynamic vines received biodynamic preparations (homeopathic-level doses of various plant, animal, and mineral composts). For the eight years of the study, soil composition, plant tissue, pruning weight and, wine composition were all measured for a variety of parameters. The overall conclusion was: very little difference, if any. Biodynamically grown fruit did, in some instances, possess higher brix, and the practice, by various measures, contributed to greater vine balance, but there proved to be almost no difference in soil health between the two practices.

The authors of the study included Jennifer Reeve, then at Washington State University, now Associate Professor of Organic and Sustainable Agriculture at Utah State;

Lynne Carpenter-Boggs and John Reganold of Washington State; the late biodynamic consultant Alan York; University of California Cooperative Extension specialist Glenn McGourty; and Leo McCloskey, founder of the wine consulting firm Enologix.

FROM JENNIFER REEVE, ET AL.,"SOIL AND WINEGRAPE QUALITY IN BIODYNAMICALLY AND ORGANICALLY MANAGED VINEYARDS"

ABSTRACT: Wines produced from biodynamically grown grapes have received increasing attention. Similar to organic agriculture, biodynamics eliminates synthetic chemical fertilizers and pesticides. The primary difference between the two farming systems is that biodynamics uses a series of soil and plant amendments, called preparations, said to stimulate the soil and enhance plant health and quality of produce. Whether these preparations actually augment soil or winegrape quality is unclear and controversial. A long-term, replicated, 4.9-ha study was initiated in 1996 on a commercial Merlot vineyard near Ukiah, California, to investigate the effects of these biodynamic preparations on soil and winegrape quality. The study consisted of two treatments, biodynamic and organic (the control), each replicated four times in a randomized, complete block design. All management practices were the same in all plots, except for the addition of the preparations to the biodynamic treatment. No differences were found in soil quality in the first six years. Nutrient analyses of leaf tissue, clusters per vine, yield per vine, cluster weight, and berry weight showed no differences. Although average pruning weights for both treatments in 2001 to 2003 fell within the optimal range of 0.3 to 0.6 kg/m for producing high-quality winegrapes, ratios of yield to pruning weight were significantly different ($p < 0.05$) and indicated that the biodynamic treatment had ideal vine balance for producing high-quality winegrapes but that the control vines were slightly overcropped. Biodynamically treated winegrapes had significantly higher ($p < 0.05$) Brix and notably higher ($p < 0.1$) total phenols and total anthocyanins in 2003. Biodynamic preparations may affect winegrape canopy and chemistry but were not shown to affect the soil parameters or tissue nutrients measured in this study.

FINALLY . . .

There's compelling evidence to suggest that—using the parlance of Michael Pollan in his groundbreaking work, *The Botany of Desire*—grapevines *selected* early man to develop the practice of viticulture, the concrete procedures required to make this life-changing and culture-changing elixir. In fact, they possess attributes that make them irresistible to agricultural practice. First and foremost, vines are remarkably good at accumulating sugar in their berries, which can be turned into alcohol with minimal effort. Alcohol might not top the list of *terroir* attributes, but without it, there is nothing to discuss. Rocks don't make wine, vines do, and without their sugar, converted into ethanol, this book would have no audience. Second, *vinifera* grapevines are monoecious: both sets of reproductive parts exist conveniently side by side in the same flower, a huge advantage for organized growing, cultivation, and vine selection over time.

In *Uncorking the Past,* wine archaeologist Patrick McGovern recounts what we know, starting in the Neolithic era (beginning about twelve thousand years ago) about the history of fermented beverages, moving from China to the Near East, Eastern Europe, Latin American and Africa. One broad thesis running through the book is that humans have craved alcoholic beverages since their first accidental discovery, when a stash of honey turned into primitive mead, or a basket of grapes fermented on its own to spectacular results. McGovern proposes that alcohol's many valuable functions—social lubrication, health benefits, tasty calories, an opening to transcendence—were a driving force in the adoption of agriculture and ultimately civilization. Settling down, after all, allowed for a steady supply of the raw material for wine, beer, and other intoxicants.

McGovern also advances what he calls "the Noah hypothesis," that winegrape cultivation began at a single source, precipitated by a single event. The following selection describes his search for these vines of origin.

FROM PATRICK MCGOVERN, *UNCORKING THE PAST*
TO THE HEADWATERS OF THE TIGRIS AND EUPHRATES

Besides securing stone and pottery samples for analysis, my trip to eastern Turkey had another goal. You might say that we were looking for the vinicultural Garden of Eden. The eastern Taurus, Caucasus, and northern Zagros mountains have long been considered the world center of the Eurasian grape: this is the area where the species shows its greatest genetic variation and consequently where it might have been first domesticated. It is also becoming increasingly clear, as we pursue our combined archaeological and chemical investigations, that the world's first wine culture—one in which viniculture, comprising both viticulture and winemaking, came to dominate the economy, religion, and society as a whole—emerged in this upland area by at least 7000 B.C.

Once established, the wine culture gradually radiated across time and space to become a dominant economic and social force throughout the region and later across Europe in the millennia to follow. The end result over the past ten thousand years or so, since the end of the Ice Age, is that the Eurasian grape now accounts for some ten thousand varieties and 99 percent of the world's wine. Even though North America and East Asia have many more native grape species, some with very high sugar contents, there is as yet, amazingly, no evidence that they were domesticated before modern times.

We were interested in finding out whether there was a unique event that precipitated domestication in this core area. This one-time, one-place proposition has been referred to as the Noah hypothesis, an allusion to the biblical tradition that the patriarch's first goal, after his ark came to rest on Mount Ararat, was to plant a vineyard and make wine (Genesis 8:4 and 9:20).

Although eastern Turkey today might not appear to be conducive to viticulture, the recent excavations painted a much different picture during the Neolithic period. Precipitation levels were higher in the period immediately succeeding the Ice Age, and because the Taurus Mountains form part of the Trans-Asiatic orogenic belt, a region of intense geologic activity today and in the past, the soils are rich in all the essential metals, minerals, and other nutrients needed by grapes as well as by numerous other wild fruits, nuts, and cereals.

The calcareous hills and valleys of this upland region are generally characterized by an iron-rich red loam known as terra rossa. This soil is often rocky, encouraging good drainage and root development, and contains enough clay to retain moisture through the dry season. Its slightly alkaline pH and low humus content are also good for grapes. Even if these conditions were ripe for exploitation, the question is whether early humans first domesticated the Eurasian grape somewhere in the Taurus Mountains and began to make wine here.

In collaboration with colleagues in Europe and the United States, we have applied modern DNA analysis to resolve this question. We sequenced specific regions (microsatellites) of the nuclear and chloroplast genome of modern wild and domesticated grape plants from Turkey, Armenia, and Georgia and carefully compared these results with those from European and Mediterranean cultivars. We have already shown that Middle Eastern grapes probably derive from common ancestors and that four Western European varieties—Chasselas, Nebbiolo, Pinot, and Syrah—are closely related to a group of Georgian cultivars and might well have some ancient Georgian ancestors. The extraction of ancient DNA from seeds or other parts of the plant, which should eventually provide more direct evidence, is being pursued.

Our search for wild vines in eastern Turkey turned out to be a high adventure. Together with my associates from the University of Neuchatel in Switzerland (Jose Vouillamoz) and the University of Ankara (Gokhan Soylemezoglu and Ali Ergul), we traveled the dusty highroads and byways in our Department of Agriculture Land Cruiser during the spring of 2004. One dramatic setting for our collecting was in a deeply cut ravine below the famous site of Nemrut Daghi, where the first-century B.C. ruler Antiochus I Epiphanes had statues of himself in the company of the gods hewn out of limestone on a mountaintop 2,150 meters high. Other promising areas were along a river valley cutting through the Taurus Mountains around Bitlis and Siirt, and along the Euphrates River north of Sanliurfa, at Halfeti.

We traveled all the way to the headwaters of the Tigris River, just downstream from Lake Hazar and the city of Elazig. Here the river cuts through one of the most metallurgically important areas in the ancient Near East, Maden (Turkish for "mine"), and an area that is still tectonically active. Maden is only about twenty-five kilometers from the important Neolithic site of Çayönü. Fortunately the Earth's crust was quiet, but in my eagerness to reach a particularly enticing grapevine clinging to a bank of the river, I almost fell into the raging torrent of the upper Tigris. But for the sure hands of my colleague Ali, I might not be telling this story.

Our risks paid off when we found a hermaphroditic wild plant which was positioned between a wild male and female vine, exactly the situation that an early viticulturalist would have needed to observe and select for. What makes the domesticated Eurasian grapevine so desirable is that it is hermaphroditic: male stamens and female pistils are located together in the same flowers on the plant, whereas for the wild variety, male and female flowers occur on separate plants. The proximity of the sexual organs on the hermaphroditic plants ensures the production of much more fruit on a predictable basis. This self-fertilizing plant could then be selected for desirable traits, such as sweeter, juicier fruit or thinner skins, and cloned by propagation of branches, buds, or roots.

The hermaphrodites, representing a mutation, account for about 5 to 7 percent of the wild-vine population. Still, it would have taken some pretty sharp eyes to pick out just those plants and domesticate them, as the sexual organs of the flowers are microscopic.

Domestication of the grapevine assumes that humans had discovered how to propagate it artificially, as growing it from seed results in unpredictable characteristics. The natural habit of the wild grapevine also doesn't lend itself to making the best wine grapes or easy harvesting. Left to itself, it grows high up in trees, shading out competing plants and producing fruit appealing to animals, especially birds who spread its seed, but not necessarily to humans, as it can be very sour.

It's not beyond the realm of possibility that Neolithic viticulturalists developed a layering method (*provenage*) to propagate the grapevine from the root and thus train it to grow up a nearby tree. The idea of propping up vines with artificial supports might well have been suggested by vines growing up trees. Early viticulturalists might have also started training the vine's height and shape, which would make gathering the fruit a lot easier. To date, our DNA studies, based on the samples we collected in eastern Turkey and the Caucasus, appear to support the Noah hypothesis, but more work is needed. Samples are needed from other parts of the Middle East, especially Azerbaijan in modern Iran, and this remains a difficult region to work in. If the hermaphroditic gene itself, a single region of the genome that accounts for the development of both male and female organs in flowers on the same plant, could be isolated, then we could target that gene for analysis in ancient and modern material. Someday soon we will have the answer.

6

WINEMAKING
The Human Element in Terroir

When we encounter the word *terroir* these days it's usually without a shred of supporting data. Instead it's recorded as impressions, observations, or sets of anecdotes presented as facts without supporting evidence. Writers, journalists, poets, and novelists the world over rhapsodize about *terroir* but can't really get a handle on what it is in any quantifiable way: physically, biochemically, sensorially. The more skeptical of modern winemakers and enologists want evidence; they want to see a flavor profile graphically presented, based on ironclad descriptive analysis and supported by whatever chemical, biochemical, or microbiological verification can be mustered. The more they know about the unique flavor fingerprint in a wine from a specific site, the better the chance of conserving or amplifying it. At the very least, winemakers talk to one another and share information; in most wine regions there's a well-established network for brain-picking, for finding out what works and what doesn't.

The human element, in short, can't be left out of the *terroir* discussion, but frequently it is. So often "nature" is credited as the sole factor shaping wine flavor, while a winemaker's stewardship of the entire process, from establishing the vineyard to bottling the finished product, is downplayed or excluded from the discussion.

After all, what lies behind a wine's distinctive flavors? Is it grape genetics alone? Is it the lay of the land, or the season's weather? Regardless of other factors, the hand of the producer as it guides and shapes the process should be an essential part of the *terroir* puzzle, and to de-emphasize its importance leads inevitably to an incomplete picture of a wine's '*terroir* gestalt.'

We start this chapter by examining the natural wine movement, a philosophy whose core tenet seeks to challenge the notion that a winemaker should have an impact upon the wine. Natural wine, after all, is ostensibly wine made with as little human intervention as is "humanly" possible, wines made without added yeasts, color, enzymes or sulfur, often resulting in wines of a certain pristine quality, and just as often, wines with a noticeable disregard for typicity, to say nothing of hygiene. Nevertheless, the movement serves as a useful paradigm with which we might address some of these lingering issues as to just where the winemaker fits into the winemaking scheme.

From there we segue to a discussion of human intervention more broadly. How much intervention is allowable in this so-called "natural" process? Is it the winemaker's job to try and stay out of the way entirely? Is it possible to make commercial and technically formulated wines, wines that emphasize a signature style or brand identity and *still* have authentic, legitimate *terroir* expression? When does the latter "break down," so to speak, or recede to the point it is not recognizable? Few winemaking texts and how-to's answer questions of this type. Instead, most winemaking texts describe making wine as a technological process, not one designed to capture *terroir*-in-a-bottle.

Winemakers will concede that microbes, yeasts and bacteria, are the real workhorses in winemaking, in partnership with all that they do. But this too is a rather slippery slope, since the yeasts and bacteria used to make wine have undergone a technological revolution themselves. "Designer" commercial yeasts can now be employed to manage any number of fermentation protocols, selected to bring out particular nuances of flavor or texture, or to survive excessive alcohol levels. They are tailor made, almost, to make any wine from any must. Is it any wonder, then, that native or "wild" yeasts are coming back into favor, in the hopes that a fermentation that occurs using (at least some) of the native yeast populations might be more true to the *terroir* of a site? Of course, native yeasts can be unpredictable, and preserving them can maintain the populations of other organisms whose odors or flavors may be perceived as faults. So which is the right direction? Are distinctive flavors in certain wines the result of careful *terroir* character conservation, or is this character merely the "sum of faults" that may creep into the wine?

The chapter closes with a section exploring where we go from here. Or, in the words of Clark Smith, should wines be spoofulated or artisanal? Inevitably, the answer depends on how you define, explore and implement your indigenous flavors, or whether you locate *terroir* markers for a given variety in your regional milieu and "reverse engineer" the wines to express that *terroir*. Which is the right path here, and is there a wrong path?

HUMAN FACTORS: THE ANTHROPIC INFLUENCE ON THE ENVIRONMENT

We start with the impact that humans have on their own surroundings, which include the vineyard and the winemaking environment. We have heard many a winemaker make the modest claims that "the vineyard makes the wine," or "less is better," "or the winemaking process should be one of benign neglect." Over time in winemaking cultures the human influence directs the evolution of style, technique, consistency. But in countless occasions in wine's long history, humans have impacted the environment in which wine is made, either physically, culturally, or socially—as well as geographically. Kyle Schlachter substitutes the term geography for *terroir* in his discussion and, by so doing, persuasively makes the case for the "anthropic influence."

FROM COLORADOWINEPRESS.COM, "THE FALLACY OF TERROIR," BY KYLE SCHLACHTER

The fallacy of *terroir* is not that it doesn't exist, it is that people keep saying there is no English-language equivalent of the concept. Many people attribute, soil, climate, and topography as the common denominator of *terroir*. According to the infallible Wikipedia, "at its core is the assumption that the land from which the grapes are grown imparts a unique quality that is specific to that growing site." Thus, *terroir* is often described as the set of special characteristics of a certain place. Depending on scale, that could be a region, a village, a vineyard, or even a specific block within a vineyard. This is exemplified by the fact that syrah wines differ from Hermitage, Barossa and Paso Robles despite being made from the same grape cultivar. Perhaps the concept of *terroir* is best epitomized in Burgundy where famed *climats* like Romanée, Romanée-Conti and Richebourg in Vosne-Romanée are only meters apart and yield distinct wines all made from pinot noir.

Yes, the soils, climate (not so much in Vosne-Romanée), and topography vary in each of these places. But one thing that is too often left out of the *terroir* discourse is the anthropic influence. After all, grape vines don't decide where they grow, harvest themselves, or stop fermentation before the product is vinegar. Some people, including myself, add the human element to the concept of *terroir*. Whether or not you include people as part of the *terroir* of a place, there is in fact an English word that covers all the definitions of the concept: geography. In English, the characteristics of a wine can be said to come from the geography of a place. Geography is more than just maps (actually, the study of maps is called cartography). Geography encompasses soil, climate, topography, geology, history, and cultural practices of a place.

In 1964, a famous geographer named William Pattison described four broad traditions in which geography, as an academic field, is based. He presented a spatial tradition, regional studies tradition, human-environment tradition, and an earth science tradition. Wine, like geography, concurrently pursues all four of these realms. A study of wine, in the geographical tradition described by Pattison and affirmed by many others more recently, can cut across all four fields because as one follows the life of a bottle of wine from sun to soil to grape to winery to bottle to consumer, every individual bottle of wine is really just geography in a bottle.

In the spatial tradition, the distribution of wine regions can be described, contrasted, and compared. Heck, that is the basic concept of *terroir*. Different vineyards impart different characteristics to different wines. Why are wine regions predominantly found between the temperate latitudes of 30° and 50° in each hemisphere? Well, that is spatial geography.

Of course the answer has to do with the climates and geology of those regions. The specific locations of vineyards and why certain vineyards produce more acclaimed wine than others is a result of the geologic, edaphic (soil), and topographic characteristics of those sites. All the basic principles of *terroir* are part of the earth science tradition of geography. I like to think that the basic tenets of *terroir* are the same as those that soil scientist Hans Jenny (pronounced "yenny") used to describe independent factors that determine the process of pedogenesis (soil formation). Jenny coined the idea of Cl.O.R.P.T. in 1941 to define the influence of regional climate (Cl), biological organisms (O), relief or topography (R), parent material or underlying geology material (P), and time (T) on current soil properties. All of those factors combined resulted in different soils depending on how the factors varied. Differences and similarities in wine is analogous to soil.

But as I said previously, wine is more than just a combination of those basic environmental factors. The historical and cultural practices of regions have greatly influenced the modern *terroir* of many places. In July 1395, Philip the Bold, Duke of Burgundy, banned the gamay cultivar from Burgundy. This left pinot noir to run unopposed for undisputed king of the region. For more on that interesting story, read Ben O'Donnell's piece in Wine Spectator. A little episode called Prohibition forever changed the wine industry in the United States of America. If you think that terrible experiment didn't affect the *terroir* of this country, just look at which wine regions were most popular and what cultivars dominated the vineyards in California in the 19th century. America wasn't known for its Napa cabernet!

Finally, perhaps the most important geographic factor in the wine industry is the human-environment tradition. Choices relating to rootstock selection, canopy management, planting density, clonal propagation, irrigation regimes, and vineyard preparation are all essential parts of modern wine growing. If you can show me a vineyard that has not been impacted by any of those factors, I'll eat my shorts. All of those human interactions can be said to accentuate the sense of place of a wine. Or they can be used to internationalize the style of a wine and negate its *terroir.* That outcome depends on the philosophy of the winemaker, the grower, and the marketing team.

Not only is *terroir* a product of all four of these geographic traditions, every place on Earth has *terroir.* However, not all *terroir* is created equal. Some places remain unplanted because the *terroir* is terrible for grape vines. Some *terroir* remains unplanted because its greatness has just not been discovered. And some *terroir* remains unplanted because the climate has yet to change enough to make it economically feasible. More so, some great *terroir* may cease to exist because of a changing climate. Not only does *terroir* exist everywhere, its expression changes everywhere. Changes may be brought about by climate change, natural disasters, political decrees, or simple cultural adaptations.

Geography may not be as catchy a term as *terroir,* but please stop spreading the fallacy that there is no English translation for *terroir.* It's geography . . .

NATURAL WINE: NATURE VS. NURTURE

Taken literally, there is no such thing as a natural wine in the world of commercially available wines, any more than there are natural woven baskets or natural oil paintings. So often nature alone is credited and nurture is excluded. But when sound wine is the desired outcome, some conservation, some level of husbandry, is required. When you look more closely at the details, you start to suspect that the words "natural" and "wine" are antithetical to each other, that wine, by its very nature, is a processed beverage, requiring interventions at several critical stages in the act of making it.

Here is just a short list of the areas where those interventions occur:

- **Reception:** Where incoming harvested grapes are received. Cleanliness, organization, efficiency, timely operating speeds, and temperature all affect the outcome of the finished product.

- **Amelioration:** *Acidulation and Chaptalization.* Acidulation refers to adding acid to the wine. Chaptalization refers to adding sugar. Both choices make up for what nature did not provide in the desired proportions.

- **Inoculation:** Refers to supplemental yeast and or bacteria in the must to conduct desired fermentations.

- **Fermentation practices:** Refers to those parts of the process that are monitored, adjusted, and controlled by humans to promote healthy microbial growth, and color and flavor extraction, especially the manipulation of temperature, the addition of air, water, and the vessel used to ferment.

- **Elevation:** The French expression *élevage* refers to the care and aging of wines during their time in the cellar.

- **Stabilization:** Refers to the use of cold stabilization, removal of proteins, or correcting microbial conditions to render wines safe and sound after bottling.

- **Clarification:** *Fining and Filtering.* Fining refers to the removal of unwanted microbes, physical solids, and tannins, which might otherwise cause faulty odors and flavors in wine, typically through the use of a binding agent. Filtering seeks the same result by use of a filter. ∎

SUCH INTERCESSIONS are critical not only to prevent the wine from spoiling, but to bring it to a finished state and therefore express its flavor potential with clarity and naturalness, and thereby its *terroir*. Attentive winemaking requires a kind of passive-alertness to each stage in the process, an enological lizard brain, jumping in when problems arise, while staying largely absent at all other times.

Natural winemaking, taken to its natural extreme, is described with fitting inevitability by chemist and Master of Wine David Bird in his book, *Understanding Wine Technology*. In this excerpt, he makes a case, in effect, for human interaction in the process by demonstrating the results of its absence.

FROM DAVID BIRD, *UNDERSTANDING WINE TECHNOLOGY:* "THE GIFT OF NATURE"

Good wine ruins the purse: bad wine ruins the stomach.

—Spanish Proverb

THE ORIGINS OF WINE

As even the most moderate of drinkers knows, wine varies enormously in quality, from the positively vile to heavenly nectar. Yet they are all the product of the fermentation of grape juice. The inquisitive wine drinker cannot help but question the diversity of quality and style, and wonder why there should be this wide range, when the raw material is simply the grape.

There can be little doubt that the first wine ever tasted by man was the result of an accident of nature. The vine is an ancient plant and has been known for millennia as the bearer of nutritious fruit. It is highly probable that grapes had long been gathered for consumption as fruit, or for the production of a delicious juice. All that is necessary for the discovery of wine was for a container of juice to have been left standing longer than usual, when the natural yeast in the atmosphere or on the skins of the grapes would start alcoholic fermentation and convert the sugars to alcohol. Initially, the juice would have been regarded as spoiled because, in its early stages, fermentation produces an odor of rot and degradation. Indeed, it is the first stage of the degradation process that reduces organic matter to its basic constituents. Only on further keeping, and after a cautious tasting, would it have been discovered that a total transformation had taken place and something had been produced with strange and wonderful properties.

This ancient wine was born without the aid of science in what would have been a very hit and miss affair—and ancient it is, going back at least 5,000 years, as witnessed by the paintings in the tombs of Egypt. And in Greek and Roman empires Hippocrates and Pliny wrote about the benefits of drinking wine. It was not until the work of Louis Pasteur (1822–1895) at the University of Lille in the middle of the nineteenth century that it was discovered that fermentation was due to the presence of microorganisma. But it was not until the last three decades of the twentieth century that scientific principles have been rigorously applied to winemaking.

The traditional way of making wine involved little science: grapes were crushed to release the juice, which was allowed to ferment with the naturally occurring micro-organisms until the juice had been converted into wine, with no temperature control and no analysis. The results were totally unpredictable, sometimes wonderful, sometimes disgusting.

THE NATURAL CYCLE

Winemaking is undoubtedly an art and the winemaker an artist, but if an understanding of the basics of the science that lies behind the transformation of grape juice into wine can be grasped, then the full potential of the grape can be realized. The pinnacle of quality can be achieved by the application of science through quality control, used in its holistic sense: controlling quality in the vineyard itself, of the vines, of the grapes, of the expressed juice, of the fermentation process and of the finished wine.

The process begins with photosynthesis, that miraculous process whereby green plants are able to synthesise sugar from carbon dioxide (CO_2) and water (H_2O) under the influence of sunlight and with the aid of chlorophyll, the green matter of plants. The sugar thus generated in the leaves is actually sucrose, which is then transported by the sap to the grapes that are the storehouse of the energy required by the pips to nurture the next generation. When it reaches the grapes, the sucrose is immediately hydrolysed by the acids to glucose and fructose.

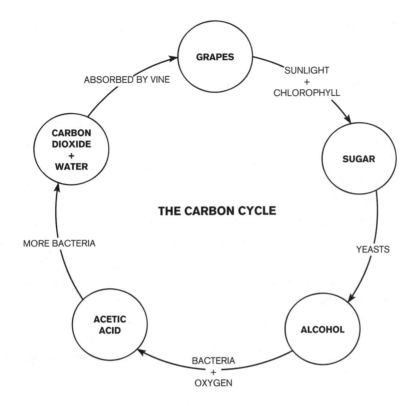

FIGURE 6.1
The Carbon Cycle.

As the grapes hang on the vine, and gradually come to full maturity with the aid of sunshine and warmth, the potential quality within the grape reaches a peak. The threat to this quality comes in the instant that man interferes and gathers the grapes. It is critically important that the quality inherent in the grapes is maintained between gathering and processing because chemical and biochemical changes occur from the moment the bunch is separated from the vine.

But more than this, maintaining the quality of the wine after it has been produced is particularly difficult because wine is a meta-stable substance; it is halfway down the slope of decomposition from grape juice to carbon dioxide and water. It is the product of the microbiological attack on grape juice, the same process that reduces all living matter to water, nitrogen, carbon dioxide in a few mineral salts. In reality, wine is a small part of one of nature's cyclic processes, the carbon cycle.

In this process green plants absorb carbon dioxide from the atmosphere and convert it to sugars, which are then used to create alcohol. This in turn gives rise to vinegar, which itself decomposes to release carbon dioxide into the atmosphere, which starts the cycle over again.

The complete cycle of detail is as follows:

1. The vine absorbs carbon dioxide (CO_2) from the atmosphere through the stomata in the leaves and water (H_2O) through its roots. With the aid of sunlight and the green

chlorophyll in the leaves, it converts these raw materials to sugars, which are stored in the grapes.

2. After the grapes have been picked and crushed, the yeasts convert the sugars to alcohol, producing wine, with some of the carbon atoms from the sugar returning to the atmosphere as carbon dioxide.

3. The next stage is the process known as oxidation, when the wine is attacked by the oxygen of the air, with the aid of bacteria, which converts the alcohol to acetic acid, producing vinegar.

4. By the action of more bacteria the vinegar is decomposed, yielding, ultimately, water and carbon dioxide.

5. The water flows into the soil and the carbon dioxide goes into the atmosphere, where they are ready for further re-cycling. ■

THE NATURAL Wine Movement has prompted many writers to contemplate just how much manipulation they are comfortable with in their wines. In Patrick Comiskey's essay, which appeared in *Bon Appetit* and was written for a lay audience, he seeks to make room for both natural wines and their less than natural counterparts. Harvey Steiman, in the Wine Spectator, seems to issue a warning that the lack of manipulation altogether can yield some fairly "funky" results.

FROM BON APPETIT: "BACK TO NATURE," BY PATRICK COMISKEY

When you uncork or unscrew a bottle of wine these days, you probably make the very reasonable assumption that your wine is a natural product, that it has more in common with orange juice than, say, Tang.

But chances are good that the wine is a long way from fresh-squeezed. In the last 60 years, numerous practices have taken winemaking further and further from its simplest, most natural form: grape juice that's been fermented by yeast and preserved in a bottle. Many of these practices amount to technological advances or solutions to long-standing problems. In the last decade, however, some have taken the form of cosmetic enhancement—nips and tucks to smooth away less than consistent expressions of the grape. More than ever before, wine is processed: in large-scale operations especially, wineries are more like factories.

All of which has led to a dogged subculture of winemakers, importers, and advocates who are devoted to a return to traditional winemaking and vine-growing, to make a product that is as pure as humanly possible. These naturalists, to varying degrees, are growing grapes without the use of herbicides, pesticides, fungicides, or any other 'cides. The fruit is nearly always hand-harvested and grown without irrigation. It is vinified using indigenous yeasts that live naturally on the surface of the grape or in the winery. New oak barrels, while not technically unnatural, are frowned upon, as is the excessive use of sulphur as a preservative. As much as possible, the winemaker allows the wine to make itself.

This type of winemaker is still rare, and the wines themselves are expensive to produce and sometimes so rustic or exotic that they're not likely to appeal to everyone's tastes. Nevertheless, natural wines are being made in sufficient numbers—and are garnering enough critical attention—to provide a real alternative to conventional wines.

When did wine become anything but natural? For more than 4,000 years, winemaking and vine-growing changed little. Technological advances amounted to the use of glass bottles and cork stoppers to reduce spoilage and preserve the wine. But in the middle of the last century, large-scale industrial farming transformed agriculture by instituting many nontraditional practices, including the widespread use of chemical pesticides and fertilizers. Europe's wine regions, decimated by war and years of economic hardship, turned increasingly to chemically enhanced farming because of its reliability. But that kind of farming reduced soil to little more than a lifeless medium for fertilizers and additives, and there were times of the year when vineyards became hazardous, even deadly, to walk through. At the same time, the global market called for production that emphasized volume, efficiency, and, above all, consistency—a trend that continues to this day.

While throughout history wine has been praised for its variability—the fact that it changes in the bottle, in the glass, and across vintages—marketplace consistency still breeds brand loyalty, whether it's to a bottle of wine, a bottle of Coke, or a bottle of dandruff shampoo. As a result, many less-than-natural processes have been developed to ensure consistency, from minor tweaks to major innovations. In the vineyard, high-performance clones and rootstocks maximize yield and machines harvest the grapes: in the winery, additives include enzymes, modified yeasts, oak chips, oak dust, tannins, sugar, and coloring agents.

Enhancements are hardly limited to mass-market wines. For boutique wines, there are costlier methods of manipulation. Does your wine taste hard or aggressively tannic? Break out a micro-oxygenator, a device that pushes tiny oxygen bubbles into the wine to soften the tannins. Grapes too ripe or too sweet? You can add water or acid to correct any imbalance. If alcohol's the problem (and it often is, with overripe grapes) simply remove it using "spinning cone" technology. Enzyme cocktails and yeast enhancers (which go by names like OptiRed and OptiWhite) can push your yeasts to over-perform. Many of these processed, or "spoofulated," wines are irresistibly delicious and often get high marks in competitions and from critics. But they can taste soulless.

A few years ago, I was interviewing a prominent Oregon winemaker who poured me an especially seductive Pinot Noir. She informed me, with obvious pride, that she adds oak balls the size of Ping-Pong balls at a certain stage of the fermentation to give the wine a creamier mouthfeel.

I was amazed at her skill, but the disclosure also seemed kind of creepy. This was Pinot Noir: a varietal famous for its ability to express vineyard character. Where was the Oregon-ness of this wine? I felt a little manipulated, as if the winemaker, like a plastic surgeon, had cosmetically altered the wine to guarantee my seduction, and sacrificed its authenticity to do so. Still, it was drop-dead delicious. Where was the fault in that?

Natural winemakers do find fault in this sort of tampering. They prefer to let the vines and the wine reflect what nature bestowed upon that vintage. This might make them wonderfully interesting, or, sometimes, just funky and weird. At the very least, they taste unique—they're like the poster children for *terroir,* the French term that refers to the way a wine expresses the place where its grapes are grown.

I still struggle with the question—natural or not?-- every time I open a bottle, but I'm grateful to have the choice, just as I am glad to have the choice between, say, a roasted heirloom potato and a Tater Tot. We all know that there are times when a Tater Tot totally hits the spot. The same is true for a ten-dollar red. It's a big world, and there's plenty of room in it for both the natural wine that proclaims difference and the satisfying sameness of a reliable conventional bottling. Your wine tastes can, and should, include both.

FROM WINESPECTATOR.COM: "IS IT ALL IN THE FUNK? HOW 'NATURAL WINES' CAN POLARIZE WINE DRINKERS," BY HARVEY STEIMAN

I love the idea of natural wines. I'm all in favor of encouraging biological diversity in soils and avoiding pesticides, something the best conventional winegrowers do, too. It's immensely appealing to think of wine fermented, aged, and bottled without any intervention. Just let the grapes ferment and stopper up the result. I admire the sense of completeness and harmony that wines from these "natural" winemakers can achieve, when all goes well.

But I keep remembering the words of the late California winemaker Andre Tchelistcheff. Left to her own devices, Mother Nature is trying to make vinegar, he liked to say. A winemaker's job is to catch it when it's wine. If things don't go exactly right, more than the vinegary pungency of volatile acidity can affect a finished wine. Mother Nature can infuse it with the barnyard smells of brettanomyces, a yeast that can proliferate post-fermentation, or the fizzy texture and sauerkraut notes of a wayward malolactic fermentation.

It takes heroic efforts on the part of the winegrower to keep such things from happening without help from technology. The best natural winemakers do, but others? Well, let's just say the array of potential flavors is much broader than many of us want to drink. The striking thing, to me, is how those who champion natural wines are willing to accept this funkiness. Not only accept it, but consider it part of what makes the wine attractive to them. Ordinary wine drinkers, who don't know that these characteristics are considered faults by the majority of winemakers around the world, often assume the funk is just part of the rhythm of the wine.

It's amazing how many people are essentially blind to brettanomyces, for example. I can't tell you how many times I've wanted to moo under my breath at the barnyard aromas wafting out of my glass while the pedigree of a famous label dazzles others around the table. Maybe I am sensitive to this characteristic because the wines I review from Oregon, Washington, and Australia seldom have it. Those accustomed to drinking Bordeaux, Rhône, or some highly acclaimed Tuscan wines, which often reflect relatively high levels of brett, may not notice.

Something similar happens in music, specifically in live performance vs. recordings. In opera, an area for which I have a special fondness, the immediacy of live performance often comes with the occasional missed note. Few want to hear a bobbled high C or a phrase that drifts sharp every time they play the recording.

When a computer application doesn't do exactly what's expected, software industry insiders deadpan, "It's not a glitch, it's a feature." I am hearing something similar from people in the natural wine movement, not as a joke but as a defense of flavors that many might define as faults.

Some fans disdain studio recordings, which can clean these bits up. They'll take the sour notes for that extra frisson you get from live performance, even if they have to put up with it every time they listen. Few, however, argue that the mistakes were a feature.

NATURE AND NURTURE

If we agree that *terroir* expression is a product of Nature and Nurture, that human intervention and husbandry are a necessary part of the successful production of a wine, that human involvement nurtures, elaborates, and even amplifies *terroir* expression, then the next obvious question is how much? When have you gone too far? When is the "nurture" component invasive and overbearing, when does it obliterate the delicate fingerprints of a place?

It's useful to remember that the recognition of *terroir* is the result of hundreds of years' worth of careful human observation and adjustment, with cultural protocols—the human factor—evolving historically, carried on by ritual, systematic practice, generational advancement, and evolution. The process of mastering the sow's ear can be difficult and protracted, but once mastered the path to the silk purse can fade into the background.

Terroir character is always conferred in hindsight, isolated after years, sometimes centuries of natural expression. Many questions remain: How does one discover one's *terroir*? How does one isolate a *terroir's* flavor markers and bring them forth? What can the winemaker do to amplify or safeguard its expression? Are there standards of input, or limits to one's vineyard and winery practices required to preserve *terroir* expression? How far is too far? Does making high quality wine mean discovering and conserving idiosyncratic elements in your wine or in your vineyard, or will doing so make your product less marketable? What if those idiosyncrasies are characterized as flaws or faults? Are you duty-bound to preserve them?

Responsible winemakers should, we think, conduct trials to get at some of these questions. Whole vintages of trial and error could follow—adopting positive practices and rejecting those that diminished the desired flavor outcome—until an optimal protocol could be designed. In the Old World this would have been done over many generations, with protocols adopted based on the gradual discovery and implementation of the best methodology.

DEGREE OF INVOLVEMENT

Scant literature exists to guide winemakers in the intricacies of "*terroir* winemaking." No textbooks exist explaining what to do and what not to do, or how to shepherd special, unique flavors into a bottle, how to craft a wine transparent enough to let the wine's unique character show through. Such practices can be applied down to the microscopic level, as endemic microflora that exists on winery surfaces and in storage vessels may become part of the winemaking landscape, and it would be incumbent upon the sensitive winemaker to enable the domestication of certain microbes while guarding against the winery equivalent of a nosocomial infection. So perhaps it comes down to a balance between "benign neglect" and "vigilant manipulation" in order to maintain an ideal environment for *terroir*.

Perhaps the best summation of the options is found in a selection from Jamie Goode's book *The Science of Wine,* in which he discusses natural winemaking vs. various levels of manipulation, in which he explores the parameters of intervention in the winery.

FROM JAMIE GOODE, *THE SCIENCE OF WINE: FROM VINE TO GLASS:* "NATURALNESS IN WINE: HOW MUCH MANIPULATION IS ACCEPTABLE?"

Wine can be made naturally; it almost makes itself. At its most simple, the process of making wine involves harvesting grapes, sticking them into some vat, crushing them a bit, and letting them ferment. When fermentation is complete, separate the solid matter from the liquid and you have wine. But winemakers almost always add things to their wine. There are several reasons for this, some of them better than others, and this leads to a thorny question that is at the heart of many of the most passionate debates in winemaking circles: just how much manipulation is acceptable?

There's no simple answer to this question. It's a grey area and any attempt to prescribe permissible levels of manipulation is a line-drawing exercise. But just because it is difficult to make these sorts of distinctions doesn't mean we shouldn't try to make them. Wine laws exist in virtually all wine-producing regions or countries which outline the type of manipulation that is acceptable and the type that is not. Some wine regulations are stricter than others. To get an understanding of the issues involved, let's consider four different positions and assess their strengths and weaknesses.

1. Add anything

Should a wine be judged purely on how it tastes? Is drinking wine just a sensory experience? Some people argue that this is, indeed, the case. If it is, then there are no real reasons to prohibit additives at all. The answer is in the glass, and if there are ways of making wine taste "better," then by rights they should be allowed. The weakness of this position is that it ignores the fact that wine is a discretionary purchase. Certainly, fine wine is something that people buy partly because it isn't manufactured—the grapes aren't just seen as the raw materials that act as a starting point in the manufacturing process. Grapes, as we have seen, have a connection to the soil as well as to individual vineyards. Part of the appeal of wine is that it is a natural product rich in culture, and its image will suffer if any kind of manipulation is allowed without scruples.

2. Add nothing

The idea of adding nothing at all to wine is an extreme position for one reason: sulfur dioxide. Sulfur dioxide (SO_2) is intrinsic to winemaking because it's hard to make good-quality wine without it. It plays a vital role as an antioxidant and also as a microbicide, preventing the growth of harmful bacteria and rogue yeasts at different stages–it is added during winemaking and at bottling. Some winemakers bravely attempt to make SO_2-free wines for the sake of naturalness or for health reasons, but it needs to be borne in mind that SO_2 is itself produced naturally during fermentation in non-negligible amounts anyway.

3. Add as little as possible

This is a laudable position for the reason mentioned earlier: that wine is perceived by consumers as a natural product, and this is part of its appeal. A sensible winemaking policy is only to add something if not adding it is going to compromise wine quality, and then only as little as possible. It's tough to make a good wine with no SO_2, but the effects of any additions can be maximized by smart use. Acid additions might be needed in warmer climates. This raises a question about other sorts of wine manipulation.

For centuries oak barrels have been used to make wine, and the use of them is uncontroversial partly because they are traditional. Use of new oak barrels certainly would count as an additive manipulation because they contributed important flavour components to the wine. Smart barrel use is a vital component of the winemaking process for the majority of fine wines, and it's hard to imagine doing without them. But consider what might happen if they had never been used for wine and someone tried introducing them now–there would probably be a bit of an outcry in certain circles. This raises the question of whether it is hypocritical to allow barrel use but exclude newer, high-tech manipulations such as micro-oxygenation and reverse osmosis (RO), each of which are given their own chapters in this book.

On balance, the case for accepting older traditional manipulations and avoiding newer ones does have a sound basis: that of preserving product integrity in the eyes of consumers. The add-as-little-as-possible school would no doubt object to newer high-tech manipulations, although at a stretch it could be argued that alcohol reduction by RO would prevent excessive alcohol levels from compromising wine quality—and thus should be allowed.

4. The Compromise

The final position in our debate would be to permit some manipulations but not those which could be deemed as "cheating." Openness and honesty are the key words here–*i.e.*, adding most things is okay as long as their use is disclosed, and wine laws don't outlaw them.

So where would the line be drawn? A strong argument could be made for banning manipulations such as non-traditional chemical flavorings, but allowing winemakers access to other techniques if they choose to use them. I can't see anyone seriously arguing that fruit flavorings or non-wine fruits should be used to make wine and the substance still be allowed to be labeled as wine. On the other hand, this more relaxed view would permit RO and micro-oxygenation. These technologies are, like any other technologies, merely tools, and if tools can be used well they can be used badly. What counts is *how well* the tools are used–not whether or not they are used.

CONCLUSION

Ultimately, a relationship of trust exists between winemakers and consumers. I feel that where manipulations are used, they should be declared, and then consumers can choose which wines they want to purchase on this basis. If they want a wine that's totally natural, with no SO_2 added, then that's fine. But if they want a wine from a producer who uses as little manipulation as possible, this is also fine. Alternatively, if they don't mind how the wine was made and only care about the taste, that's their choice also. I suspect that many consumers would be surprised by the degree of manipulation that does take place with some wines, because the popular conception is that wine is a relatively "natural," additive-free product.

As stated above, the various means for manipulating wines are just tools; as such they can be used wisely, used badly, or not at all. Whether or not the use of these tools is justified is a decision that can't be made globally and enshrined in legislation. Would you rather have a flawed natural wine when a simple manipulation would have eliminated the fault? It's a difficult, multilayered question (for example, what is "flawed" in the context of a wine?). That's why I would advocate a policy of freedom on the part of the winemaker, coupled with honesty about disclosing to the consumer the sort of "manipulation," if any, that is used.

It is possible to argue that different categories of wines should be treated differently. While manipulation might be necessary to help out a commodity wine made from less than perfect grapes, there is less of a case for using more manipulation than is absolutely necessary for fine wines. Rules are important to preserve wine integrity and protect consumers from fakes, but they should be implemented locally rather than globally. Some manipulations, such as adding chemicals as flavouring agents (whether they occur naturally in the wine or not), are indefensible, and should be banned altogether.

THE FRENCH METHOD

In a telling passage in Hugh Johnson and James Halliday's important book *The Vintner's Art,* the authors share an anecdote from Chateau Margaux in Bordeaux. In the late seventies, André Mentzelopoulos, the new owner of the Chateau, called the winery one harvest weekend to see how the harvest was progressing. To his surprise no one answered the phone. So he chartered a plane and flew to Bordeaux to see if he could get some answers and found the Chateau largely deserted. "He thought it was inadmissible," wrote the authors, "that employees closed on Saturday and came back on Monday, during fermentation, which lasts only three weeks of the year." Turns out he needed some additional lessons on French winemaking, and on being French.

In France, though, the non-interventionist approach extends right through the winemaking process. Natural yeasts for both primary fermentation and malolactic fermentation are regarded as essential: they are an extension of the expression of *terroir,* providing subtle but palpable complexity in the wine, unlike the one-dimensional and "foreign' character of a single cultured yeast (or so the French believe).

■   ■   ■

If all this paints a picture of France as a land of stubborn traditionalists, it only tells half the story. The emblem and embodiment of scientific rationalism is Bordeaux's famous Professor of Oenology, Emile Peynaud. Peynaud has been characterized as hero and as villain, as savior and destroyer. His critics tell the (probably apocryphal) story of the expert given an unknown wine to comment on, and his response was: *"I can't tell you what the wine is or where it came from, but I can tell you it was made by Professor Peynaud."* (Other countries have their equivalents: in Australia Brian Croser and Dr. Tony Jordan have been accused of a similar Svengali-like influence.) The supporters of Professor Peynaud would simply say that all he sought to do was give the winemaker a better understanding of, and greater control over, all aspects of winemaking: if chemical and bacterial reactions take place in the course of making the wine (and they do) those reactions should be planned and their conse-quences understood. If there is a problem with this approach it is that it takes much of the mystique out of winemaking, and exposes impotence or incompetence for what it is. And by eliminating the chance consequences of that bacterial contamination, oxidation, acetification, or whatever, it is perfectly true that the wines made under Peynaud's control exhibit a degree of family resemblance: they are devoid of major technical faults.

Fully understood, science does not mean the end of individuality nor enforce the making of sterile, squeaky-clean wines. Giotto proved his skill by drawing a perfect freehand circle; Picasso showed in his early realist period that he had the ability to portray nature as precisely as any artist. Once winemakers have mastered the basic skills and techniques, then of course they may eschew them: it is an entirely different thing to ignore technique simply because it is not available in the first place–or because you do not understand it. ■

IN THE following passage, Oz Clarke and Margaret Rand puzzle over how ephemeral the French notion of *terroir* is, particularly in the country's hallmark region for its expression, Burgundy.

FROM OZ CLARKE (WITH MARGARET RAND), *GRAPES & WINE*
WHERE GRAPES GROW

If climate and latitude alone determined where vines grew, life would be so much simpler. You would simply look at the sunshine, the temperature, the rainfall and the chance of frost, and work out what to plant where. Hey, presto: great wine, every time. Unfortunately, it's not quite that simple. Weather is only one of many factors that determine the quality and style of a wine. When considering where grapes grow, it's worth remembering that European vineyards were seldom planted after close analysis of the weather or the soil. If by chance they turned out to be great vineyards, then people tried to work out why. It's been a long process of discovery, and it's not over yet.

Winemakers have been studying Burgundy's Côte d'Or for years, but we still don't know precisely what it is about this little stretch of French vineyard that produces such marvelous Pinot Noir and Chardonnay. And if people can't agree on that, it's not surprising that they also can't agree on which of the attributes of the Côte d'Or you should try to imitate if you want to make great Pinot elsewhere, or how closely you should imitate them. Should you find somewhere that mimics the Burgundian climate? Or is the climate in fact a disadvantage? If you think that, then you'll seek somewhere warmer and drier. Should you be trying to copy the soil? And if so, should you be looking at its structure, its mineral content, or what?

Vineyard dirt is receiving intense study from winemakers everywhere, even in the New World, where they used to be much more interested in climate. Yes, climate is vital too–but Australian winemakers are now saying that the greatest advances in quality will come from greater understanding of the soil; and they're probably right.

Terroir

This is still a relatively poorly understood concept—but it is important to realize that *terroir* does not mean "soil." The *terroir* of a vineyard is the sum of all its parts: its geology, its climate, its topology, its water-holding ability and the amount of sun it receives, and the effect of man. Without human intervention there wouldn't be vineyards in the first place. So the soil, both topsoil and subsoil, is important, as are the mineral components of the soil. How fertile or infertile it is, and its depth and structure, which affects how well or poorly drained it is, are also factors. Altitude, steepness of slope, and exposure to the sun all matter, as does the meso-climate, or climate particular to that vineyard. From the French point of view (and it is most of all a French concept), it is the *terroir* that makes each vineyard different. It underpins the Appellation Controllée system not least because, as the underlying factor behind wine quality and style, it should show in the wine no matter who the winemaker is or what he or she does to the wine. Winemakers come and go; the *terroir* remains.

However, good viticulture and winemaking can permit the expression of the *terroir* while bad viticulture and winemaking can mask it. And since good vineyard practice can mean installing drainage where necessary, and since good winemaking in Northern Europe does not exclude chaptalization (the addition of sugar before or during fermentation to increase a wine's alcoholic strength) growers are in practice not absolute slaves to what their *terroir* dictates.

WINEMAKING STEPS

Terroir winemaking discussion is absent from winemaking textbooks. Though there are discussions of elements of style (from tools used, aging techniques, blending, etc.) there are no instructions per se on how to recognize or preserve flavors related to site. We have never seen a diagram showing all the "junctures of decision" in the winemaking process as they relate to *terroir* expression. There are certainly many decisions along the way which can result in a flavor developing one way or another.

What if you were given one hectare of historic vineyard land planted to Pinot Noir in the Côte de Nuits that had been used to make fabled wines since time immemorial? What would you do? Firstly, the cultural forces (read: local advice) would be overpowering, maybe stifling. You'd probably be compelled to follow the accepted protocol . . . peer pressure might see to that, but it's worth noting that straying too far might risk an AOC compliance audit (your wines may not get the approval to use the AOC on the label).

The classic enology professors all preached: "You can make bad wine from good grapes, but you can't make good wine from bad grapes." We never doubted the truth of the first clause. As for the second, while maybe true, we took as a challenge. If the site always yielded bad grapes, you'd eventually replant or abandon; if the weather or seasonal conditions gave you bad grapes, you'd have to really earn your pay as an enologist to achieve decent or better quality and keep the inventory alive.

Traditional enological training has always involved mechanics. Reliable lab methodology evolved to analyze and help head off potential problems (especially sub-threshold ones) and to let winemakers know when more diligent sanitation might be required. Perhaps the best description for effective winemaking technique isn't merely benign neglect, but rather vigilant benign neglect.

Although many excellent winemaking texts exist in all wine cultures, Old World and New, they are all concerned with physical enology and processing. There is no enological "Strunk and White" for making *terroir*-inflected wine. Below, we'll excerpt some winemaking texts and hear from several winemakers as to what they say they do.

One of the classics is the winemaking text from Boulton, et al., *Principles and Practices of Winemaking*, from 1996. Here they describe a few of the rationales applied to winemaking, and deliver a complicated but failsafe schematic for determining the praxis for nearly every winemaking contingency:

FROM ROGER BOULTON ET AL., *PRINCIPLES AND PRACTICES OF WINEMAKING*
GENERAL SEQUENCE OF OPERATIONS IN WINEMAKING

These operations interact and their nature, timing, and sequence are key in producing different wine types and styles. Furthermore, failure to apply each operation optimally increases costs, is likely to change quality, and can lead to failure.

Wines differ so much that processing by rote is possible only for mass-produced, relatively nondescript wines. Such wines may be inexpensive and pleasant, but they are out of favor with consumers and are likely to become more so since interest has focused on premium types with diverse and specific characteristics. Of course, quality relative to others available is still vital among wines

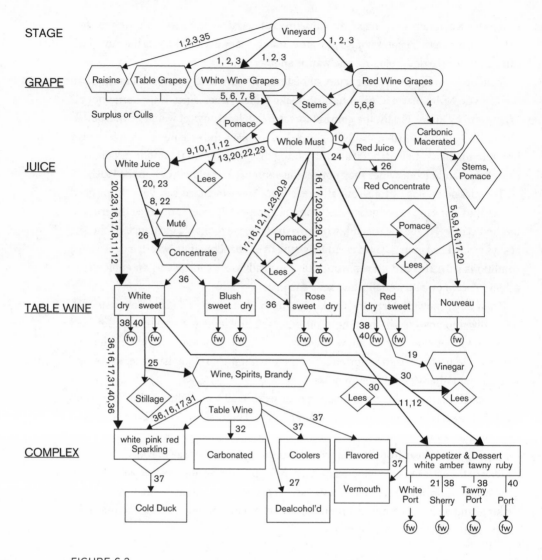

FIGURE 6.2
An amplified outline scheme of making various wines, alternate products, by-products, and associated wastes.

competing primarily on the basis of low price. The U.S. market, and increasingly others also, will not buy poor wine regardless of low price.

To some degree, the nature and sequence of winemaking operations are obvious. The grapes must be obtained and fermented. The young wine must be clarified, processed, and distributed to the consumer. The winemaker must consider that wine may be held a long time before actually being consumed. Some operations may not be necessary for all wines (sterile filtration of microbially stable dry wines, for example) and others are applicable only where a specific style is sought (carbonic maceration, sur lies, etc.). The number of times a process such as centrifugation fining or filtration may be applied affects cost and ultimate quality owing to the potential for loss of one volume, loss of volatiles, exposure to air, etc. Often more experience and restraint can give better but less costly wine if the objective desired is firmly in mind. Some wine might be filtered half a dozen times and others not at all or once. The latter is frequently better and certainly cheaper, barring special considerations. It may be fatuous to advise the novice winemaker "don't just do something, sit there," but contemplation of the ramifications should precede action in the well-managed winery. Figure 1 (*see Figure 6.2—eds.*) attempts to amplify the general sequence of operations to cover the full panoply of winemaking practices. It is recommended that it be consulted each time a new wine type is considered. ■

IN HIS landmark 2006 lecture on *terroir* delivered at the University of California Davis, Warren Moran delineated six facets of *terroir* that impact a wine. He termed the winemaking facet "vini-terroir," and in the lecture he described the ways in which winemakers were responsible, to varying degrees, for stylistic overlay.

FROM "'YOU SAID *TERROIR?*' APPROACHES, SCIENCES, AND EXPLANATIONS," BY WARREN MORAN

6. VINI-TERROIR

As Denis Dubourdieu is fond of saying, terroir is about overcoming the disadvantages of the natural environment where you find yourself growing grapes for wine. Vinification and *assemblage* are one means of ameliorating, or even capitalising on, such disadvantages, especially those of climate. The region Champagne is the ultimate cool climate success story and extreme example of this point.

Champagne, especially in its real home in France, is an innovative method of vinifying and assembling Pinot Noir, Chardonnay, and Pinot Meunier to make a distinctive wine. It capitalises on the high acidity and aromas of these grapes that in many years would make relatively ordinary still wine. Part of its success lies in local people and enterprises selecting and developing over centuries cultivars, pruning methods, and canopy management that suit the Champagne style. Vines in Champagne yield at levels over four times higher than the same varieties grown in regions of France, where still wines of quality are made from them, although often different clones, of Chardonnay and Pinot Noir.

Growers from different communes or groupings of communes in Champagne have gradually specialised in the varieties that suit their local environment and the demands of those vinifying their grapes, whether it be a cooperative or an enterprise as large as Möet and Chandon that operates across many communes. Despite this specialisation by commune, Gislain de Montgolfier of Bollinger (pers. com.) describes colourfully how on adjacent parcels of land with virtually the same soils and climate, but with different owners, the qualities in the grapes produced are quite different. Vinification in small parcels, as practised by some enterprises, allows the personalities of the place and the people to emerge even when later masked (or should I say enhanced?) by the *assemblage*.

Sauternes is another obvious example where an apparent disadvantage (moist atmospheric conditions prior to vintage) has been transformed into a wine of great distinction. Like everywhere else, the atmospheric conditions vary from vintage to vintage in Barsac-Sauterne. That vignerons are able to achieve reasonable quality in many seasons, and outstanding in some, is a considerable achievement in understanding how to manage the vine and its crop to ensure the highest probability that the berries are in fine physiological health when the weather conditions are ideal for the so-called "rot" to be noble.

Frequently, we do not know what causes the distinctive flavours of varieties or cultivars grown on particular pieces of land. One has to go no further than the New Zealand experience in Marlborough to emphasise this point. When Montana (now Allied Domecq or Pernod-Ricard) planted its first grapes in Marlborough in 1973 nobody had the faintest idea that Sauvignon Blanc grapes grown in this environment would result in a wine with a distinctive array of aromas that is almost instantly recognisable and appealing to consumers in most wine drinking countries. Now other New World regions are closer to a reasonable imitation of it, even though the nuances of their natural environments are different. Such a convergence must have much to do with winemaking as well as viticulture. People in other parts of the world such as South Africa and parts of Australia have learned— sometimes through New Zealand mentors—how to tease out the particular qualities that provide the Marlborough style in the vinification and to some extent the vineyard. ∎

IN 2014 author Natalie Berkowitz published *The Winemaker's Hand,* a collection of interviews with winemakers all over the world in which they discuss their practice and their methodologies. Inevitably, the issue of *terroir* crept into nearly every discussion, with each winemaker describing what to do—and what not to do—to preserve its expression in a developing wine. Here is a small set of excerpts.

FROM *THE WINEMAKER'S HAND,* NATALIE BERKOWITZ, ED.

Joel Burt makes the still wines at the sparkling wine house Domaine Carneros. In this excerpt, Burt enumerates the number of methods by which he and his viticulturalist facilitate terroir expression in the vineyard, and he also sheds light on some of the ways in which terroir is maximized, or isolated, or selected into existence, using techniques that are meticulous and attentive.

JOEL BURT, DOMAINE CARNEROS

I work with the viticulturalist to get the fruit I want. Factors like the number of shoots, irrigation, and stress on the vines are decisions that affect the amount and consistent quality of fruit. The vines are exposed to a lot of sunshine to achieve thick skins and higher tannins than more delicate sparkling wine needs. For still wine we start in the vineyard, looking for low yields of small berry clusters that grow from one or two shoots per vine.

■   ■   ■

Wines made in big tanks are innocuous. The best Pinot noir is made in small fermentation lots so we have homogenous lots to blend. Small batches of Pinot noir, Chardonnay, and Pinot Meunier give me complexity, richness, and fruity flavors. Pinot Meunier is a grape that adds a little spiciness and a bit of tannin to the wine. Our Carneros Chardonnay has characteristics similar to those of Burgundian white wine, since Burgundy is where my soul lies.

We hand pick at night. The fruit is picked at 40°F so when it hits the tank it's ready for a pretty cool fermentation. In 2012, Chandon introduced an optical sorting machine that takes high-speed photos to scan individual berries. It's the same machine Tom Tiburzi uses. The operator sets the parameters the winemaker is looking for, and the machine checks the shape and color of the berries as they ride down the conveyor belt. An air cannon shoots undesirable berries to a different belt and then into different bins. The sorter can check sixty tons rather than sixteen tons when grapes are sorted by hand. It's a game-changer. Grapes are delivered straight into stainless steel tanks. We can ferment less desirable grapes and sell that wine on the bulk market. It's particularly useful for Pinot and is a place where technology is good. It's certainly better to start with great fruit than to modify it with additives to get better wine.

Some winemakers guide and herd grapes for high extraction and then use more oak than the wine can handle. Every step along the way decides a wine's fate. I learned to wait and see rather than hurry an event that changes the wine. I want its character to develop by itself. ∎

IT SEEMS clear that Cathy Corison has a sensibility that can embrace the uniqueness of her vineyard in the Napa Valley, a very particular stretch of the valley floor from Rutherford to St. Helena. To be sure, her wines' style is dictated by that place, but it is also influenced by the winemaker's sensitivity to and experience with that place. This despite the fact that all around her in the Napa Valley winemakers practice terroir-defying techniques in pursuit of ripe, sweet, buxom, high-alcohol wines—a trend that Corison seems intent on avoiding altogether.

CATHY CORISON, CORISON WINERY

Terroir is undoubtedly a key factor, but it comes down to a winemaker's vision of how to work with the soil. Truly great, complex wine is a combination of a unique place meets the hand of an individual winemaker with passion and a personal vision. In the best case, the winemaker's hand becomes an integral part of the terroir. Wine is interesting to think about, beyond its use for washing down food. A consumer should become familiar with a recognizable, individual wine style. Someone who has never seen a Gauguin painting will always recognize his paintings, even though they may have only seen one. The same experience should be true of a particular wine. Good wine can be made by committee, but often too many hands, too many conflicting ideas muddle the results. Fortunately, there are occasions when, at large wineries, vintners are given the opportunity to make wine that reflects their ideas and attitudes . . . And then there are also times when a capable wine technician makes flawless wine without character and soul.

Terroir is emphasized today as an antidote to the current international standardized style of winemaking. Terroir makes sense when it is a consistent, unique factor in each vintage, as it has been for generations in Burgundy. But even there, house style can be so strong it often trumps the vineyard. It's why I needed to choose the right terroir to achieve the style I wanted before I made the first drop. Some of my favorite Cabernet Sauvignons come from the gravelly, loamy alluvial soil of

Rutherford bench between Rutherford and St. Helena, a terroir that is perfect for Cabernet grapes. There's copious sun that ripens them fully and the cool nights that produce dark color, complex flavor, and adequate natural acidity. The best vintages occur when the cooling fog rolls in each evening, clearing up by 9 or 10 A.M.

▪ ▪ ▪

Terroir is part of my hand. Together, we deliver a sense of place, power, and elegance, a wine that graces the table, one that achieves a balance between food and wine, each making the other better. I try to balance between a vine's green growth and its fruit. It's a struggle, but it's important to me to keep the alcohol level under 14 percent. Grapes picked too early have green flavors, but if picked too ripe, the wine loses the deep red, blue, and purple fruit characteristics I love in Cabernet. The result will have unsatisfying black and pruney flavors with high alcohol and low acidity. Grapes grown and picked correctly from the Rutherford Bench exhibit soft, ripe tannins that originate in the vineyard. It's not possible to remove bad tannins surgically in the winery, leaving only the good ones behind. Nine out of ten years, our tannins feel like velvet.

▪ ▪ ▪

I hope I've gotten better at what I do. I've become more intuitive about the way I make wine. The miracle of winemaking is underlined by experience. I don't need to measure sugars anymore. I'm not interested in what's "in," like the current trend of boozy, very ripe, low-acid, sweet wine. I keep my radar focused through changes in fashion. I stick to my style because I've seen fashions come and go. ▪

DAVID STEVENS is a winemaking consultant who has worked with many wineries, most notably a long stint at Bouchaine. He makes the case here for winemaking's discipline, its rigor, and its adherence to scientific ground rules, and judiciously employs the tools needed to bring the *terroir* out of the wine.

DAVID STEVENS, CONSULTANT, CALIFORNIA

Winemaking is attention to detail. I think of winemaking as a sport with rigid rules determined by science. Winemaking is a science that requires constant planning. The right staff is crucial, the growers have to be on board, and the materials all lined up. My goal as winemaker is to play the hand nature deals me the best way I can. Terroir plays a big role in a wine's character, so the deck is stacked for and sometimes against the winemaker. A winemaker has to do the homework—picking cleanly without leaves, choosing the brix level, punching down the skins to keep the cap wet, adding sulfur dioxide, and choosing the right barrels. Finding a good cooper is as difficult as finding a wife. Yeasts are crucial because natural yeasts found on grapes don't do the job.

I was trained to believe diversity is the key to complex and high-quality wines. Sometimes wine is about a specific place. Some of my colleagues are excited when cases roll off the line, or when they see their label in a restaurant. Not me. I'm totally turned on by the creative process. Using grapes from several local vineyards, and selecting new versus used wood helps to achieve desirable diversity. Decisions about early or late blending also make a huge difference. If wines are blended, I blend early, so the wines get to know each other in the barrel. Plus it gives me more time to blend later if needed.

* * *

Winemaking is an accumulation of a million small choices about twenty big ones. It boils down not just to the critical decisions about when to harvest, how fermentation temperature affects sugar to convert to alcohol, when to press off, how to get extraction from the skins, pulp, and seeds. It's all a miraculous change. Tasting a young wine is not even remotely like what it will be like in two or ten years. It's like a long pass in football or like flying by the seat of your pants. It pays to remember Nature bats last. ∎

JOHN WILLIAMS is one of the most dedicated and articulate terroirists in the Napa Valley, one of the great believers in soil expression, particularly his prime patch of valley floor in Rutherford known as the Rutherford Dust. His relationship to that Dust, as you'll see, is deeply personal, philosophical, and even spiritual, and it defines his winemaking practice, which he characterizes as much by what he doesn't do as by what he does.

JOHN WILLIAMS, FROG'S LEAP WINERY

I believe terroir has the power to transport you back into time and place. Our Rutherford Cabernet Sauvignon is an example of the best qualities in an appellation noted for the quality of its cabs. My most noble goal is to make a wine that smells and tastes like where it comes from. Wines should be prized for their elegance, balance, ability to age, and most of all, to reveal the truth of the place where they were grown.

Since I settled in Rutherford, I've tried to unravel its terroir, the sense of how this unique place is part of our wine. It's always exciting when I taste Rutherford Dust in other wines from our appellation and find it disappointing when I don't. The dust has a unique earthy, dusty smell, and a distinctive mouth-feel that is like rubbing velvet against its nap. Rutherford is an integral part of my body. It's in the water I drink, the food I eat, the wine I make and drink. Every winemaker should want to feel those intense qualities. I'm tuned in to aromas and flavors. Olfactory experience, the powerful chemistry of smell, can be an out-of-body experience, in the same way certain odors remind me of Christmas candy. My Sauvignon blanc reminds me of the aroma of rain on a slate roof in Ithaca, New York. It's the place in my head where I go for aromas of Sauvignon blanc. I summon up the Rutherford terroir in the same way.

* * *

Vines adapt to many conditions, so a winemaker has to decide whether to be a master manipulator or a co-conspirator. There needs to be a correct balance between our needs and nature's requirements. It's part of an ongoing process that starts in the vineyard as a series of thoughts and actions that deepen the relationship between grapes and wine. We can submit vines and grapes to our will, but our goal should maintain the vines' natural balance with proper trellising, correct row alignment, and protection from insect pests and predatory animals.

I believe the winemaker's duty is to stand aside and let Mother Nature determine the natural beauty of the grapes. One single precept motivates me beyond all others and leads me to follow a natural path to winemaking. It is the Taoist principle that expresses the practice of *not-doing.* Standing back sounds easy to do, but in fact, it is the most difficult way to make wine. I prefer to work within nature's patterns and rhythms rather than to take actions that follow man-made paths and conventions. I never undertake a procedure merely for its own sake. Decisions about winemaking should respect Nature's natural order rather than a human time schedule. I don't believe in forcing fermentation to fit my schedule, or racking based on time rather than taste. We use oak barrel aging to subtly enhance flavors in some wines, rather than to disguise or overwhelm the wine. Current wine styles favor overextracted wines produced by winemakers with big egos, who push for 16 percent alcohol using all sorts of manipulative techniques. The Taoist concept makes you deal with what is given to you to work with. Otherwise, the natural process is stifled. My goal is to work with purity of purpose to craft fine wines in a more traditional style. Accepting this premise takes me down all sorts of remarkable winemaking paths. ∎

PHILIPPE DELFAUT and Alexander van Beek are both Bordeaux winemakers, a place that, like Napa, can be thought of as a bit insensitive to the vagaries of *terroir,* more driven by style than by place. Not so these two winemakers, who both make a case for guiding the *terroir* by keen observation of what's been presented, then getting out of its way.

PHILIPPE DELFAUT, CHÂTEAU KIRWAN

My personal philosophy is to work more by intuition, innovation, and experience. Transforming juice into wine requires more feeling than technology. My job is to guarantee a sense of terroir in Kirwan's wine. I start from scratch with the land and work toward wine in the bottle. I have a stylistic goal and reach it with whatever nature gives me. I prefer the term *wine-helping* to *winemaking,* which I think refers too much to operating through a technical, scientific approach.

The human aspect of winemaking plays the most important role. I am the decision maker. Wine is a living product and therefore each year requires new decisions that can't be determined in advance. I guide Kirwan's entire process to produce fresh, complex, and balanced wines. . . . My worst fear is a natural accident that is impossible to control. The vagaries of weather in Bordeaux make winemaking as difficult as walking a tightrope.

ALEXANDER VAN BEEK, CHÂTEAU DU TERTRE, CHÂTEAU GISCOURS, AND CAIAROSSA

At our two châteaux we are the opposite of industrial producers whose winemakers never step into the vineyard to check the fruit. Winemakers who deal in quantity produce bland wines without specific terroir character. Think of Kraft cheese-makers who never see a cow. Good wine requires contact with the vintner. A winemaker is the concertmaster who pulls together disparate elements from difficult conditions of poor soil and seasonal vagaries of weather to make a symphony in the bottle. It is why wines vary from one winemaker to another and from season to season. Each vintage requires different choices and decisions. Every year we ask questions about how to do it better without extreme change. It's better to work with knowledge about the personality of every parcel of our vineyards. Time helps us know the potential of individual plots and how richness in wine develops slowly. Clarity of purpose adds to our vision. We are meticulous, always looking for finesse, elegance, richness, and femininity in our wines. ∎

RAIMOND PRÜM is one the world's great riesling practitioners. Here he outlines some of the interesting added parameters to *terroir* expression in a grape whose wine bottlings are composed of an infinite number of must weights, shades, and nuances, many of these based on picking practices and winemaking decisions, though all of the finished wines express *terroir,* despite not resembling each other in the slightest.

RAIMUND PRÜM, S.A. PRÜM

It's exciting to produce wines. Each season presents a different scenario. We start to prepare each vineyard in the summer, working to educate our grape producers about better growing practices. Grapes for our wines must be handpicked even though machines are now able to harvest on our steep slopes. At harvest, we collect what we judge will make good wine. Each harvest has fantastic potential, but 1980, 1984, and 1987 were disasters because of cool summers, late bud break, little sunshine during the summer, and cold weather at harvest. We recently served the '87. It was still young. The acidity wasn't harmonious, but it was interesting and will change during another thirty years in the bottle. Unlike many other white wines, Rieslings can age from ten to fifty years, developing different characteristics in the bottle that are a joy to discover.

I say Riesling requires a warm foot and a cool head. Making wine depends on whether you listen to your experience to make great wine or follow the demands of consumers and the market. Following the market is tricky because it changes faster than the ripeness of the grapes.

The start of harvest is one of the most important events of each vintage. The quality of grapes is dramatically affected by weather, especially in our climate. We check sugar content, acidity, pH levels, and other factors, but it is difficult to wait for perfect ripeness. It is imperative that our wines maintain their style. I believe ripeness is a more valid way to describe Rieslings. For us, ripeness develops great concentration of sugars, acidity, and balance.

It is the winemaker who is the main influence on which course to take in vinification. Versatile Riesling has the capacity to be vinified in an extraordinary range of styles that have a very wide range of sweetness that defines different styles . . . from bone dry to unctuously sweet and ambrosial.

MICROBIAL *TERROIR*

It stands to reason that if *terroir* exists at all, it must be conveyed by some means to the finished product. For the past two decades, researchers have been locating the potential microbial sources for its particular sensory properties. At least some of these findings suggest that if you define a well-tended vineyard as a biome, the microflora that exists within that biome contains a population particular to that place, in those conditions, and relating to the cultivar that predominates. "Microbial *terroirs*" are measurable and identifiable, and at least in some cases seem to reflect a kind of fingerprint for a given place, what Nicholas Bokulich and his team have defined as a "biogeography."

Furthermore, a number of researchers have been able to isolate a biome within the winery walls, on its surfaces, in its barrels, on its ceilings and floors. What they've learned about this particular tiny population is surprising and even a bit confounding. Clearly, further research is necessary.

FROM "MICROBIAL BIOGEOGRAPHY OF WINE GRAPES AS CONDITIONED BY CULTIVAR, VINTAGE, AND CLIMATE," BY NICHOLAS A. BOKULICH, JOHN H. THORNGATE, PAUL M. RICHARDSON, AND DAVID A. MILLS

ABSTRACT

Wine grapes present a unique biogeography model, wherein microbial biodiversity patterns across viticultural zones not only answer questions of dispersal and community maintenance, they are also an inherent component of the quality, consumer acceptance, and economic appreciation of a culturally important food product. On their journey from the vineyard to the wine bottle, grapes are transformed to wine through microbial activity, with indisputable consequences for wine quality parameters. Wine grapes harbor a wide range of microbes originating from the surrounding environment, many of which are recognized for their role in grapevine health and wine quality. However, determinants of regional wine characteristics have not been identified, but are frequently assumed to stem from viticultural or geological factors alone. This study used a high-throughput, short-amplicon sequencing approach to demonstrate that regional, site-specific, and grape-variety factors shape the fungal and bacterial consortia inhabiting wine-grape surfaces. Furthermore, these microbial assemblages are correlated to specific climatic features, suggesting a link between vineyard environmental conditions and microbial inhabitation patterns. Taken together, these factors shape the unique microbial inputs to regional wine fermentations, posing the existence of nonrandom "microbial *terroir*" as a determining factor in regional variation among wine grapes. ∎

FORTUNATELY FOR us, Nicholas Wade of the *New York Times* wrote an incisive, limpid summary of Bokulich and Mills's paper, which encapsulates their findings in a much less technical format.

FROM NICHOLAS WADE, "MICROBES MAY ADD SPECIAL SOMETHING TO WINES"

Terroir is a concept at the heart of French winemaking, but one so mysterious that the word has no English counterpart. It denotes the holistic combination of soil, geology, climate, and local grape-growing practices that make each region's wine unique.

There must be something to *terroir,* given that expert wine tasters can often identify the region from which a wine comes. But American wine growers have long expressed varying degrees of skepticism about this ineffable concept, some dismissing it as unfathomable mysticism and others regarding it as a shrewd marketing ploy to protect the cachet of French wines.

Now American researchers may have penetrated the veil that hides the landscape of terroir from clear view, at least in part. They have seized on a plausible aspect of terroir that can be scientifically measured—the fungi and bacteria that grow on the surface of the wine grape.

These microbes certainly affect the health of grapes as they grow—several of them adversely—and they are also incorporated into the must, the mashed grapes that are the starting material of winemaking. Several of the natural fungi that live on grapes have yeastlike properties, and they and other microbes could affect the metabolism of the ensuing fermentation. (Several species of microbes are available commercially for inoculation along with yeast into wine fermentations.)

But are the microbial communities that grow on the grapes of a given region stable enough to contribute consistently to wine quality, and hence able to explain or contribute to its terroir?

Such a question would have been hard or impossible to address until the development of two techniques that allow the mass identification of species. One is DNA bar coding, based on the finding that most species can be identified by analyzing a short stretch of their genome, some 250 DNA units in length. The other is the availability of machines that can analyze prodigious amounts of DNA data at a reasonable cost.

Armed with these new tools for studying microbial ecology, a research team led by David A. Mills and Nicholas A. Bokulich of the University of California, Davis, has sampled grape musts from vineyards across California. Grape varieties from various wine-growing regions carry distinctive patterns of fungi and bacteria, they reported Monday in The Proceedings of the National Academy of Sciences.

They found, for instance, that one set of microbes is associated with chardonnay musts from the Napa Valley, another set with those of a must in Central Valley and a third grouping with musts from Sonoma. They noticed a similarly distinctive pattern of microbes in cabernet sauvignon musts from the north San Joaquin Valley, the Central Coast, Sonoma, and Napa.

The discovery of stable but differing patterns of microbial communities from one region's vineyards to another means that microbes could explain, at least in part, why one region's zinfandel, say, tastes different from another's. The links between microbes and wine-growing regions "provide compelling support for the role of grape-surface microbial communities in regional wine characteristics," the researchers conclude.

"The reason I love this study is that it starts to walk down a path to something we could actually measure," Dr. Mills said. "There are high-end courses on terroir, which I think are bunk. Someone has to prove that something about terroir makes it to the bottle, and no one has done that yet."

Microbes are deposited on the grape surface by wind, insects and people, and may fail or flourish because of specific local conditions such as the way the grape vines are trained. And there may be genetic affinities between particular microbial species and each variety of grape, the researchers say.

Even if Napa's chardonnay grapes, say, carry a distinctive pattern of fungal and bacterial species, the Davis scientists need still to prove that these microbes affect the quality of the wine. Microbes could exert an influence both during the lifetime of the grape and during fermentation, when they may add particular ingredients to the wine. "We will look at how overall microbial communities correlate with quality traits in the wine, and whether you can predict quality from the microbes present," Mr. Bokulich said.

Thomas Henick-Kling, a professor of oenology at Washington State University, said it was plausible that microbes are a component of terroir. "Unripe grapes taste the same the world over," he said. It is known that single strains of yeast can have a strong effect on a varietal's flavor, he continued, "so it's likely that microbes play a larger role than presently known and are probably a part of the regional differences that we recognize."

While Dr. Mills said that "I make fun of terroir all the time," he believes that regional distinctions between vineyards do exist and that microbes have a role in creating them. If the specific links between microbes and the sensory properties of wine can be identified, growers will be able to take a savoir-faire attitude to terroir instead of a je ne sais quoi shrug.

On the other hand, he added, pinning the qualities of wine on bacteria and fungi may spoil that frisson of enchantment for some connoisseurs. "Many people don't want this figured out," he said, "because it demystifies the wonderful mystery of wine." ∎

MUCH OF this research has quite rightly focused on the populations that bring grape juice to its miraculous end, without succumbing to alcohol poisoning, without petering out as the food source diminishes to dryness. We're talking about yeast, more than one, but in particular the remarkable yeast *Saccharomyces cerevisiae,* the one workhorse microbe that transforms food to this grand elixir. A number of papers in recent years have explored its origins and its function, even its location, on the fruit and in the winery, parts of which are reproduced here.

FROM ALESSANDRO MARTINI, "ORIGIN AND DOMESTICATION OF THE WINE YEAST *SACCHAROMYCES CEREVISIAE*"

The term "natural" is conventionally used in wine microbiology to designate a grape-must fermentation initiated by the yeast flora "naturally" present in the environment, while a "guided" process requires the use of a selected yeast fermentation starter. In freshly pressed grapes, several microbial groups can be found, in addition to the predominant yeast flora. Bacteria, filamentous fungi and (sporadically) protozoa are normally present, and may take over in musts obtained by hail-damaged grapes. Grape must, however, is in itself a differential growth medium on account of its low pH value (3.5)—which prevents the growth of most bacteria—and because its high sugar content (often over 25%, w/v) only permits the growth of yeasts; and, among them, only of those able to ferment the substrate; and among the fermenting ones, only of those capable of producing and tolerating high concentrations of ethanol. As a result, spontaneous fermentation of grape-juices definitely should be regarded as an enrichment culture.

The first ecological survey on naturally fermenting musts revealed the presence of a definite pattern: the initial phase of the fermentation process is consistently characterized by the presence of a single, definite yeast species possessing lemon-shaped (apiculate) cells (Kl. apiculata). After 3–4 days, the "apiculate" cells are rapidly replaced by oval-shaped, slightly elongated cells (Sacch. cerevisiae). Only a few hours after their appearance, exclusively "elliptical" cells are isolated from fermenting musts.

Muller-Thurgau (1896) performed the classical experiment involving the formation of sterile grape-must by Kl. apiculata and Sacch. ellipsoideus (now Sacch. cerevisiae) separately as well as with a mixed culture of both yeasts. Chemical analysis of the fermented musts showed for the first time that these two species possess greatly different technological properties. The "apiculate" yeast can produce only small amounts of ethanol (up to 4–5%, v/v) and is characterized by the production of considerable quantities of secondary compounds, such as acetic acid. In contrast, the "elliptical" yeast manifests better technological properties, because, besides being able to produce high amounts of ethanol, it produces very low quantities of secondary products during fermentation and is endowed with a much higher growth rate. The above observations were soon transferred into the practice of operating grape-must fermentations with only one yeast species, selected beforehand for its superior technological characteristics.

Later, the ecological surveys of Martinand (1909) on the Mosel area revealed the intervention of yet another species characterized by small, globose, multilaterally budding cells, subsequently classified in the new taxon *Torulaspora rosei* (presently *Torulaspora delbrueckii*) by Guilliermond (1913). Its tolerance to ethanol and growth rate were higher than those of *Kl. apiculata* and lower in relation to *Sacch. ellipsoideus*, but the amount of acetic acid formed during fermentation was almost nil.

These ecological and technological studies gave rise to two separate research trends: (i) The definition of the applied aspects of winemaking, aiming at the optimisation of process parameters in relation to the micro-organisms responsible for the fermentation; (ii) research was essentially centered on the "ecology of wine yeasts," intended as the identification of yeasts living on grapes, in fermenting musts, and in freshly obtained wines from as many different regions, areas, zones, and microclimates as possible. In other words, a monumental isolation exercise was mounted, ambitiously aiming at obtaining evidence on the effect of indigenous yeast strains, such as those isolated from a small hill slope of the Chianti or Bordeaux area, on the quality and specific organoleptic characteristics of the wine obtained. In addition, there were efforts to discover a yeast culture endowed with superior oenological properties to use as a superselected yeast starter in winemaking.

From the redundant and monotonous body of information accumulated over the first part of this century on the yeast flora associated with the fermentation of grape musts in all the nations of the Mediterranean basin or in geographical areas as different as Japan, India, Brazil, and Australia and reviewed by Kunkee and Amerine (1970), the following situation, already evident from the first ecological surveys, was largely confirmed: (i) the species *Kl. apiculata* and *Sacch. cerevisiae* are the main inhabitants of fermenting grape-musts; (ii) the species *Torulaspora delbrueckii* (formerly *Torulaspora rosei*) may, in some cases, compete with *Sacch. cerevisiae* during the second phase of natural fermentation; (iii) *M. pulcherima* is relatively frequent in Italian and Spanish wine areas, while *Candida stellate* (ex *Torulopsis bacillaris*) is widespread in the Bordeaux region of France; and (iv) all the remaining species (about 150) isolated from fermenting musts are occasionally present. ■

THE AUTHORS go on to evaluate yeasts found in the winery, on winery surfaces and storage vessels. To their surprise, they can locate just a single strain: *Sacch. cerevisiae*, dominating the environment. To their conclusion:

As a consequence of the above evidences, it must be concluded that the wine yeast *"par excellence" Sacch. cerevisiae* does not live in nature at all, but can only be found in the winery environment.

It is commonly believed that the wine yeasts that colonize a microzone, such as a vineyard or hill slope, are specific to the area; fully adapted to the pedoclimatic environment; fully adapted to the must to be fermented; and responsible, at least partially, for the unique characters of the wine obtained. Also commonly accepted is the idea that the making of great wines does not benefit from the use of selected starters. Some enologists admit, however, that good results also can be obtained with selected yeast starters originating from the micro-area where these great wines are produced, though the majority believes that starters are only good for producing ordinary wines.

From the information presented in this review on the ecology of wine yeasts in nature, any beneficial role in winemaking of the exceedingly few cells of *Sacch. cerevisiae* dwelling in vineyard soil or on grape surfaces must be excluded. However, the idea cannot be excluded that a local, individual, specific fermenting yeast flora, selected through the years in each microclimatic area, may be targeted to the winery environment, where we have seen before that the variability necessary for the selection of a winery-specific strain is always guaranteed by the presence of exceedingly many cells and generations of *Sacch. cerevisiae.* A selective pressure on the population of *Sacch. cerevisiae* in the winery may be operating through limiting environmental factors, such as ethanol concentration, in favor of high fermenting power strains; sugar concentration, in favor of strains capable of fermenting in adverse osmotic conditions; or high concentrations of sulfur dioxide in favor of strains resistant to its action. It is conceptually impossible, however, to justify the presence of a selective pressure causing the enrichment of strains capable of producing volatile compounds that characterize wine organoleptically, because such a property cannot possibly endow the strain possessing it with a clear advantage over the others.

In any case, there's already some preliminary evidence to show that the strains of *Sacch. cerevisiae* isolated from the winery environment possess technological properties on the average comparable with those of commercial selected starters. ■

LINDA BISSON, in the *American Journal of Enology and Viticulture,* has further studied how strains of *Saccharomyces* have morphed and evolved into site-specific, phenotypically distinct strains depending on where they're grown and presumably influenced by the environment they find themselves in. The abstract for her long paper is here:

FROM "GEOGRAPHIC ORIGIN AND DIVERSITY OF WINE STRAINS OF *SACCHAROMYCES*," BY LINDA BISSON

The availability of genome sequence information from a large collection of strains of *Saccharomyces* isolated from a variety of geographic regions and ecological niches has enabled a detailed analysis of genome composition and phenotype evolution, the two components of strain diversity. These analyses have also provided a relatively complete depiction of the origins of wine strains. In population genomic analysis, wine strains of *S. cerevisiae* cluster as a highly related group, but one that shows a greater level of phenotypic differentiation than would be predicted based on the level of genomic similarity. Natural and human selection and genetic drift have played roles in the evolution of wine strain diversity. Phenotypic diversity is so extensive that no one strain accurately represents all wine strains with respect to biological properties and fermentation performance. In addition, both commercial and native isolates have been found to carry introgressions, regions of DNA derived from nonhomologous organisms, suggestive of cell fusion events with yeast of different genera and species. Comparative sequence analysis has thus refined our knowledge of yeast lineages and offers an explanation for the evolution of phenotypic diversity observed in winery and vineyard populations. ∎

ALL OF these populations to one degree or another are managed, regulated, and contained by the judicious use of sulphur dioxide in the winemaking environment. Jamie Goode, in his book *The Science of Wine*, does a particularly good job of summarizing its use in the winery environment, and stresses its importance, even as he explores the consequences of not using it.

FROM JAMIE GOODE, *THE SCIENCE OF WINE: FROM VINE TO GLASS*, CH. 15: "SULPHUR DIOXIDE"

Sulphur dioxide: it's one of the most frequently discussed, yet simultaneously one of the most frequently misunderstood issues in winemaking. Winemakers, merchants, writers, and even consumers talk about it constantly, but with the exception of the first group mentioned above, I suspect that most don't have a clear understanding of the issues involved. It's undoubtedly a technical sort of subject that fits firmly into the category of the chemistry of winemaking, but I'm going to try and keep this chapter readable and interesting without sacrificing depth of content. Sulphur dioxide is an important subject, so it's a good idea I have a decent grasp of the issues relating to its use.

■   ■   ■

The key to understanding the effects of SO_2 is the ratio between its free and bound forms. When SO_2 is added to a wine, it dissolves, and some of it reacts with other chemical components in the wine to become "bound." This bound fraction is effectively lost to the winemaker (at least temporarily) because it has insignificant antioxidant and antimicrobial properties. Various compounds present in wine, such as ethanal (acetaldehyde), ketonic acids, sugars, and dicarbonyl group molecules, are responsible for this reaction.

Winemakers routinely measure total SO_2 and free SO_2, with the difference between the two being the amount existing as the bound form. Importantly, an equilibrium occurs between the free and bound forms, so that as free SO_2 is used up, some more may be released from the bound fraction. It's slightly more complicated than this, however. Some of the bound SO_2 is locked in irreversibly; the remainder is releasable. And of the free portion, most of it exists as the relatively inactive bisulfite anion (HSO_3-) with just a small amount left as active molecular SO_2 . . .

THE IMPORTANCE OF pH

One of the key factors affecting the function of SO_2 is pH. For the benefit of those who have long forgotten their school chemistry lessons, pH is a measure of how acidic or alkaline a solution is (technically it relates to the concentration of hydrogen ions in solution). A pH of seven is neutral, and below and above this figure, the solution is progressively more acidic or alkaline, respectively. Thus a wine with a lower pH is more acidic. All wines are acidic (with a pH of less than seven), but some are more acidic than others. The pH is important here in two respects. Firstly, at higher pH levels, more total SO_2 is needed to get the same level of free SO_2. Secondly SO_2 is more effective—that is, it actually works better—at a lower pH, so as well as having more of a useful free form for the same addition, what you have works better as well. It's a double benefit.

HOW SULPHUR DIOXIDE WORKS IN WINE

The most useful attribute of the wonder molecule is that it protects wine against oxidation. Professor Roger Boulton of UC Davis explains, "as a wine is exposed to oxygen, the key initial reaction is the oxidation of monomeric phenols with a special reactive group to form hydrogen peroxide. The peroxide can be consumed by a number of other reactions, either being quenched by tannins and other phenols or the formation of acetyldehyde by reaction with ethanol." Boulton points out that the assertion that SO_2 is "protecting" wine from oxidation is technically incorrect. "There is a general misconception that SO_2 will protect against oxidation," he explains. "Its rate of reaction with oxygen is so slow that it cannot compete for the oxygen and stop the phenol oxidation. While it does compete for the peroxide formed, its main role is binding up the aldehyde formed, so that we do not smell the oxidation product."

The second type of oxidation that occurs in wine is caused by enzymes known as oxidases, which speed up oxidation reactions drastically. These enzymes are present in damaged or rotten grapes, so where these are likely to be present it is especially important to use sufficient SO_2. Winemakers should therefore use grapes that are as clean as possible, with the absolute minimum of fungal damage. It follows that sweet wines made from botrytized grapes need substantially higher levels of SO_2 to protect them against oxidation. Significantly, botrytized wines are also very high in compounds which bind free SO_2, with the result that winemakers can end up adding enormous levels and still not have significant free SO_2.

White wines generally need higher levels of SO_2 than reds to protect them from oxidation. This is because red wines are richer in polyphenolic compounds, which give the wine a natural level of defense against oxidation. White wines that have been handled reductively (that is, protected against oxygen exposure through the use of stainless steel and inert gases in the winemaking process) are especially vulnerable to oxidation, and need careful protecting.

The aging of wine is what is known as a "reductive" process. It works properly in the absence of oxygen, which is why a good tight seal by the closure, whether a cork or a screwcap, is important. While there's some debate about whether tiny traces of oxygen might be needed to ensure optimum evolution of wine in the bottle, it is universally recognized that any significant influx of oxygen will rapidly oxidize the wine—that is, oxygen will combine chemically with compounds present, negatively affecting the flavour . . .

But SO_2 is also microbicidal. It prevents the growth of—and, at high enough concentrations, kills—fungi (yeasts) and bacteria. Usefully, SO_2 is more active against bacteria than yeasts, so by getting the concentration right, winemakers can inhibit growth of bad bugs while allowing good yeasts to do their work. SO_2 is usually still added to the crushed grapes in wild yeast fermentations;

while it kills some of the natural yeasts present on grape skins, the stronger strains survive, and thus are selected preferentially. Sweet wines and unfiltered red wines are at higher risk of rogue microbial growth, so with these it is especially important that correct SO_2 addition is practiced.

It follows from all this that if you don't use enough SO_2 in your winemaking, you run the dual risks of oxidized wine and off-flavours and aromas from unwanted microbial growth, together with potentially considerable bottle variation.

CONCLUDING THOUGHTS

Winemakers in the modern world, old or new, must find their own way through intelligent trial and error to achieve their conceptual flavor goals. Based on multiple vintages with a particular site, they can learn the techniques the site teaches them about its own expression, learning eventually to steer towards a consistent, desirable flavor endpoint that reflects the *terroir*. Depending on what nature throws at them, they can either stand back and do nothing, or react, dodge, facilitate, ameliorate, and do what it takes to coax their grape must towards positive *terroir* flavor output.

Modern curricula in enology and viticulture provide students with a scientific background in the physical and biological sciences, including food science and microbiology, as well as process engineering. The emphasis on parameters and mechanics and on how to do what you do (or fail to do) ought to give students the tools to recognize and isolate *terroir* expression. That ability will only grow the longer they're in the business or interacting with a single expressive site, resulting in a unique skill set, a toolbox of viticultural practices—of the sort that John Williams and Phillippe Delfaut suggested—which maximize a site's expressivity, in the near and far term. Students may learn some of the ways to subtly persuade wine toward desired quality endpoints but ultimately, in the real world, this is an interpretation of the existing wine culture, and the understanding required to steer your business practices toward this enlightened and authentic outcome as the ultimate goal.

And that, in the end, is a measure of authenticity, to which notions of *terroir* inevitably contribute. So we conclude with two mildly conflicting views of the same coin, one by Clark Smith, whose pot-stirring 2014 book *Postmodern Winemaking* brought up several compelling issues with respect to the role of winemaker in an age where wine musts and the very notion of *terroir* can be successfully manipulated, for good or ill, the process of which he called *spoofulation*. And finally, a short editorial opinion from Jamie Goode and Sam Harrop MW, taken from the introduction to their book *Authentic Wine*.

FROM CLARK SMITH, *POSTMODERN WINEMAKING*, CHAPTER 20: "ARTISANAL OR SPOOFULATED?"

A common claim among today's scribes is that technology makes wines taste the same. Wine technology fall guy that I am, I can scarcely have a conversation with anybody in the trade without the issue emerging. I think my own wines prove otherwise. But when I'm open about the new tools I've employed, concerns immediately surface about eradicating distinctive terroir expression.

In a way, they are right. Technology has certainly robbed us of the spoiled wines we regularly encountered in the 70s. These were certainly (gag) more distinctive and varied: high VA, stuck fermentations, malolactic in the bottle, geranium tone, and aldehydes were well-known benchmarks we almost never see today. Darn.

Believe it or not, there are some very vocal proponents of imbalance in the name of naturalness. Alice Feiring comments in a recent *New York Times* post, "Call me a silly girl, but if it was a hot year I want to taste the heat. A wet one? I want to taste it. High acidity? Low acidity? Give me the best a winemaker can do. A fine winemaker can always make something fascinating. Vintage subtleties are part of the wine passion. I do not want 'corrected' wine."

I ran into Alice recently at the First Annual Qvevri Symposium in the Khakheti region of Georgia and I asked what she thought of a particular Chinuri done in qvevri. She said she found it "a little hot." She was right, but the observation surprised me. If you advocate for, say, no pesticides, you need to go beyond ignoring the apples with spots; you need to actively look for them at the store, pick them out, buy and eat them, leaving more perfect ones for others less enlightened. If you advocate against smoothing vintage to vintage and claim to want to taste what nature provides, then you should overlook, even treasure, minor imperfections of balance.

In the real world, of course, this doesn't happen. Quality in the bottle comes first, Natural comes second. Any winemaker needing to feed her family had better do both. In practice, this means that uncompromising Natural Wine is made mostly by trust-funders, and in tiny quantities. The way to support wine diversity is to get in your Prius and go find your local vintner, whether you're in Sandusky, Ithaca, Charlottesville, or Des Moines. I'm afraid that more the norm is to hit the organic section at Dean and Deluca, then go home and grouse about sameness on the Internet.

Eric Asimov also takes it for granted that less is more when it comes to artisanal wines. As he puts it, "The conflict . . . comes with winemakers who claim to believe that their wares are art, who say they believe in terroir and all the associations that go along with wines that convey a sense of place. These wines ought to be made naturally, without major technological reshaping. They are not intended to appeal to the broad populace, but to be distinctive."

My favorite corollary of Murphy's Law states that "nothing is impossible for the man who doesn't have to do it himself." The real truth is that wines, and I mean all wines, become distinctive through artifice. That's what winemakers do, don't you know. Asimov's words simply have no meaning for a winemaker. How are we to know what constitutes Natural Wine-making and what qualifies as "major technological reshaping?"

Winemakers must not abandon the moral high ground to nonpractitioners. Rules from the uninitiated are more cynical than reverent. These men and women who actually give their lives to making wine are my colleagues, my clients, and my friends, and they deserve more respect from their supposed fans. They generally agonize, quite privately, a lot more about process than armchair critics are in any position to appreciate.

Like a pop singer, you gotta balance self-expression against what sells. Screwballs like me buck the system, trying to make a go of it with fringe experiments like Faux Chablis and sulfite-free Roman

Syrah—fortunately not my day job. Winemakers are not acquisitive souls, but they do have bills to pay, and generally have dug themselves sizable financial holes. How to make a small fortune in the wine business? Start with a large one.

To some extent, we all end up chickening out. Just to pay the bills, we are sometimes reduced to kicking out on the side of a White Zinfandel or the basic Napa Cabernet that tastes just like the rest. The reality is that winemaking choices are ruled by a democracy in which people vote with dollars.

Heads up, ye geeks: hard-core wine buffs are disadvantaged in this context because they have no product loyalty. They like to sample a different wine every day, so they don't support brands very well even when they love them. I am one of those people, so I speak from experience. Sip, smile, and onto the next thing. If you want your voice to count, you must enroll your less adventuresome friends who just want to buy good wine by the case.

Even at its leanest, the transformation from grapes to grand vin is hands down the most manipulative of cooking processes. Like any chef, the artisanal winemaker is charged with transforming raw groceries into an offering that displays their special attributes to best advantage. Distinctive flavor expression comes first, but the outcome, not the tools, are the issue, provided the winemaker knows what he's doing.

In my view, until we focus on ends instead of means, we won't make much progress toward naturalness. Banning the steady-cam won't make film more artistic. Banning black mascara from the shelves at Walgreen's wouldn't convince women to present their natural, unpainted selves in public. The real enemy is artlessness, not evil machines and methods. Lackluster wine is just bad cooking.

When Alan Goldfarb came to interview me, he articulated an interesting distinction. I showed him my Faux Chablis, a French-style Chardonnay bottled at 12.9% alcohol with good minerality and distinctive aromatic expression. I also showed him the wine prior to alcohol adjustment at 14.8% alcohol–bitter, hot, and aromatically null. He remarked quite rightly that the alcohol-reduced wine had truer terroir expression but that the unadjusted wine, which he did not prefer (few do), was more authentic. Bingo.

This is the heart of the Natural Wine quandary. What matters most, terroir expression or authenticity? If authentic wines were actually jumping off the shelves, winemakers would be more inclined to sleep in and let the wine make itself. But artlessness is not, by itself, a turn-on. I find for the most part that consumers are more concerned with what's in the bottle than with the winemaker's methods. But okay, there is an element that does care about process, and these folks deserve the facts.

FROM JAMIE GOODE AND SAM HARROP MW, *AUTHENTIC WINE: TOWARD NATURAL AND SUSTAINABLE WINEMAKING*

The issue of naturalness and authenticity is one of the key current debates in the world of wine, and it is likely to become more heated over the next few years. Why? Because wine is now at a metaphorical fork in the road, and from here it can go one of two ways. The first is to continue down the road taken by New World branded wines: huge volumes, a reliance on technology and marketing, reliability at the cost of individuality, an emphasis on sweet fruit flavours, and a loss of terroir (the possession by wines of a sense of place). The destination? Wine would gradually become indistinguishable from other drinks, and grapes would be seen simply as the raw ingredient in a manufacturing process. It's easy to see how wine is being pushed down this road by changes in retailing practices and demand for branded, homogenous wine. Marketplace-driven consolidation has hit the wine industry. Players who can't manage large volumes with low margins are in danger of being forced to retreat to the heavily saturated and competitive fine wine niche or to bow out completely. The middle ground, once flush with diversity, has rapidly eroded, and those still in the game are seeing their access to market dry up. This is a real concern because many of the most interesting wines have come from this middle ground: midsized producers with perhaps dozens of hectares, rather than hundreds, who make the sorts of wines that we fell in love with and that persuaded us that wine is interesting in its own right. Nowadays, a small group of large drink companies dominates the world wine market. The accountants and managers rule the roost. Their products hit price points, are made in huge volumes, and don't offend anyone, but they do not excite. They are consistent from vintage to vintage, made to reflect the style rather than a sense of place.

For a vision of where the wine industry might currently be heading, it is worth looking at what has happened to the beer industry in recent years. The big companies and suits (business executives) moved in. The marketers realized the product quality wasn't the selling point, and instead, they focused on building brands and selling the concepts underlying the brands to consumers rather than talking about the taste of the beer. The result was product homogenisation. Does the wine industry want to tread the same path? There's a real danger that if wine is treated solely as a manufactured product, blended and tweaked to fit the preferences of specially convened panels of "average" consumers, the wine industry will become moribund as a sector. Diversity based on regional, cultural, and winemaking differences will be lost, and any sense of continuity with the past may vanish forever.

The other road involves a retracing of steps and a celebration of what has made wine different and special: a respect for tradition, a sense of place, and an acknowledgment that diversity is valuable and not just an inconvenience. Wine is embedded in the deeper culture. The destination of this road is the rediscovery of "natural," authentic wine. This is wine with a vital connection to the vineyard it came from, wine that is unique to a particular distinguished site. "I believe in the concept of 'naturalness,' as it is at the core of the concept of terroir," says renowned Australian winemaker and wine scientist Brian Croser. "Terroir is at the core of the fine wine endeavor and ethic, as it defines the quality factor which is enduring and cannot be competed away by technology. I maintain that the finest, best-balanced, and most unique wines will be made naturally from great expressive terroirs. Not only will the absolute quality across many vintages and tasters aggregate to the best (compared to manufactured wine), but the very ethic itself adds a halo that is in accord with the human spirit trying to reconnect to nature in a largely disconnected life. The spiritual and intellectual needs are in accord with the satisfaction derived from the personality and quality of fine wine." ∎

TO THE extent that the world's greatest winemakers exist to safeguard this legacy, there is reason enough to respect their relation to *terroir* expression.

7

SENSORY
Validating Terroir

If you believed that certain places yielded wines with distinctive character, wouldn't you run right out and organize some tastings to prove your point?

The ultimate validation of a *terroir's* existence has to come through tasting, through the use of human sensory receptors. After all the geological prospecting, soil surveying, topographical mapping, meteorological forecasting and clonal inventorying, the fundamental question remains: What are the flavors in the wines from Place X that make them taste like wines from no other place else on earth? Because ultimately we're more interested in the taste of *terroir* than in the *terroir* itself.

Historically, all but a thimbleful of *terroir* commentary is anecdotal. Such validation is hardly without value; it has everything to do with the joys of individual wine consumption, and the fact that there's a mountain of it definitely leads to the hunch that there's something bona fide going on. But anecdotes by themselves don't prove anything, any more than alleged Bigfoot sightings.

If *terroir* is such a fundamental element of great wine, surely there must be volume upon volume of reports from blind tastings confirming its reality, its indisputable existence? In fact, hardly any exist. The identification of *terroir* influences has only recently been subjected to rigorous sensory methodology. Demonstrable, reproducible evidence is long overdue. We're neither satisfied with nor convinced by "I smell, I taste, I think, therefore it is."

Without reliable results from organized tastings, claims of *terroir* expression start to sound hollow. As *Wine & Spirits* editor Joshua Greene put it in his introduction to a special issue on *terroir* called "Hi-Fidelity Terroir" in 2012, "Scientists can trace the DNA of a vine, giving a factual basis to buying and selling wine by variety. But there are no widely accepted scientific tools to measure terroir expression in wine, so many consider the discussion of such matters bogus. And to the degree that much of the discussion of terroir is based on supposition rather than fact, they may well be correct."

Fortunately, some of the missing methodological tools are at hand, or at least under development, and researchers are also getting clearer about the very elements of a given *terroir* that can be tasted and tested. It leads us to believe that this particular chapter of the *terroir* story may ultimately have a happy ending.

We start by noting the absence of tips on tasting for *terroir* in standard wine reference works, and then present what was probably the first attempt to deploy rigorous sensory methodology to wine, a 1984 piece by Ann Noble—in which the experts failed miserably.

We then take up the dubious concept of minerality, the notion that certain wines convey directly to the glass the character of vineyard soils. A long excerpt from an article by geologist Alex Maltman challenges the widespread assertion that minerality, in all its forms, is detectable in wine at all, no matter what critics and wine writers assert—he argues that a literal "taste of the vineyard" is impossible.

The next section examines three emerging methodologies which might be used to quantify, characterize, or otherwise "map" *terroir* character: Descriptive Analysis, Natural Terroir Units, and a less accurate but visually compelling practice called Napping.

The chapter closes with an evocation of the frustration many people feel when they can't quite find the minerality that's supposed to be in their glasses, by one of this volume's co-authors.

HOW DO YOU TASTE FOR *TERROIR*?

In the modern era *terroir* has enjoyed a largely positive reputation as a concept, with connotations that suggest it is one of the most valued aspects of wine. Yet, hardly anything has been written about exactly how to define it, or how a wine drinker could know which sensations produced in the nose and mouth actually constituted "*terroir*" elements. While claims were made about certain "*terroirs*" or "*climats*," or "*lieu dits*" and their sensory signatures, these were more or less the stuff of ongoing legend. The identifiers of uniqueness were often idiosyncrasies, expressed as some flavor derived from rocks or soil. It was tacitly assumed that *terroir* flavor profiles existed and that with experience one could learn to recognize the *originalité* or "area character" in a wine of *terroir*—whatever it was.

Very little early sensory data, not even the vaguest impressions, seem to exist. Through the twentieth century, studies reported the effect of a given viticultural practice or winemaking technique on wine flavor, or they identified the volatile chemical compounds responsible for this or that aroma, but until very recently attempts to move past anecdotes to tie specific wine aroma and flavor profiles to particular places were rare. The vast majority of the work is yet to be done.

REFERENCE BOOKS WITHOUT REFERENCES

When we look for help with *terroir* from the teachers of tasting, very often the instruction comes up short. We get introduced to the mechanics of tasting; sometimes the sources of flavors and aromas are explored in considerable depth, but when it comes to *terroir*, it rarely gets more than a passing nod. Here, summarized, are the approaches of six standard texts on the question of *terroir*.

Wines: Their Sensory Evaluation, by Maynard Amerine and Edward Roessler, 1976. A classic in its time with meticulous instruction on composition of wines. But the book does little to define flavor profiles, and the closest it comes to exploring *terroir* is this parenthetical, passing reference: "Wines differ from each other because of the composition of the grapes used (color, maturity, region where grown, climatic factors, diseases, etc.), fermentation processes used." Thereafter the matter is dropped.

The Taste of Wine, Emile Peynaud, 1976. The Bible of its time for wine aficionados, from one of France's most esteemed enologists. Here one finds *terroir* mentioned in the index with a single page reference, but the reference is no more than a typical definition of orthodox usage, establishing its place in the hierarchy of wine pedigree, alongside such terms as *"cru"* and *"climat."*

How To Taste, Jancis Robinson, 2000. Three references to *terroir* are made only in passing, for example, "the local physical environment that the French call *terroir*. . . ." No specific flavors are associated with any particular environment.

How to Taste Wine, Pierre Casamayor, 2002. There are only two short mentions in this instructional volume; one of these is a little more specific and satisfying than what we've seen to date: "Riesling grapes represent a striking example of the exchange between grape variety and *terroir*. Depending on the soil, they may express delicate floral aromas, or heavier scents of naphtha or truffles."

Tasting & Grading Wine, Clive Michelsen, 2005. One finds only passing references to Old vs. New World flavors, mineral character, and some of the specific positive and negative odors common in the world's wines, but no mention of *terroir*, or how components might come together consistently to define a flavor or aroma relating to site.

Wine Tasting: A Professional Handbook, Ronald Jackson, 2009. A top-notch tasting textbook with plenty of discussion of the chemical nature of flavors and odors. Here the focus is on grape varieties rather than *terroir*. The word itself is only mentioned in a couple of places and even then, the author is dismissive: ". . . often misused in an attempt to justify the supposedly unique quality of wines from certain vineyard sites."

If the taste of *terroir* is so obvious, why haven't the experts picked up on it?

THE OTHER FAMOUS BORDEAUX TASTING

Most serious wine lovers sooner or later learn about the famous 1976 "Judgment of Paris" tasting, in which wine merchant Stephen Spurrier put the best of Bordeaux and Burgundy up against wines of the Napa Valley, and the panel of French judges rated the California interlopers higher than the wines of their own beloved country. A few years later, a much more sophisticated comparative tasting was conducted on the wines of Bordeaux, one which revolutionized the application of sensory analysis methods to wine.

During a research stint in England in 1982, University of California at Davis sensory scientist Ann Noble—who would go on to create the famous Wine Aroma Wheel—worked with two co-authors to organize a highly structured tasting of vintage 1976 wines from four Bordeaux communes by two different panels, and then parsed the results with sophisticated statistical methods. They called that method Descriptive Analysis, and it represents the first major application of modern sensory analysis to wine; this remains one of the most frequently cited papers in the field.

The first important pillar of this approach is its use of objective descriptors, terms that can be learned and used consistently—"strawberry," "rotten egg," "highly acidic"—rather than comparative, evaluative, and metaphorical terms—"this one is better," or "this wine is more elegant." This move from the subjective to the objective represents a huge step forward in sensory studies.

Second, this experiment did not require its tasters to be experts; its methods were carried out by non-professional, carefully-trained lay people. The training is explained in the selection below.

As for the results, major differences between wines did not appear between communes, and the major attributes of difference—whether green bean/green olive character or elements of fruit—appeared not to be driven so much by place as by other factors in winemaking that may influence bitterness and astringency, such as extraction, pressing off, and oak treatment. Perhaps more striking, neither the Masters of Wine nor the trained tasters could tell the difference between communes in any statistically significant sense.

The excerpt below does not include the extensive discussion of statistical methods; for that, consult the original article.

FROM ANN NOBLE, ANTHONY WILLIAMS, AND STEPHEN LANGRON, "DESCRIPTIVE ANALYSIS AND QUALITY RATINGS OF 1976 WINES FROM FOUR BORDEAUX COMMUNES"

ABSTRACT

Twenty-four wines from four communes in Bordeaux were evaluated by descriptive analysis by trained assessors. The same 1976 wines were assigned quality ratings by Masters of Wine (MW). The major aroma difference between the wines was attributed to variation in the intensity of the "green bean/green olive" character by canonical variates analysis (CVA) of the aroma descriptor ratings across wines. The CVA of the flavour by mouth ratings showed the wines to be discriminated primarily on the basis of astringency and bitterness. By multivariate analysis of variance across regions, and by examination of the configurations derived from the CVA across wines, it was shown that the wines did not vary significantly between communes. No significant difference between the wines in quality ratings of the MWs was found.

1. Introduction

Many far reaching claims about the uniqueness of wines from different Bordeaux communes have been made by wine writers. However, despite the extensive literature on the chemistry of Bordeaux wines, no sensory analytical evaluations have been reported.

To profile flavour analytically, the technique of descriptive analysis has been applied to a variety of beverages, including wines, cider, beer, and whisky. Success in the use of this technique depends on the selection of appropriate, well defined and consistently used flavour attributes. To define the terms rated in beer, Mecredy *et al.* prepared standards in several different brands of beer, so that specific flavour notes were recognisable against different backgrounds. Similarly, panelists were trained in the consistent use of descriptors in wine descriptive analysis by Schmidt using reference standards prepared by adulteration of a base wine. To train panelists for evaluating flavours of cider and perry [pear cider], Williams defined flavour terms by addition of specific compounds or essences to paraffin wax.

In this paper the descriptive analysis of wines from four Bordeaux communes by trained assessors is presented and compared with their quality ratings by experts.

2. Experimental

2.1. Wines   Five wines, selected from each of four Bordeaux communes, varied in quality designation from unclassed to second growths. Four additional regional Bordeaux wines were included for comparison. Details of the 24 1976 wines are provided in Table 1. The wine used for preparation of all reference standards was a Carignane made from French grapes at Long Ashton Research Station.

2.2. Descriptive analysis panel training   A 17-member panel was selected from available personnel at Long Ashton Research Station (nine men and eight women, aged between 22 and 50 years). The majority of the judges had participated in previous descriptive sensory tests, but only a few had extensive wine tasting experience.

Following two orientation sessions in which terms were generated by the assessors from individual assessment of two wines, six sessions were held, in each of which a different set of three of the 1976 Bordeaux wines were presented with 15 to 20 reference standards. The aroma reference standards were prepared by adding to the base red wine food products, flavours, or chemicals to define terms resulting from discussions in the session. No standards were provided for the flavour by mouth terms. At each session assessors smelt the reference wines, and then rated the intensities of the terms in each wine. The appropriateness of each term for the wines was then discussed, both to select important descriptors and achieve a consensus as to the meaning of each term. From the discussions in the training sessions, a final set of ten aroma and five flavour by mouth terms was selected for use in the formal descriptive analysis. The compositions of these 10 aroma standards are listed in Table 2.

TABLE 7.1 Commune or district of origin, Château and growth designation of 1976 Bordeaux wines[a]

Code	Commune or district of origin	Château	Growth
1	St Estèphe	Ch. Houissant	
2		Ch. Montrose	2nd
3		Ch. Calon Segur	3rd
4		Ch. de Pez	
5		Ch. Haut Marbuzet	
6	St Julien	Ch. Lagrange	3rd
7		Ch. Ducru Beaucaillou	2nd
8		Ch. Gloria	
9		Ch. Talbot	4th
10		Ch. Léoville Lascases	2nd
11	Margaux	Ch. du Tertre	5th
12		Ch. Brane Cantenac	2nd
13		Ch. Malescot St Exupéry	3rd
14		Ch. Giscours	3rd
15		Ch. Rauzan Gassies	2nd
16	St Émilion	Ch. Roudier	
17		Ch. Grand-Corbin-Despagne	Grand Cru Classé
18		Ch. Fombrauge	
19		Ch. L'Angelus	Grand Cru Classé
20		Ch. Canon la Gaffelière	Grand Cru Classé
21	Haut Médoc	Ch. Cissac	
22	Haut Médoc	Ch. La Tour St Joseph	
23	Médoc	Ch. la Cardonne	
24	Bordeaux	Ch. du Pradeau	

[a] According to 1855 Medoc and 1954 St Émilion classification.

Because of the lack of intensity of most attributes in the wines, seven additional training sessions were held prior to the formal testing, using the same protocol, wines, and conditions as those used in the formal sessions.

2.3. Descriptive analysis protocol and design   Using a partially balanced incomplete block design to permit duplicate evaluation of each of the 24 wines, three wines were presented at each of 16 sessions. No wine was presented in combination with any other more than once. At each session, assessors smelt the 10 aroma reference standards before evaluating the test wines. The intensity of aroma attributes was scored on each of the wines in the randomised order in which they were presented, after which the intensities of the flavour by mouth terms were rated. The intensity of each attribute was scored on a 10 point scale, where 0 = not present, 1 = low, and 9 = high intensity.

TABLE 7.2 Aroma terms selected for descriptive evaluation and composition of the reference standards created to define them

	Term	Composition of reference standard*
1.	Berry (blackberry/raspberry)	10 ml liquor from canned blackberries
		2–3 (thawed) frozen raspberries
		5–6 g strawberry jam
		5–6 g raspberry jam
2.	Black currant (canned/'Ribena')	7–10 ml liquor from canned black currants
		2–4 ml Sainsbury's black currant drink
3.	Synthetic fruit	5 ml cherryade (Corona drinks)
		1 pear drop
4.	Green bean/green olive	4–5 ml liquor from canned green olives
		8–10 ml liquor from canned green beans
5.	Black pepper	4 particles black pepper (fine ground)
6.	Raisin	10 raisins
7.	Soy/"Marmite"	0.5 ml soy sauce
		0.5–1 g "Marmite" yeast extract
8.	Vanilla	0.25 ml vanilla flavouring essence
9.	Phenolic/spicy	5–10 µl 4-ethyl guaiacol
10.	Ethanolic	5 ml (950 ml litre^{-1}) ethanol

*In 30 ml Carignane wine.

All wines were presented in coded, standard, tulip-shaped, clear 215 ml wine glasses and evaluated at 16–22°C in isolated booths illuminated by fluorescent lighting. A 25 ml sample of wine was poured into each glass and covered with a watch glass at least 30 min prior to testing. Distilled water was provided for cleaning the palate between wines, and all samples were spat out.

2.4. Quality rating by wine experts Ten MWs evaluated the overall quality of the wines in one session. In an incomplete block design, each MW rated 16 of the 24 wines, in blocks of four, providing 6–8 replicate assessments of each wine. The MWs rated overall quality of the wines on a nine-point scale, where one was defined as an unacceptable, defective 1976 Bordeaux wine, five was a standard 1976 Bordeaux with no defects and nine was an excellent 1976 Bordeaux. The wines were rated on two criteria: "if immediately consumed" and "when ready to drink." In addition, for each wine the MWs were requested to describe briefly the attributes of the wines which influenced their assignment of these overall quality ratings.

Wines were presented in the same coded, covered glasses as used for descriptive analysis. Judges were seated at individual tables in a room (20°C) illuminated with northwest light.

2.5. Data analysis Individual analyses of variance (ANOVA) were run on each term rated by the trained assessors and on the two overall quality ratings of the experts. The first nine aroma terms

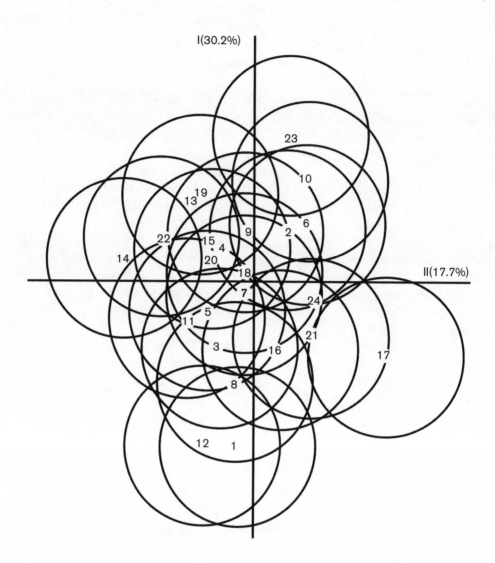

FIGURE 7.1

Canonical variates analysis of nine aroma attributes across 24 Bordeaux wines. Canonical variate scores and their 95% confidence intervals.

■ ■ ■

and the five flavour by mouth attributes, respectively, were then analysed by multivariate analysis of variance (MANOVA) and canonical variate analysis (CVA) using a Genstat program; (the ethanol term was not included in the data analysis since many of the panelists did not rate it).

3. Results

The largest difference among the aromas of the wines is due to variation along the first dimension in the "green bean/green olive" attribute, although differences in the "vanilla" and "soy/Marmite" characters also contribute. The second dimension is primarily a contrast between the "black currant" and "phenolic/spicy" notes. Wines 1 and 12, highest in the "green bean/green olive" aroma and lowest in the "vanilla" character, were from St Estèphe and Margaux, respectively. In contrast, those lowest in "vegetative" character and highest in "vanilla" were from St Julien (10) and the Médoc (23). A St Emilion wine (17) was higher in the "phenolic/spicy" character and lower in "black currant" than all but five wines. Examination of Figure 1 (*Figure 7.1 here—eds.*) further confirms the lack of significance found across regions by MANOVA. No apparent clustering of the wines by commune occurs, nor is there any partitioning based on growth designation.

∎   ∎   ∎

The significant dimension in the flavour by mouth data represents a variation in astringency and bitterness among the wines. The most astringent (and bitter) wines, 2, 14 and 21, were from St Estèphe, Margaux and Haut Médoc, respectively. The least astringent (and bitter) wine (23) was from the Médoc. As shown by non-significance in the MANOVA across regions and by the widely scattered wine means in Figure 2 (*not reproduced here—eds.*), the wine flavours (by mouth) were not significantly different between regions.

3.2. Masters of Wine: Quality Ratings   The overall quality of the wines was rated significantly higher when "ready to drink" than "if consumed immediately" (t = 13.38, d.f. = 23, P 0.001). There was no significant difference among the wines for either quality rating, as shown by the summaries of the ANOVA of the quality ratings in Table 9 (not reproduced here—eds). This is due to differences existing among the assessors assigned quality ratings over and above systematic differences accounted for by ANOVA. The various sensory aspects of the wines were obviously weighted differently by the assessors in arriving at an overall quality rating. As an example of these differences in interpretation of quality among the MWs, comments on one wine ranged from "rich and concentrated fruitiness," "pleasant, but cheap fruit taste," and "not much depth," to "totally defective, nose appalling."

In contrast to their different methods of integrating the wines' attributes and estimating quality, the MWs were often consistent in their specific descriptions of the wines. In descriptive analysis of the wines by the trained assessors, wines 1 and 10 were rated as having very intense "green bean/green olive" and "berry" characters, respectively. The MWs described wine 1 as having the "smell of rotting vegetation" and "pronounced vegetal (sic) aroma, bell pepper." Six of the eight assessors evaluating wine 10 described it as "fruity," with the more specific descriptions, "plummy," "raisins," and "black currant" also being used.

4. Discussion

For this set of 1976 Bordeaux wines, the differences in flavour were primarily due to variations in the intensities of the vegetative (green bean/green olive) and fruit (black currant) aromas and astringency and bitterness. Greater differences were observed among wines than among communes or growth designations. In contrast to the analytical profiles of the wine flavours provided by the descriptive analysis, the overall quality ratings made by the MWs yield little information. Quality is a composite response to the sensory properties of a wine based on the assessor's expectations and

TABLE 7.3   Mean intensity ratings for wines 22 and 23 and corresponding least significant differences for descriptive analyses

| | Mean ratings | | |
	Wine 22	Wine 23	L.s.d. (0.05)
Aroma			
Berry	3.60	3.85	
Black currant	5.15	3.65	1.008
Synthetic fruit	1.50	1.85	
Green bean/green olive	1.80	1.15	0.897
Black pepper	1.15	0.70	
Soy/"Marmite"	0.80	0.50	
Phenolic/spicy	1.45	0.75	
Vanilla	1.05	0.95	
Raisin	1.05	0.95	
Flavour by mouth			
Sour	3.41	3.35	
Bitter	3.44	2.68	0.648
Astringent	3.88	3.06	0.631
Berry by mouth	3.24	3.59	
Black pepper by mouth	2.32	1.77	

NOTE: For aroma terms, n = 20 (10 assessors × 2 reps); for flavour by mouth terms, n = 34 (17 assessors × 2 reps).

hence is an individual response based on preferences and experiences. Accordingly, although the wines were shown to have different flavours, no significant differences in quality were found, because of the differences in preferences of the expert judges. These results are consistent with those of others. In contrast to the reproducible and consistent performance of trained assessors in descriptive analysis of systems varying in colour, viscosity or sweetness, Trant *et al.* found considerable variation in hedonic or preference responses to the system made by the same assessors. Further, many of the individuals changed their preference over time.

Although they were not rated as being significantly different, wines 22 and 23 were rated the highest and lowest respectively for both quality "now" and "when ready to be consumed." The sensory characteristics of these wines can be defined from the mean intensity ratings for the attributes in Table 10. Wine 22 was rated significantly higher in the intensities of black currant aroma, bitterness and astringency. Although not significantly so, wine 23 was lower in "green bean/green olive," "phenolic/spicy," and "black pepper" and higher in synthetic fruit aroma. Examining Figures 1 and 2, it is apparent that neither axis corresponds to a "hedonic" dimension, significant to the MWs in separating the two wines. Wine 23 was one of the lowest in the vegetative character and astringency, but 22 had only mid-range intensity values for these attributes.

Ideally, by descriptive analysis, the sensory characteristics of wines can be defined and used to interpret the patterns in quality ratings by experts or in preference of consumers. Only by combining the two evaluations can the reasons for quality assessments be interpreted in terms of variation in specific sensory attributes.

CAN YOU TASTE THE VINEYARD?

Wines of *terroir* should, by definition, taste different from each other; that's the whole point. And yet some common elements pop up again and again in descriptions, formal and informal, of wines considered to be expressive of *terroir*. Most of them suggest that in some form or other, the wine in the glass conveys a taste of the vineyard from which the grapes hailed. Wines that are "Old World" or "authentic" or that show "area character" more often than not are also alleged to carry traces of rocks, soil, earth—a literal link between place and wine: stony, steely, flinty, chalky, and most of all, mineral-laden, possessing minerality.

In the 2012 Terroir issue of *Wines & Spirits,* a number of well-established wine writers were asked how they identified *terroir* expression in their tastings, and aside from some vague and ethereal ruminations, the body of comments is laden with chalk, minerals, granite, "pencil shavings" (implying both cedar and graphite smells), slate, volcanic, flint, stones, crystals, sandstone, limestone, clay, and primary rock. While each of these descriptors has its own aromatic twist, they can all be seen as variations on the core term, minerality, the Holy Grail of *terroir* lovers and the bane of skeptics.

To state the obvious, invocations of minerality inevitably suggest trace markings of rocks, dust, dirt, gravel or soil to the attentive taster. And yet it has long been known that grapevines are insignificant conduits for minerals, and furthermore, that minerals are neither particularly volatile nor aromatic. Still the word is used to evoke the smell of the vineyard site, whether it be chalky, flinty, claylike, dusty, gravelly, granitic, and so on.

Chances are that the wine experts who go on and on about minerality are picking up something; but what, exactly? Could it be a vineyard's mineral contents? Is that even physically possible?

Professor Alex Maltman of the Institute of Geography and Earth Sciences at Aberystwyth University in Wales offers his explanation in a classic paper from 2013. The heart of his case lies in challenging the notion that vineyard minerals can or do manifest themselves in the glass; this, he argues, is impossible. Later in the essay, Maltman offers possible explanations as to just what tasters are detecting when they report finding minerality.

FROM ALEX MALTMAN, "MINERALITY IN WINE: A GEOLOGICAL PERSPECTIVE"

Tasting "minerality" in wine is suddenly highly fashionable. And unusually for a wine-taste descriptor, the term is very often taken to imply a genesis: the sensation is the taste of minerals in the wine that were transported through the vine from the vineyard rocks and soils. However, there is an array of reasons why this cannot be. The minerals in wine are nutrient *elements*—typically metallic cations—and only distantly related to vineyard geological minerals, which are complex crystalline *compounds.* The mineral nutrients in wine normally have minuscule concentrations and they lack flavour anyway. Although attempts to explain the perception of minerality involve allusions to geological materials, these are irrelevant to its origin. Whatever minerality is, it cannot literally be the taste of minerals in the vineyard rocks and soils.

INTRODUCTION

Minerality is a word currently much used by populist wine writers to describe a perceived taste in wine. The growth in its usage has been phenomenal, from virtual non-existence just a decade or so ago to near ubiquity today. Yet there is much debate about what the term actually means: it lacks any agreed definition. Its scientific basis is at best conjectural; indeed flavour scientists remain skeptical even about its validity. A further remarkable aspect about minerality is that unusually among common tasting words it is often accompanied by at least an implication of its origin. And in one way or another, this is taken to involve the vineyard geology. It is these supposed geological aspects of minerality that are explored in this article.

■ ■ ■

MINERALITY AS THE FLAVOUR OF MINERALS

It seems self-evident that the terms normally used to describe wine-tasting sensations are metaphorical, a way of attempting to put a flavour impression into words. No one thinks that a wine perceived as smelling of, say, tropical fruits or new-mown hay, or tasting of spice or leather has actually involved those materials in its production. But minerality is different. So often, reports of minerality in wine are accompanied by something to the effect that the sensation is the flavour of minerals actually present in the wine. Because people know that wine—like other foodstuffs—contains minerals it does seem a straightforward proposition. And even if it is not put explicitly, descriptions of wines being "mineral-rich," "mineral-loaded," "packed with minerals" and so on, clearly signal that these wines are thought to contain unusually high amounts of tastable minerals.

Moreover, most wine tasters probably know that essential to vine growth are the minerals derived from the rocks and soils of the vineyard. (In fact this dependence is so often emphasized that it almost seems that some still view vines as being largely made from minerals in the ground, a legacy from before the discovery of photosynthesis.) It would seem logical, therefore, to infer that minerality is the taste of the minerals that were originally taken up by the vine roots, transmitted through the vine to the berries and ultimately to the finished wine. Thus, a direct connection between the perceived minerality of the wine and the specific vineyard geology would seem entirely plausible.

The latest trend is to embellish the term with a reference to some specific mineral (as in a quartz, gypsum, or graphite minerality) or rock (as in a chalky, slaty or granite minerality), as though specific geological materials conferred particular kinds of minerality. (Entering into a computer search-engine the words mineral or minerality along with the name of a well-known mineral or rock returns plentiful examples.) The idea is simple, romantic, and manifestly a powerful marketing device in terms of giving a wine a specific provenance. In fact, this way of linking a wine with the much-revered "sense of place" may be part of the reason for the explosive growth in the use of minerality.

The purpose of this article is to marshal the arguments why this idea, although attractive in its simplicity, has to be a misconception. Any connection between a sensation of minerality in a wine and vineyard geology cannot be literal and direct, but has to be complex and circuitous.

CONFUSION BETWEEN NUTRIENT MINERALS AND GEOLOGICAL MINERALS

Directly relating minerals in wine with those in the vineyard soils implies that they are the same things. However, although ultimately linked, they are not the same. When we talk about minerals in foodstuffs such as wine, we usually mean single elements, chiefly metallic elements such as magnesium, zinc, or iron. They are minerals *in the nutrient sense.* If they are in solution, as in vine sap, grape juice, and wine, these nutrient elements exist in ionic form, as cations, e.g. K+, Ca++, and Mg++. But minerals in the vineyard bedrock, stones, and the physical framework of the soils—minerals *in the geological sense*—are almost all compounds, and usually complex and insoluble ones at that. Of course, the nutrient minerals in vines and wine are very largely derived from the geological minerals (unless there is contamination of some kind) but by processes that are complex, protracted, and constantly changing, being subject to a host of evolving variables. In other words, there is a major disconnect between the two kinds of minerals, even within the vineyard itself let alone through to the finished wine.

As an example, take the most common geological mineral in the outer part of the Earth, and presumably the most widespread in the world's vineyards: feldspar. This is actually a family of minerals with various permutations of calcium, potassium, sodium, aluminium, silicon and oxygen ionically and covalently bonded into a crystalline lattice that gives a grain of feldspar strength and rigidity.

Not only that, but normally the feldspar particle will be bonded together with a host of other mineral grains to give the solid, rigid aggregate we call rock (or, if fragmented, a stone). To be accessible to vine roots, therefore, these elements have to somehow become detached and abstracted from the outer boundaries of this aggregate of crystal lattices. This may utilise mycorrhizae and other microbiota, but these function mainly in rich, shallow level soils that are usually not taken to produce wines with minerality. The other chief mechanism involves interchanging cations between either humus, also found mainly at shallow levels in the soil, or the surfaces of certain geological minerals and those in adjacent pore water.

Feldspar, and for that matter the other geological minerals common in vineyards such as calcite and quartz, has virtually no cation exchange capacity (CEC). Some clay minerals, however, with their large surface-volume ratios and electrostatically negative surface charges, are able on their extremities to loosely hold cations that can be interchanged with others in the adjacent pore water. Weathering of feldspar to clays involves the water-driven rearrangement of the constituent elements into forms such sericite or kaolinite, depending on circumstances. Both forms also have a relatively low CEC so an improvement in cation availability requires further reactions to produce higher CEC clays, such as illite or montmorillonite. The particular reaction routes taken by the degrading feldspar and the subsequent clays depend on a host of factors such as chemical environment, time, temperature, moisture content, and pH. Additionally, the extent to which the CEC is actually utilised depends on the vine metabolism, allowing the expulsion of exchangeable protons and establishing appropriate gradients. Ionic transport from the clay surfaces to the vine roots, involving advection or diffusion through those pore-throats in the soil that are interconnected, also depends on variable chemical and hydrostatic gradients, as does the extent to which the transported ions actually pass into the vine roots.

The above is a terse outline of one example, but it hints at the complexities and the variables that are involved in making a constituent element of a geological mineral available as a nutrient mineral to the vine roots, and hence creating a detachment between the two different kinds of minerals.

DIFFERING PROPORTIONS OF IONS IN GEOLOGICAL MINERALS AND WINE

This disassociation in the vineyard between the geological and nutrient minerals increases within the vine. Various transporter proteins, lipid bilayers in membranes, hydrophobic deposits in cell walls, etc., determine how much of the nutrient ions absorbed by the roots are actually loaded into the vine xylem. The vascular system then apportions nutrients differentially around the various components of the vine. Even within the berries themselves, differing ratios of nutrients reach the skins, seeds, and juice. Hence in normal circumstances the inorganic chemical profile of the grape juice bears only a distant and indirect relationship with the vineyard geochemistry.

This disconnect grows yet further during vinification. Fermentation removes from the must certain mineral nutrients, such as zinc, copper and barium, while adding others, such as aluminium, calcium and iron. Fining and/or filtering can remove yet more (and where geological materials such as bentonite are used, cations may be leached from them and actually added to the must). For example, Ruzic and Puskas found that anywhere between 20% and 50% of a wine's copper content was removed by filtration, whereas Tatár, Mihucz, Virág, Rácz, and Záray found that fining with bentonite could increase the rare earth element content by up to 830%. Ageing can increase copper, iron and manganese whereas calcium, aluminium and chromium are removed along with precipitates such as potassium tartrate.

Consequently, the proportions of mineral nutrients in the finished wine bear only a complex, indirect and distant relationship with the geological minerals in the vineyard. This is why, incidentally, it has proved so difficult to find a reliable chemical way of using the inorganic constituents of a wine to detect adulteration or to fingerprint its provenance. Attempts have had to resort to trace elements, isotopes, sophisticated statistics and so forth, and although most conclude with some "potentially promising" correlations, wines subject to counterfeit still rely in practice on diagnostic packaging devices. In summary, the complex relationships summarised above undermine the idea that minerality is simply the taste of vineyard minerals in the wine.

MINUSCULE CONCENTRATIONS OF MINERAL NUTRIENTS IN WINE

The above discussion concerned the relative *proportions* of the mineral elements; the fact is that their *actual concentrations* in the finished wine are typically minuscule. Potassium, being the primary mineral nutrient of vines, is an exception but even this rarely exceeds around 1000 ppm, i.e., roughly 0.1% of the wine, and in the wines analysed by Sauvage, Frank, Stearne, and Millikan it averaged only 577 ppm (0.06%). In fact the *total* inorganic content of wines typically ranges between only 0.15% and 0.4%, according to Coombe and Dry.

The other main mineral nutrients present in wine are calcium and magnesium. Illustrative concentrations for calcium are 30–120 ppm, 50 ppm, and 10–15 ppm. Castiñera Gómez et al. reported magnesium concentrations between 60 and 80 ppm.

Eschnauer and Neeb detected a total of 50 different inorganic components in wine. However, about 25 of these were trace elements with concentrations 1–100 ppb (parts per billion) at most, and about 20 were ultra-trace elements at concentrations measured in parts per trillion. Such concentrations are so low that they barely exceed detection limits, even with modern analytical techniques. Cobalt, cadmium, nickel and selenium contents, for example, were for Cox, Eitenmiller, and Powers at the very levels of detectability (0.3, 0.05, 0.3 and 0.02 ppm, respectively).

These tiny concentrations conflict with the popular assertions that various wines are "mineral crammed," "mineral laden," "brimming with minerals" and the like. However, minute though these numbers are, the real point is that these mineral elements practically have no flavour anyway. There is a whole range of organic metabolites in wine that we can sense in such low concentrations, but not so with inorganic cations. Some cations may be detectable on the palate above a certain threshold, but usually giving an unpleasant sensation. In other words, coupling these minuscule concentrations with the fact that almost all minerals are flavourless seems fatal to the idea of minerality in wine being simply the taste of minerals.

INABILITY TO TASTE MINERALS

With very few exceptions, minerals—in both geological and nutrient senses—lack flavour. Regarding taste, our mouths' gustatory organs can only deal with liquids. However, the geological minerals relevant to vineyards are solid and for all practical purposes are insoluble. Of the minerals common in vineyards, only calcite (the constituent of limestone) has a significant solubility but even here it is no more than about 47 ppm maximum in ordinary water; the values for feldspar and quartz are much less and for the clay minerals virtually zero. In other words, practically all geological minerals are tasteless. The only significant exception is the halide mineral called halite (sodium chloride, salt), which, of course, gives the sensation of saltiness on the tongue. Salinity, however, is to be avoided in vineyard soils, as is saltiness in wines. Licking a mineral or rock surface gives a tactile sensation but this is not a taste. Freshly polished surfaces of rocks (say limestone, granite, or slate) or of minerals (say quartz, calcite, or feldspar) cannot be distinguished by the tongue or by smell.

Aroma (= odour or smell, correctly olfaction), with taste the other component of flavour, is perceived in the olfactory bulb of the nose. In order to reach the organoleptic receptors located there, a substance has to volatilise (become vapour). Rocks and minerals cannot do this. The tendency for a substance to volatilise is indicated by its vapour pressure, and this is considerable for many of the metabolites such as esters, ketones, and higher alcohols found in wine (which is why they are collectively called aromatic molecules). Their vapour pressure is measured in tens of *kilo*pascals and more. In contrast, the vapour pressure of geological minerals and almost all nutrient elements is measured in a few tens of pascals at the most. A few metals show some tendency to sublimate (change directly from solid to vapour) but they are so unstable as elements that they barely exist in nature. Moreover, they typically have unpleasant odours. Consequently, apart from the taste of saltiness, flavour is not a property listed in catalogues of metal or mineral properties. (The non-metallic element sulphur sublimates, and its smell is familiar in some vineyards in volcanic areas. However, it is widely used in viticulture and vinification anyway, and, though not as an element, is ubiquitous in wine.)

There have been attempts, mainly in the context of impurities in drinking water, to establish human "detection thresholds" for some inorganic elements. Such values can only be very approximate as there are so many confounding factors, especially the variability between individual tasters and the nature of the accompanying ions and any other substances. Even so, the numbers demonstrate several points that are very relevant here. Most importantly, in most cases even the lowest values for the detection thresholds are considerably higher than the concentrations normally found in wine.

For example, Companys, Naval-Sanchez, Marinez-Micaelo, Puy, and Galceran found zinc contents in red wines from Raimat, Spain, averaged 3 ppb and 4–5 ppb in white wines, yet the World Health Organisation puts the taste detection threshold for zinc (in tap water) at 4 ppm, that is, a concentration a thousand times greater than that in the wines. Zacarías et al. determined a taste detection threshold for copper in water of 2.4–3.8 ppm and Epke and Lawless an odour threshold (in the presence of certain anions) of between 7.8 and 24.6 ppm; concentrations in Puglian red and white wines ranged only (despite copper applications in the vineyards) from 116 to 462 ppb, and the range of copper contents of some Croatian red wines peaks at 1.1 ppm. Sauvage et al. found calcium in south Australia wines at concentrations from 30 to 120 ppm; Lockhart, Tucker, and Merritt reported a taste detection threshold in tap water of 125 ppm.

Note that these values are the minimum concentrations for the presence of something to be sensed; moreover they are literally thresholds of detection and not identification. That is, the presence of some flavour might be sensed but humans are not capable of recognising the kind of taint it is. In fact at values much greater than threshold levels, for many metals tasters report disagreeable sensations even if they cannot be identified. For instance, copper levels in water above about 4 ppm give a detectable bad taste and zinc "imparts an undesirable astringent taste to water." These are hardly desirable attributes for a wine.

Furthermore, these are thresholds for tasting in water. Here, there are few competing flavour compounds (in tap water) and none in de-ionised water. Most bottled waters, incidentally, are drawn from a well or spring having resided in an aquifer for a long period, and hence can have high levels of directly dissolved solutes. The detection thresholds must be vastly higher in wine, because of the presence of both sensory nonvolatile compounds and, especially, the hosts of aromatic vinous compounds that give wine its flavour. For example, although Sauvignon Blanc wines are often said to have marked minerality, the compounds particularly responsible for the characteristic flavour of this varietal are now known to be various mercaptans and methoxypyrazines. And, in marked contrast with inorganic minerals, humans can detect and recognise these at extremely low concentrations. For example, 2-methoxy-3-isobutyl pyrazine can be sensed "even at low parts per trillion levels" and 4-mercapto-4-methyl-4-pentanone has an odour threshold of a mere 0.0001–0.005 ppb. Moreover, because—unlike inorganic ions—such metabolites can offer both gustatory and olfactory sensations, we may be able to sense their flavours even at the subthreshold values for taste or smell when measured independently. Any semblance of flavour that the tiny amounts of inorganic elements might have in this environment will simply be swamped and lost.

There are wines that have anomalously high concentrations of mineral elements, almost always due to some form of contamination such as from agro-chemicals, traffic pollution, and winery plumbing. These present problems for the winemaker. Not only do they taste disagreeable, some elements (such as copper, iron, and aluminium) present haze and colouration difficulties, and there may be a risk to public health. In fact it is telling to point out that the obvious question regarding the detection thresholds of inorganic elements actually *in wine* seems unresearched. The reason may be the potential health issue: the concentrations needed to be added in order to bring the minerals up to detectable levels would almost certainly make the wine toxic.

OTHER FLAVORS OF MINERALITY

Maltman goes on to subdivide several of the so-called mineral elements common in wine tasters' depictions of minerality in a tasting note or description. It's rarer than you'd think: Calling a wine "fruity" is often followed by details, like "black cherry," "fig," or "cassis." Not so, minerals: no one isolates "manganese" or "molybdenum" when describing a wine's mineral character.

Sometimes, though, attempts are made to describe certain traits. Here is Maltman's explanation for some of these characteristics.

FLINTY TASTE/SMELL, OR FLINTINESS

Minerality is often related to a perceived flinty taste in a wine. Geologically, flint is rather loosely defined and sometimes used interchangeably with terms such as silica, quartz, and chert. However, all these are forms of silicon dioxide — silica. The silicon and oxygen atoms are locked in an efficient three-dimensional crystalline framework that makes all these materials unusually stable (non-vapourising), tough, insoluble and virtually inert. Consequently, they lack any taste or odour. Indeed, it is because of these properties that silica is used for glass, and hence the very bottles and glasses that contain wine. Consequently, the notion of a wine that was stored in a glass bottle (silica) and tasted in a drinking glass (silica) having the flavour or smell of flint (silica) somehow derived from the vineyard is something of an oxymoron.

However, another repercussion of the efficient crystalline framework of silica is its lack of cleavage planes, instead inducing conchoidal fracture, a tendency for the material to break with irregular, concave surfaces. Intersections of these fractures give sharp edges and points, leading to the well-known archaeological applications of flint and the like as cutting tools. Thus flint is also mentally associated with edges and sharpness, and hence, evidently, metaphorically with very dry, acid wines. It seems that instead of reporting a sharpness or tartness in wine, some writers like to refer instead to a flintiness. And for some, it is even more preferable to use instead the French equivalent of flint: *silex*. Curiously, although most geological and wine-tasting terms have their equivalents in the other European languages, of these the French word for flint virtually alone finds its way into tasting notes.

Vine roots cannot take up the inert, insoluble compound silica. So if sharpness or metaphorical flintiness is seen as at least some component of minerality, it cannot be derived from flint or related materials in the vineyard. This is illustrated, for example, by Chablis wines, to many the epitome of flintiness—coming from vineyards that are calcareous in composition and lacking flint or similar siliceous material.

GUN-FLINT AROMA/STRUCK FLINT AND MATCHES

Also frequently mentioned in attempts to elucidate what minerality means is an allusion to the aroma of gun-flint, or the smell of a struck flint or match. These odours arise from the property of pyrophoricity. Many solids might, if struck forcibly enough, expel some of the percussive energy as a spark and this may have some burning smell. However, certain substances in the presence of oxygen are capable of spontaneously bursting into flames—auto-igniting. That is, they are pyrophoric. Sodium, potassium and calcium, for example, are extremely pyrophoric and hence not found as uncombined elements in nature. The phosphorus used in match heads needs the addition of a little heat, such as from the friction of striking a match. A few metals, such as iron, aluminium and magnesium, can be pulverised such that the surface area of each tiny particle becomes exposed to sufficient oxygen for it to auto-ignite, making a spark. A spark is a speck of burning material, usually producing an associated smell as it vaporises. Tiny fragments of the geological mineral pyrite—iron sulphide—can be pyrophoric, giving a distinct sulphur-tinged smell.

People long ago discovered that fragments of iron could be induced to spark, especially by striking the metal sharply against a fine-grained, tough material such as flint. This later became the basis of the flintlock mechanism in early firearms — and the smell of gun-flint. The aroma comes not from the flint, which is acting purely as an anvil, but from the burning particles of pyrophoric iron. (Modern firearms and lighters, incidentally, employ 'flints' made of a synthetic alloy of cerium and iron.) These well-known smells associated with sparking and striking matches may be useful comparators for minerality but clearly their pyrophoric genesis has no connection with processes in vineyards or vines.

EARTHY SMELL

An earthy smell, with which minerality is frequently compared, cannot be due to the inorganic components of rocks and soils, because they themselves are flavourless, as explained above. Rather, it appears to arise from organic compounds common on vines and in wineries, such as 2-methylisoborneol derived from algae, and a terpene known as geosmin (trans-1,10-dimethyl-trans-9-decalol) due to bacteria and moulds. Both these compounds have aromas that arise when the earth is being tilled. Moreover they have astonishingly low sensory thresholds, down to a few parts per trillion. In fact levels higher than this in wine can lead to it being regarded as tainted.

SMELL OF WARM/WET STONES

Similarly, the well-known aroma of stone on a hot summer day or after a shower of rain is not due to the geological material itself but the release of the organic oils mentioned in the previous section together with what Bear and Thomas called petrichor. As mentioned earlier, a freshly fractured geological surface has no flavour but on natural exposure to air it rapidly becomes filmed with volatile compounds present in the atmosphere from the decomposition of animal and vegetable matter. On warming, wetting, or when the relative humidity of the atmosphere approaches saturation, these volatile compounds are released to give the familiar petrichor smell. The substances include lipids, terpenes, carotenoids and, according to Bear and Kranzs, various fatty acids.

SEASHELLS AND FOSSILISED SHELLS

Some writers relate their perception of minerality to seashells. The link must really be with associated marine things because the shells themselves, being composed very largely of the (geological) minerals calcite and aragonite, have no taste or smell. More often though, the connection is made not with modern seashells but with their fossilised ancestors, which happen to be conspicuous in the bedrock of a number of the world's vineyard regions. But equally, such fossilised shells have no flavour.

On dying, organisms in nature soon disappear, through scavenging and decay. Any hard parts, such as teeth, bones and shells, will survive longer and, if circumstances are right may become fossilised (irrespective of whether they are still intact or broken up). Normally this comes about either by internal rearrangements and replacements to give a durable crystalline structure, or by dissolution leaving an imprint in the host sediment, which eventually becomes rock. Either way, the fossil is a replica, normally with none of the original organism remaining, composed of exactly the same geological minerals that make rocks and stones (most commonly calcite and quartz). Hence the materials are flavourless, for the reasons discussed earlier. Seeing fossil seashells in vineyard soils may prompt us to think of seafood and things maritime, but for the vines fossils are indistinguishable from any other piece of stone. Fossils in a vineyard bring nothing different to the nutrition of the vines or the composition of the resulting wine.

METALLIC SMELL

Some people may recognise a "metallic" smell, for example, the aroma we associate with handling coins and metal implements. However, as discussed above, metallic minerals lack flavour. The odour arises not from the metals themselves but through our having touched them, and the rapid reaction between the metal and skin chemicals to give highly volatile compounds. For example, Glindemann, Dietrich, Staerk, and Kuschk found that an odour described as metallic and mushroom-like in vapours next to skin touching iron was due to the ketone 1-octen-3-one, detectable by humans at very low concentrations.

In ways such as these, odours involving geological materials can be created and may make helpful comparators for explaining the sensation of minerality, but the processes are not relevant to the growth of vines. The rocks and minerals themselves remain flavourless.

■　■　■

CONCLUSIONS

The notion that minerality in wine is the taste of vineyard minerals leads to a contradiction about what kind of vineyard situation promotes it. Logically, those soils that are able to yield most nutrients to the vine would seem the most likely to imbue the wine with a high (nutrient) mineral content. That is, greater minerality would arise from the most fertile soils—the opposite of what is commonly believed. Most vine nutrition takes place just below the surface where some combination of high CEC clays, humus, and mycorrhizae will be relatively plentiful. However, wine mineral reality is most frequently associated with infertile soils: those that are particularly stony, or where vine roots have to probe deeply into bedrock. Here it is water that the roots are seeking, and at those depths (as well as with stony soils), organic material will be sparse and the transformations needed to convert rock into minerals with high CEC will have progressed little. The rock will have undergone only minor weathering: the water will have little solute.

This is not to say, however, that the anecdotal belief of minerality arising from unproductive soils is unfounded, but rather that any connection must be indirect. For example, it may be that the low nitrogen content of infertile soils leads to grape musts in which the yeast has to metabolise sulphur instead of nitrogen: there has been much speculation that minerality may involve sulphur-bearing compounds. Furthermore, it is well established that very small amounts of metallic elements can influence the course and progress of a host of metabolic reactions in the vine and in wine. These include acid buffering, yeast activation, polyphenol oxidation, co-factors in enzyme metabolism, and wine stabilization.

In fact, it may turn out with further research that the nutrient minerals of geological origin in vines and wines — minuscule in concentration and virtually flavourless though they may be themselves — are pivotal in determining wine character and flavour. However, this would have to take place in complex and circuitous ways. Thus perceiving minerality in wine would not involve tasting minerals but permutations of complex organic compounds whose production has been influenced by inorganic cations. Future research will no doubt evaluate this speculation. In any event, for all the reasons explained here, minerality in wine—whatever that perception is–cannot be in any literal, direct way, the flavour of minerals derived from vineyard rocks and soils.

WHAT IS MINERALITY, REALLY?

Alex Maltman's scientific polemic about minerality in wine is sharply drawn, but its main arguments—that vineyard minerals are not taken up as bits of slate or limestone, but as elemental ions; that they make it to the glass in sub-threshold concentrations; and that minerals do not have aromatic properties—are almost universally shared by geologists and plant physiologists. And yet wine tasters keep reporting minerality as if it is some substance inhabiting the glass, carried directly from the vineyard.

Those claiming to find minerality, among the best-honed palates on the planet, are certainly tasting *something*. Researchers in several countries have taken the lead in trying to capture what's behind the perception of minerality, using some ingenious research designs to parse out some tentative answers. Here we reprint portions of three studies published in 2013.

We start with a study done by a French team headed by Jordi Ballester of the University of Burgundy. The study employed a panel consisting of both trained tasters and wine

COULD TERROIRISTS BE SYNESTHETES?

What if the sensory perception of minerality and *terroir* is all in our heads? What if the flavors are not based on real molecules, but rather are strong mental associations arising when we smell and taste a unique assortment of aromatic attributes, and what if these are linked to impressions built into our sensory memory?

Experienced wine tasters are familiar with the flavors of a particular place and have often visited actual vineyard sites. Is it possible that these sensory impressions—of slate, flint, limestone, etc.—are catalyzed by vivid vineyard or winery surroundings—in view of slate terraces, granitic slopes, flinty riverbanks, calcareous swales? And isn't it possible that the memory of such places adheres to the wine we've tasted? It's not uncommon to hear spontaneous associations: One "smells" the sea breeze in the wine after walking an ocean-facing vineyard, for example, or one tastes chalk in a wine after hearing the crunch of fossilized rocks underfoot on a vineyard tour; how can you not smell *garrigue* herbs in a southern French wine after picking your way through the hills there, or likewise detect the smoky remnants of volcanic vapors in an Etna Rosso, which may or may not have wafted over vine rows for centuries.

It seems possible that such inter-sensorial experiences are fused in our memories. Neuroscientists have a term for it—synesthesia—describing how one cognitive pathway crosses over to another; perceptions in one sensory modality can elicit responses from another. In our everyday working vocabulary we may describe colors as soft, sounds as sharp, ideas as hot. These are well-recognized linguistic mix-and-matches. Why not a phenomenon that mingles powerful experiential input with sensory expectations, a clue embedded with a memory?

One recognized form of this phenomenon is called "associative synesthesia" and might explain the strong, almost involuntary connection between the memory of place, or familiarity with vineyard geology, and the flavors in a particular wine. Again, most experienced tasters have been persuaded of the typicity of a wine based on a memory of where it's from—or merely a linguistic allusion.

Alex Maltman, for example, mentions the "flavour of liquorice" in wines from Priorat but explains that the odor comes not from geology, but rather from the biosynthesis of anethole (anise camphor) in leaves or needles—a phenomenon that doesn't occur in grapevines, nor, really, in Priorat. This suggests an unfounded mental association between the alleged perception between liquorice and "llicorella," the name of the area's soils, a thing that could not possibly translate into the glass. "It seems in the Priorat case that the whole connection might be based on a misunderstanding," writes Maltman. "Llicorella is an ancient, very localised name in the local dialect of the Catalan language for these stony soils. It has no connection with what in English is called liquorice."

Elsewhere, he cites the famous impression of iodine, or the smell of the ocean, in the wines of Chablis. It's said to be an expression of the chalky soils the vineyards rest upon, a white limestone containing marine sedimentary bedrock. That, to Maltman, is highly unlikely: "Many of the references to iodine are allusional, though few people can be familiar with the smell of iodine, which textbooks list as pungent and irritating."

experts. The most interesting wrinkle in the study design was that the experts weren't told, at least initially, that they should look for minerality. We will follow the expert panel only in this excerpt, which demonstrates that while they could agree on a definition of minerality, they could not agree on which wines exhibited it.

FROM JORDI BALLESTER ET AL., "EXPLORING MINERALITY OF BURGUNDY CHARDONNAY WINES"

ABSTRACT

Background and Aims: The use of minerality as a wine descriptor has increased in the last few years. Minerality always suggests high quality and evokes a link between wine and the soil. The sensory meaning of minerality, however, is not yet clearly understood. The present study was designed to understand how wine experts conceptualise minerality and to explore whether they can judge wine minerality in a consensual way.

Methods and Results: Experts carried out an orthonasal free sorting task on 16 Chardonnay wines. Afterwards, they rated their mineral character according to two conditions: orthonasally and on the palate while wearing a nose-clip. The experts also answered a questionnaire in which they defined minerality. A trained panel independently performed a sensory description of the samples. The wine experts showed strong disagreement in their minerality judgements under both conditions. Three groups of experts emerged for each condition. Each group considered as mineral wines with quite different sensory characteristics which prevents any generalisation concerning the sensory meaning of minerality. Surprisingly, definitions of minerality by the experts showed some commonality despite the use of idiosyncratic terms.

Conclusions: Minerality is an ill-defined sensory concept, despite the apparent consistency emerging from verbal definitions by the experts.

Significance of the Study: Minerality is nowadays a popular term in wine marketing. Some attempts to understand its chemical origin have been made; however, this study has shown that a sensory definition of minerality should first be developed.

MATERIALS AND METHODS

A panel of wine experts and a trained panel participated in this study. The wine expert panel included 34 wine professionals (10 women and 24 men, mean age = 42 years, standard deviation (SD) = 10.7 years). All of them were winemakers, wine researchers, and/or wine teachers. The trained panel consisted of 33 panelists (18 women and 15 men, mean age = 37.6 years, SD = 19.4 years). Twelve of them had previous experience of sensory evaluation of wines.

Sixteen commercial Chardonnay wines from Burgundy were used in the study. The wines came from three subregions: Chablisien (CH), Côte Chalonnaise (CC), and Mâconais (MA) and from nine wineries. The prices ranged from 5 to 15 euros. The wines were selected according to descriptions

TABLE 7.4　Characteristics and codes of the wines selected for the study

Region/subarea	Appellation	Vintage	Producer	Code
Burgundy/Côte Châlonnaise	Montagny villages	2008	Buxinoise	CC1
Burgundy/Côte Châlonnaise	Montagny 1er Cru les Coeres	2008	Buxinoise	CC2
Burgundy/Côte Châlonnaise	Bourgogne Côte Chalonnaise	2009	Buxinoise	CC3
Burgundy/Chablis	Chablis La Sereine'	2008	La Chablisienne	CH1
Burgundy/Chablis	Petit Chablis Pas si petit'	2007	La Chablisienne	CH2
Burgundy/Chablis	Petit Chablis Pas si petit'	2009	La Chablisienne	CH5
Burgundy/Chablis	Petit Chablis	2009	Bichot	CH8
Burgundy/Chablis	Chablis	2009	Goissot	CH4
Burgundy/Chablis	Chablis Pierres Prehy	2009	Brocard	CH3
Burgundy/Chablis	Chablis 1er Cru Vaucoupin	2008	Brocard	CH6
Burgundy	Bourgogne Kimmeridgien	2009	Brocard	CH7
Burgundy/Mâcon	Mâcon Villages	2007	SCV Lugny	MA1
Burgundy/Mâcon	Mâcon-Fuissé	2009	Maison Aegerter	MA5
Burgundy/Mâcon	Viré-Clessé	2009	Cave de Viré	MA4
Burgundy/Mâcon	St Véran 'Classic'	2009	Poncetys	MA3
Burgundy/Mâcon	St Véran Clos des 'Poncetys'	2009	Poncetys	MA2

found in popular wine guides or given by their respective producers so as to potentially represent a wide range in perceived minerality. The wines are presented in Table 1 with their codes, origin and vintage.

Tasks performed by the expert panel. Free sorting task by nose only. The free sorting task was carried out in one sitting. The experts were asked first to smell all the wines from left to right following the proposed order and then to sort them according to their odour similarity. During sorting, experts were allowed to smell the wines as often as needed. They were allowed to make as many groups as they wanted and to put as many wines in each group as they wished. Upon completion, they were asked to provide some attributes to describe each group of wines. They were not allowed to taste the wines. At this point in the experiment, experts were not aware that the goal of the experiment was to study minerality.

Minerality judgement. After a 10-minute break, fresh samples were presented to the experts in a different presentation order and with random codes different than those used during the sorting task. The experts were asked to smell the wines again from left to right and to evaluate their minerality on a 10-point scale going from "absence" to "very strong." The wine samples were evaluated monadically.

After a 10-minute break, the same wines were presented one last time in a different order and with different random codes. The wine experts were then asked to assess the minerality on the palate using the same scale and the same procedure. They had to wear a nose-clip in order to prevent retronasal aroma perception.

Questionnaire. At the end of the session, experts filled out a questionnaire in which experts were asked to indicate how they assess minerality (nose, palate, or both) and to provide a short definition of minerality.

DISCUSSION

The goal of our study was to understand better the sensory meaning of wine minerality. Based on the definitions of the experts, and in agreement with the scarce literature, it appears that wine minerality involves a combination of both olfaction and mouthfeel sensations. The definitions of the experts included the main characteristics described in wine manuals as being typical subdimensions of minerality: stone-related odours (gunflint, wet stone), seashore-related odours (shellfish, iodine, and saltiness), acidity, and freshness.

The apparent consensus among experts, however, from declarative data is not confirmed from a behavioural point of view, because we have shown that the experts showed strong disagreement when judging wine minerality both on the nose and in the palate. Moreover, few mentions of minerality or minerality-related terms were generated at the end of the sorting task, which suggests that without knowing the target of our study, the experts did not tend to use minerality-related descriptors to describe their groups. Instead, they used terms globally related to reduction (reductive, cabbage, sulfur, cardboard), fruity and lactic odours for cluster 1, fruity and oxidised for cluster 2 and very fruity and intense for cluster 3.

This rough description of the sample set obtained in the sorting task is validated by an independent description made by the trained panel. As the experts, the trained panellists did not use many minerality-related terms. They showed the same opposition between wines dominated by fruity and flowery odours versus another group of wines composed of exactly the same Chablis wines present in cluster 1 (Figure 1 *[not reproduced here—eds.]*) and characterised by some reductive notes like wet dog, wet mop, and undergrowth. Some of these Chablis wines were among the most mineral wines for the experts from OMG-2 and OMG-3.

It appears that when the experts are not explicitly asked to search for minerality, they spontaneously use reduction-related terms. Another explanation is that mineral aromas could be confused with some forms of reduction. The earlier mentioned opposition between fruity aromas and minerality is supported by the low frequency of citation of fruity aromas and anecdotally, the citation of less fruit by four of the experts in the definition task. With this respect, the minerality judgement of the experts from the orthonasal group 1, who judged more mineral the wines described as fruity and floral, is quite difficult to explain. In order to better understand their mental representations of wine minerality, qualitative techniques like deep interviews may be more informative.

According to the judgments of palate minerality, three groups of experts showing different behaviour were obtained. Two groups gave minerality ratings significantly related to acidity (palate expert group 1) and bitterness (palate expert group 3), respectively, while the third did not show a significant difference in minerality among the samples (palate expert group 2). The suggested link between acidity and bitterness with minerality is consistent with the experts' definitions and the literature. A recent study suggests that saltiness could be related to minerality. Unfortunately, the trained panel did not rate saltiness; therefore we cannot determine how this taste relates to minerality. Future studies on wine minerality should take saltiness into account.

The lack of significance of the correlation coefficient between orthonasal and on the palate average ratings suggests that both judgments are quite independent. Most of the experts, however, indicated in the questionnaire that they used both nose and palate evaluation to assess minerality. It is difficult at this point to know whether two different minerality concepts (on the nose and on the palate) exist or if there is one unique concept, which is a weighted combination of olfactory, taste, and mouthfeel sensations. Moreover, because all participants of the study were French and experts on Burgundy wines, they have probably built their own minerality representation through exposure to Burgundy wines. This type of cultural effect has been previously shown by Langlois between wine experts from two different regions when assessing the potential for ageing of red wines. It is therefore to be expected that wine experts from a different wine region will develop different minerality representations.

CONCLUSION

Despite the increasing occurrence and marketing importance of the descriptor minerality in the past few years, there is no global consensus between wine experts about the sensory meaning of this concept. Instead, several ways of understanding wine minerality emerged. A possible link between minerality and reductive notes was pointed out. Because the latter can be considered as a wine taint, further research is needed to better elucidate the differences and similarities between reductive and mineral notes. The link between minerality and wine acidity and bitterness was confirmed for two groups of experts. Moreover, further studies on wine minerality should include also global minerality assessments (i.e. allowing the perception of the aroma, taste, and mouthfeel at the same time) in order to better understand how the different senses interact when judging wine minerality. ∎

ANOTHER STUDY, by UC Davis researcher Hildegarde Heymann and members of her lab zeroes in on the association of minerality as a descriptor with the presence of high acidity and sulfur compounds. Below is the abstract from a recent paper, again using both expert and trained panels, and also examining several parameters of wine chemistry. The wines, said to show minerality in the press, included examples of Chardonnay, Pinot Gris, Riesling and Sauvignon Blanc from California, Oregon, Washington, France, Italy, Germany, South Africa, and New Zealand.

FROM HILDEGARDE HEYMANN, HELEN HOPFER, AND DWAYNE BERSHAW, "AN EXPLORATION OF THE PERCEPTION OF MINERALITY IN WHITE WINES"
ABSTRACT

Minerality is a way for wine writers to associate wines with their *terroir*. Little research has been carried out on the concept of minerality. In this study white wines were compared by projective mapping (PM) performed by wine industry professionals to a standard sensory descriptive analysis (DA) by trained judges. The PM found minerality to be positively correlated with acid taste and citrus, fresh, wet stone, and chemical aromas, and negatively correlated to butter, butterscotch, vanilla, and oak aromas. The PM panel minerality was associated with both aroma and taste perception. The DA found minerality to be positively correlated with reduced, chalky, and grassy aromas and bitter taste, and negatively correlated with barrel, caramel, honey, juicy fruit, musty, and cat pee aromas. Wine groupings were similar between the two panels. Minerality was highly associated with malic acid, tartaric acid, and titratable acidity, and moderately associated with free and total sulfur dioxide. ■

HEYMAN'S ARTICLE goes on to depict a sophisticated experiment involving experienced tasters who were asked to describe an array of white wines. The word "minerality" crept into their descriptions readily. Regardless of variety, the wines most often described as "mineral" were correlated with high levels of malic acid, of high titratable acidity, and tartrate levels.

Additionally, the panel "found mineral to be positively correlated with reduced, chalky, and grassy aromas and bitter taste, and negatively correlated with barrel, caramel, honey, juicy fruit, musty, and cat pee aromas." The authors remained troubled by the vagueness of the term, and concluded that "defining and limiting the scope of the minerality descriptor would result in greater universal understanding and would thus limit overuse. It may even be necessary to split the aroma of minerality into various stone, chalk, and metal aromas, as well as further defining the bitter, acid, and salty tastes associated with wines described as mineral." Further on in the study, they add: "Innovative sensory research may someday elevate the minerality term from vague discussions of terroir and soil properties into a standardized group of aroma and/or taste compounds."

FROM "THE NATURE OF PERCEIVED MINERALITY IN WHITE WINE" (NEW ZEALAND WINEGROWERS)

New Zealand Winegrowers recently funded the New Zealand component of a project aimed at shedding light on the somewhat elusive wine characteristic "mineral." Over the last 18 months, an international and interdisciplinary collaboration between researchers from New Zealand (Lincoln University; Plant & Food Research) and France (University of Burgundy; University of Bordeaux II) has been investigating perceived minerality in white wine. More specifically, we aimed to delineate the nature of perceived minerality in Sauvignon Blanc wines from New Zealand (Marlborough) and from France (Sancerre/ Loire, Bordeaux, and Burgundy) by undertaking both sensory and physico-chemical analyses on wines from both countries.

MINERALITY IN WINE: WHAT IS IT?

Perception of minerality in wine is currently a hot topic in the industry. Wine producers and wine writers/critics appear increasingly interested in the term, pointing out its ill-defined nature and listing terms implicated in describing minerality in wine (e.g., flinty; smoky; oyster shell; chalky; calcareous; silex). As well, wine consumers are indirectly affected by the term in that perceived minerality in wine is linked with high quality wine by being described as a component of some of the world's more expensive wines (e.g., white Burgundies; Sancerre wines). Despite such industry interest, there is a distinct lack of scientific work addressing the topic, with little clarity as to precisely what the term means to wine professionals or consumers, whether minerality is solely a taste and/or mouth-feel (trigeminal nerve) sensation or can be smelled, and which wine components are implicated (e.g., acidity) in perception of "mineral" in a wine.

Much of the interest in perceived minerality in wine centres around attempts at understanding sources of perceived minerality, in particular the possible links with terroir and/or acidity in wine. For example, in a recently published article a graduate of UC Davis in California points out that the association of chalky or limestone vineyard sites with perception of minerality in the wine from such sites could involve a mediating variable, namely acidity. This is because limestone and chalky soils are alkaline, producing wines with higher total acidity and lower pH.

More recently, a new issue has become important. The relation between perceived minerality and reductive characteristics in wine (thiol/disulphide compounds) has raised its head, gaining momentum since the introduction of inert wine-bottle closures (e.g., many brands of screw-cap closure) in New World wine-producing countries. The concurrent increased usage of the descriptor "mineral" and the increased usage of screw-cap, bottle closures has not gone unnoticed by several wine writers. Various hypotheses have been put forward by wine writers, often indirectly, to suggest that increased perception of minerality in wines from New World countries such as New Zealand could have its basis not in factors considered important in Europe (e.g., soil profiles; qualitative and quantitative aspects of acidity), but in factors such as (i) sulphide reduction, or (ii) in the sensory context created by unripe fruit and therefore relative absence of fruity flavours in the finished wine (i.e., low concentrations of volatile thiols & esters). In the study reported below, we investigated several of these hypotheses.

EMPIRICAL COMPONENT: INVESTIGATION OF PERCEIVED MINERALITY IN SAUVIGNON WINES

Sauvignon Blanc wines from central France (i.e., Sancerre and Pouilly sur Loire) have historically been described in terms of their minerality. Words such as "silex" and "chalky" are frequently employed in descriptions of their sensory profiles, an underlying assumption being that the particular variant of descriptor used to describe the perceived minerality also reflects the source (i.e., type of soil in the vineyard).

On the other hand, Marlborough Sauvignons have until recently seldom been described in terms of their minerality, but as "fruit-driven" wines, with fruity and green notes dominating their aroma and flavour profiles. Several authors of the current article have previously published the little research to date that has provided data concerning perception of minerality in New Zealand Sauvignons. Initial scientific articles on this topic reported that perception of "mineral" in the NZ wines was not only low relative to that in French wines, but was associated negatively with Marlborough typicality and with liking (2004 & 2005 vintages).

Interestingly, more recent data (wines from 2007 vintage that were evaluated in June 2008) suggest not only an increased perception of minerality in the Marlborough wines, but also a change in conceptualisation of the term by New Zealand wine professionals. That is, the more recent data show that some wines perceived high in minerality were also judged high in Marlborough typicality (i.e., as "good examples" of Marlborough Sauvignon), a result that was not found in the earlier work. Further, these recent data demonstrate that although wines from France and New Zealand were perceived similarly in terms of intensity of mineral characteristics, the chemical compositions of the wines from France and New Zealand differed significantly.

The funding from NZWG allowed us to investigate scientifically several hypotheses that to date have been based largely on anecdotal evidence. Our major aim was to delineate the nature of perceived "minerality" in 100% Sauvignon blanc wines from four sub-regions of Marlborough (Awatere Valley; Rapaura; Wairau Lowlands; Southern Valleys) and from four regions of France (Sancerre; Loire; Burgundy (Saint Bris); Bordeaux). To do this we developed and executed methodologies that allowed us to:

Associate via multivariate analyses the umbrella term "mineral" with (i) assumed sub-components of perceived minerality (e.g., flinty/smoky; chalky/calcareous) and (ii) perceived reductive characteristics (e.g., sulphide; burnt rubber).

Investigate cultural differences in perception of the concept of mineral by French Oenologists in comparison with New Zealand Wine Professionals/Oenologists.

Investigate whether minerality is a smell, a taste, or tactile (mouth-feel), or a combination of several of these sensations.

Investigate relations between wine composition and perceived characteristics as a function of wine country of origin (France or NZ). This involves associating via multivariate analyses the sensory data concerning perceived minerality with selected chemical and physical measures.

MODERN METHODOLOGY AND THE PATH FORWARD

For the concept of *terroir* to gain validation and legitimacy in the wine world, it needs to move past anecdotal accounting and into something more concrete and genuine—and the best way to do that is to quantify it in some fashion. That means some sort of sensory analysis is a critical part of the program, since the smells and flavors of wine are the very things we want to explain and better appreciate.

We've already explored one indispensable methodology, Descriptive Analysis (see Noble et al.), with its emphasis on rigorous objective evaluations as opposed to subjective and anecdotal characterizations.

But even before the first glass is tasted by the first trained panelist, someone has to define the *terroir* being tested. Asking anyone what the wines of Spain taste like is an impossible question; Spain is not a unit of *terroir*. Asking anyone to identify the distinctive aromas of California's Sonoma Coast is equally futile, because that American Viticultural Area was defined largely as a marketing tool, not by geology and climate. There's

TASTING THE VINEYARD—BUT NOT THE SOIL

Some vineyards possess a *terroir* signature that extends beyond soils and beyond dirt—the environment leaves its mark on the wines in other ways. For decades, one of the prime examples of California *terroir* was the distinctive taste of the Heitz Cellar Martha's Vineyard Cabernet Sauvignon. The Heitz Martha's Vineyard always conveyed a hint of something vaguely minty in its aromas and flavors, a characteristic derived from the towering eucalyptus trees framing the vineyard, annually depositing volatile oils on the vines below.

Another example comes from wine merchant Kermit Lynch in his memoir, *Adventures on the Wine Route,* in which he recounts a visit with René Loyau, a *négociant* wine-buyer and blender from the Loire Valley with ties all over France. Loyau recounted for Lynch the time when a new vintage of the Gevrey-Chambertin from a particular grower in Burgundy suddenly lacked the wild currant aromas that had always been there. Without ever having visited the vineyard, Loyau asked the grower when he had taken out the wild currant bushes; and sure enough, a large patch on a neighboring property right beside the vineyard had been ripped out that year. Mystery solved; *incroyable!*[1]

More recently, wildfires have left their mark on wines from at least two continents. For more than a decade, fires near parts of the Australian winegrowing country have blanketed vineyards with smoke at crucial times in the growing season, leaving a residue in the grapes and affecting the taste of the wines. In 2008, wildfires in California's North Coast wine counties forced vintners to seek methods to remove smoke taint.

These examples demonstrate the power of organic matter to affect the sensory character of grapes and wines. Organic matter in the soil contributes nutrients to the vines, but any aromatic properties it might have are not exported to the glass. On the other hand, organic matter deposited on the *skins* of grapes can certainly make its way into the bottle and can have a substantial impact on the wine.

---

1. Kermit Lynch, *Adventures on the Wine Route: A Wine Buyer's Tour of France* (New York: The Noonday Press, 1990), 36–37.

---

no point to rigorous tasting without a process of rigorously delimiting the *terroir* you wish to isolate. Introducing: Natural *Terroir* Units.

NATURAL *TERROIR* UNITS

European appellations of origin, US AVAs, and similar systems throughout the wine world are designed to be rough approximations of *terroir* spatial definitions, and most have at least a thread of soil or climate or winemaking tradition running through them. But a good number of such defined regions, including some critically important ones—Bordeaux comes to mind—are either the products of political infighting, reflections of the interests of large producers, excessively large and thus non-homogeneous, or otherwise groundless.

Over the past two decades, traditional appellations are being further defined by a metric of Natural *Terroir* Units (NTUs), or in the French order of things UTNs, in an attempt to up the game of definition with more thorough analysis of physical features of different areas and the use of more sophisticated technologies like geographical information systems (GIS) that integrate hardware, software, and data to capture and visualize the characteristics of a given area. Here is the abstract of the 1993 article in which French Geologist Pierre Laville coined the term and explained its value.

FROM PIERRE LAVILLE, "NATURAL TERROIR UNITS AND TERROIR: A NECESSARY DISTINCTION TO GIVE MORE COHERENCE TO THE SYSTEM OF APPELLATION OF ORIGIN"

At present the variety of meanings given to the word "terroir" contributes to the popularization of an old, relatively unmistakable concept. In France it was practically synonymous with "appellation d'origine." Deliberately upheld by supporters of a free economy, standardized agricultural production, and an exclusive trademark protection, this confusion is harmful to local specific agriculture, which can be protected by appellation and which is well adapted to a market of specific, high quality products.

■   ■   ■

Focusing on the idea of terroir, its composition and identification, allows a new basic concept to be put forward, independent of agronomic usage: "l'unité de terroir naturel" (natural terroir unit) of UTN. An operational method for determining it accompanies this concept. The idea springs from the work of French teams who have demonstrated the specific roles of the various "terroir units" in the presence of features in the products from this terroir. At the same time the main issues of these ideas are presented. This attempt to clarify terroir and the system of appellation of origin is needed to show that other alternative forms of agriculture exist, which are more local and also more profitable than those based on a maximum production even when of good, wholesome quality. In the western world the local or foreign consumer is ready to pay for these differences based on the variations shown by tasting. Allowing the concept of terroir to lie fallow, to give it up to the agro-foodstuffs industry, is to lose a basic capital for the identity and motivation of the farmers. It also destroys the sole link which ties them to the consumer. If today the CAP (Common Agricultural Policy) shows signs of big setbacks it is because neither politicians nor farmers through their local products were able to picture the shift to intensive, standardized, agricultural production that now has to be reduced following overstimulation. ■

THE SOUTH AFRICAN wine industry and its academic brain trust have been particularly active in promoting and defining NTUs in their country, as this paper from the journal *WineLand,* the South African Journal of Enology and Viticulture, demonstrates.

FROM VICTORIA CAREY, EBEN ARCHER, AND DAWID SAAYMAN, "NATURAL TERROIR UNITS; WHAT ARE THEY? HOW CAN THEY HELP THE WINE FARMER?"

Terroir is not a new concept. We know that the Egyptians (3000 B.C.) had an understanding of the importance of the interaction between the environment and the vine as they built artificial hills in the flat Nile-Delta and divided their wines into five categories, partially based on the origin of the product. Georgic authors (200 B.C.–200 A.D.) underlined the role played by the environment in viticulture both at a macro and microscale and the importance of choosing the site according to the cultivar to be planted. This concept has formed the base of many geographical indication systems, not the least being the Wine of Origin System in South Africa.

A *natural terroir unit* (NTU) is a unit of the earth's surface that is characterised by relatively homogenous patterns of topography, climate, geology and soil. It has an agronomic potential that is reflected in the characteristics of its products, resulting finally in the concept of terroir. A *terroir,* therefore, is defined as a complex of natural environmental factors that cannot be easily modified by the producer. With the aid of various management decisions, this complex is expressed in the final product, resulting in distinctive wines with an identifiable origin. The terroir cannot be viewed in isolation from management and cultivation practices, although they do not form part of the intrinsic definition.

The above definition of a terroir determines the way that we study and identify viticultural terroirs. No single environmental component of the terroir system can be studied in isolation, rather the full complex of factors must be taken into account. There will always be two steps for a terroir study. Firstly, all the relevant natural factors (such as aspect, altitude, geology, soil type, effective soil depth, water supply to the vine, *etc.*) must be identified and characterised in order to identify relatively homogenous NTU's. Secondly, the response of the vine on the NTU's and the organoleptic properties of the wine originating from these units must be determined over a period of time so that the NTU's that result in a similar product can be grouped into viticultural terroirs.

Terroir studies usually focus on wine character or style rather than quality, but a good terroir is considered to be one that ensures a slow but complete maturation of grapes with a certain regularity in quality of the product from vintage to vintage.

DESCRIPTIVE ANALYSIS OF REGIONAL *TERROIR* UNITS: A SENSORY WORK IN PROGRESS

The cutting edge of current *terroir* work lies in the characterization of the aromas, flavors and other characteristics of wines from well-defined natural *terroir* units through rigorous Descriptive Analysis. When characteristic flavors from the wines of a place are thought to exist, or a specific and reproducible flavor profile, the powerful technique of DA may be used to "fingerprint" those flavors. Theoretically, one can show through testing that they are reproducible and homogeneous within that area and are measurably different from the wines of other nearby areas. The sensory descriptors and parameters for each area can then become an essential part of its identity and its certification as a protected geographical name and place.

This process is underway in the European Union and, to a lesser extent. other wine-growing countries, but is still in its infancy. Millions of acres under vine may eventually lay claim to an official designation. In the meantime, a few abstracts from research journal articles reprinted here give a sense of what is being done.

FROM JAMES SCHLOSSER, ANDREW REYNOLDS, MARJORIE KING, AND MARGARET CLIFF, "CANADIAN TERROIR: SENSORY CHARACTERIZATION OF CHARDONNAY IN THE NIAGARA PENINSULA"

Twenty four VQA [Vintners Quality Alliance] Chardonnay wines from three regions ("Bench," "Lakeshore Plain," and "Lakeshore") of the Niagara Peninsula appellation (Ontario, Canada), plus three international Chardonnays, were used to investigate the effect of sites using chemical and sensory analyses. Descriptive analysis revealed that "Bench" wines had more apple, citrus and melon aromas and flavours compared to those produced from the "Lakeshore Plain," and lowest grassy and earthy aromas compared to those produced from both the "Lakeshore Plain" and "Lakeshore" regions. Differences were not observed for other aroma (oak, butter), flavour (oak, butter, toast, grassy), taste (perceived acidity, bitterness), and mouthfeel (astringency, body) terms. Wines from the "Lakeshore" were golden yellow in colour, while those from the "Bench" were pale yellow in colour; "Lakeshore Plain" wines were intermediary between these two shades. Chemically, "Bench" wines were most unique, exhibiting the lowest pH values and highest concentrations of ethanol, titratable acidity, and phenols.

FROM ULRICH FISCHER, DIRK ROTH, AND MONIKA CHRISTMANN, "THE IMPACT OF GEOGRAPHIC ORIGIN, VINTAGE, AND WINE ESTATE ON SENSORY PROPERTIES OF VITIS VINIFERA CV. RIESLING WINES"

The official quality designation for German wines is solely based on the degree of ripeness in grapes, while most other wine producing countries in Europe rely on a geographic classification system. A descriptive analysis (DA) investigated the sensory properties of commercial Riesling wines from two vintages, five wine estates, and six vineyard designations within the viticulture region Rheingau. Based on the number of significant F-ratios among 10 odor and 4 orally perceived attributes, vintage and wine estate proved to have a similar impact as vineyard designation. Principal component analysis revealed substantial variations within the same vineyard designation and demonstrated the strong impact of the individual wine estate and vintage. Hence, a classification system focusing on geographic origin alone would be rather confusing for consumers, because wines differ substantially regarding their sensory appearance within the same vineyard designation.

FROM LUND *ET AL.,* "NEW ZEALAND SAUVIGNON BLANC DISTINCT FLAVOR CHARACTERISTICS: SENSORY, CHEMICAL, AND CONSUMER ASPECTS"

A trained sensory panel (n = 14) identified key flavors in Sauvignon blanc wines from Australia, France, New Zealand, Spain, South Africa, and the United States. Sixteen characteristics were identified and measured: sweet sweaty passion fruit, capsicum, passion fruit skin/stalk, boxwood/cat urine, grassy, mineral/flinty, citrus, bourbon, apple lolly/candy, tropical, mint, fresh asparagus, canned asparagus, stone fruit, apple, and snow pea. Principal component analysis was used to describe differences among regions and countries. Sauvignon blanc wines from Marlborough, New Zealand, were described by tropical and sweet sweaty passion fruit characteristics, while French and South African Sauvignon blanc wines were described as having flinty/mineral and bourbon-like flavors. Chemical analyses of these wines also showed that wines from Marlborough had more methoxypyrazine and thiol compounds. A consumer study (n = 105) showed that New Zealanders significantly prefer New Zealand-style Sauvignon blanc.

FROM ELIZABETH TOMASINO, ROLAND HARRISON, RICHARD SEDCOLE, AND ANDY FROST, "REGIONAL DIFFERENTIATION OF NEW ZEALAND PINOT NOIR WINE BY WINE PROFESSIONALS USING CANONICAL VARIATE ANALYSIS"

Pinot noir is the most widely planted red grape variety in New Zealand and is considered a premium product based on the price per volume. To date no studies have attempted to characterize the different styles of the four main Pinot noir producing regions: Central Otago, Marlborough, Martinborough, and Waipara. The intensities of aroma, flavor, and mouthfeel attributes of commercial regional wines from two vintages were investigated. Descriptive analysis was carried out by a panel consisting of experienced but untrained wine professionals. Canonical variate analysis showed that the four wine regions were differentiated according to aroma (barnyard, black cherry, herbal, raspberry, red cherry, oak, spice, and violet), in-mouth flavor (fruit density/concentration and red fruit), and mouthfeel (balance, body, and finish length) attributes. Results show that Pinot noir wines from the four regions of New Zealand are stylistically different and that experienced but untrained wine professionals produce reliable results for this type of sensory analysis.

FROM ROBINSON *ET AL.,* "INFLUENCE OF GEOGRAPHIC ORIGIN ON THE SENSORY CHARACTERISTICS AND WINE COMPOSITION OF *VITIS VINIFERA* CV. CABERNET SAUVIGNON WINES FROM AUSTRALIA"

The current study explores the relationship between sensory characteristics and wine composition of Cabernet Sauvignon wines in relation to Australian geographical indications (GIs). Descriptive sensory analysis was conducted to characterize the sensory attributes of commercially produced Cabernet Sauvignon wines from the Barossa Valley, Clare Valley, Coonawarra, Frankland River, Langhorne Creek, Mount Barker, Margaret River, McLaren Vale, Padthaway, and Wrattonbully GIs.

Canonical variate analysis using the significant sensory attributes demonstrated that each GI could be distinguished from the others. A recently developed analytical method was used to analyze over 350 volatile compounds in the wines assessed, and measures of the major nonvolatile components were also determined. . . . Results demonstrate that Australian Cabernet Sauvignon wines have common sensory attributes related to geographic origin. The work also identifies a number of candidate components that are related to individual GIs which warrant further investigation. The study is the first to explore the concept of regionality in Cabernet Sauvignon wines from Australia.

NAPPING

Recently a simpler, useful method for "mapping" sensory data has been developed by French researchers called "napping" (after the French word for tablecloth, "nappe"), in which comparative sensory assessments are made on a flat plane and mapped, with the resulting pattern assessed, often by some method of descriptive analysis. Excerpts from a recent paper on the subject by lead researcher Helen Holt at the Australian Wine Research Institute are reproduced here.

As you'll discover, it is a far less robust process than either method explored above, but it seems to have some value as an initial assessment tool, which can steer or frame Descriptive Analysis.

FROM HELEN HOLT, WES PEARSON, AND LEIGH FRANCIS, "NAPPING—A RAPID METHOD FOR SENSORY ANALYSIS OF WINES"

The method of quantitative descriptive sensory analysis has been used for many years to discover the most important sensory characters for a group of wines. This method involves a trained sensory panel rating defined characters for their intensity in each wine, so that the presence and intensity of wine descriptors can become paired to give an overall picture of the similarities and differences among a group of wines. This is a very sensitive and reproducible method and gives a great deal of information about the individual wines and how they compare to others in the group tested. The data can also be compared to wine chemical measures or preference and quality data. The method can be used, for example, to see the sensory effect of a new yeast strain for winemaking, or to examine the effect of grape ripeness on wine sensory properties.

However, to apply this method a highly trained panel of sensory assessors is required, with specific training for each group of wines the panel assesses. This can be both time-consuming and expensive. A number of alternative methods have been developed which can be used to assess a set of wines more rapidly. Several papers have been written about the different methods, their similarities and their differences, their weaknesses and their strengths. One such recently developed method is "napping." The name "napping" comes from "nappe" or tablecloth in French, a term used originally by French researchers working with this method. It is a type of projective mapping which may also include sorting tasks.

Napping is a way of evaluating a group of samples, with individual assessors arranging them in a two-dimensional space according to how similar or different they consider the samples to be from one another. Samples which are most similar are placed close together and samples which are very different are placed well apart. This sample placement can be made on a sheet of paper on a flat surface like a table or bench (hence "nappe") or on a computer screen, using the horizontal and vertical axes on the computer screen in the same way as an assessor would use the two dimensions of a paper sheet on the bench.

▪   ▪   ▪

Assessors may be instructed to assess the wine separately by appearance, aroma, or palate (partial napping), or overall, using all aspects (global napping). Assessors use their own criteria and their own experience to decide whether samples are similar or different. In this way each assessor places the wines according to what is important to them, not according to a group consensus. Each assessor receives their coded wines in a different, randomised order, like any other sensory assessment, removing the effective tasting order. In general only a single replicate of each wine is used but there have been some explorations regarding the use of replication in napping, with either whole sample sets or selected samples within a set. The number of samples that can be assessed is still a matter for debate, but a set of up to 18 samples has been found to be practical using the AWRI's facilities. This is an important constraint as it limits the number of samples which can be assessed in a single study.

To gain more information, assessors can be asked to write a few words describing each wine or group of wines. This can be performed using any words the assessor chooses (free-choice) or by choosing words from a prepared list.

▪   ▪   ▪

Napping provides a rapid snapshot of a group of samples—how similar or how different they are. With the use of descriptive terms, information about the nature of the differences can be gained. It provides a picture—a consensus map—of how wines sit in a two dimensional space, using the data from all assessors to make this picture.

▪   ▪   ▪

In summary, this method, like any other sensory method, has advantages for specific types of study depending on the aim of the project. It is particularly useful in screening studies. It is rapid, less expensive, and generally easier to use than conventional sensory methods. It is very helpful for projects, including contract or commercial studies, where timeframes are tight and budgets are limited. Overall, it is a valuable new addition to the suite of sensory analyses at the AWRI.

ON THE OTHER HAND . . .

Some people sip a glass of wine and immediately point to the "minerality" in the glass; others at the same table, drinking the same wine, wonder, what mineral? Molybdenum? This selection from *Wine Enthusiast* describes the travails of one seeker of minerality, this volume's coauthor, Tim Patterson.

FROM TIM PATTERSON, "ROCKS IN MY HEAD"

Nothing in winedom is more prized than the miracle of minerality. This elusive characteristic, found only in certain wines and discernible only to selected palates, carries a potent symbolic charge: tasting minerality is tasting the living soil that gave birth to the grapes.

There's a good chance, of course, that minerality is mostly hooey.

When I was first being socialized into winespeak, I found minerality everywhere, even in $5 Trebbianos. Soon I realized I was likely picking up acidity, not true grit. Still, I knew it was out there somewhere, because everybody said it was. I resolved to try harder.

I fell in love with German Riesling and mastered the distinctions between regions, vineyards, even vineyard blocks; I came to revel in the shades of difference between the Bernkasteler Doktor and the Bernkasteler Lay. I attended seminars illustrated with geological timelines and vineyard cross-sections and soil samples passed from hand to knowing hand. More than once I watched Terry Theise, the charismatic avatar of German wine *terroir* in the American market, swirl a glass of wine in front of a hushed room, close his eyes, and whisper, "Schlossböckelheim Felsenberg," or words to that effect.

When I got home, I made the mistake of checking the science behind all this, and it wasn't encouraging. First of all, people have the darndest time agreeing on what "minerality" is. Is it a flavor, an aroma, a texture? And exactly which minerals are we talking about? There's no trace of minerality on the UC Davis Aroma Wheel—nobody could come up with a reference standard.

Then there's the little problem that rocks mostly don't taste or smell at all. Clean off a rock and there's precious little, organoleptically speaking, to write home about.

My mineral-centric buddies explained, with some condescension, that wet rocks clearly do have a smell—even a hosed-down sidewalk gives off a scent. I soon found out there's a word for this phenomenon, *petrichor,* the smell of rain on dry ground. Showers after an arid spell activate oily substances given off by the surrounding vegetation; in other words, the smell of wet rocks is the smell of plants.

It gets worse. Modern plant biology holds that flavors and aromas are manufactured within the grapes through photosynthesis, not transported up from the ground into the berries. Peach and raspberry flavors don't come from fruit in the ground, and there's no evidence that little molecules of schist or limestone or clay work their way into your glass, either.

More likely, according to state-of-the-art research, what gets called minerality is some combination of acidity and sulfur compounds. So, maybe I wasn't so far off with that cheap Trebbiano, after all.

On my dining room sideboard, I now keep a little dish full of pieces of red slate from famous Mosel vineyards. The rocks, alas, have no smell. But because there was extra room, I added some fragments of fossilized dinosaur poop from Montana. Talk about ancient soils. They don't smell either, but they provoke some very interesting conversations.

8

MARKETING
Terroir *for Sale*

The most revered *terroirs* of the world differ wildly in soil types, topography, climate patterns, in grape varieties, and winemaking traditions, but they have one essential thing in common: they are all famous. Which means that someone got the word out.

At its heart, marketing hype seems completely antithetical to *terroir* in all its essential purity. But without it—whether it's a publicist's spin, a lyrical terrestrial description, a seductive tasting note, an artful slide show, a passing reference to a famous neighboring vineyard, a pile of rocks on a tablecloth, a jar of dirt passed around in a seminar—no one would know to ask whether *terroir* existed. From the classics to the latest sensation, a great wine's greatness is inevitably accounted for by way of its *terroir,* of the hallowed ground from which it is derived.

That process, of course, is never entirely tidy nor utterly objective. *Terroir* is rarely a "eureka" moment; it does not emerge as the result of an experimental finding by lab-coated teams of tasters issuing thick reports. More often than not, *terroir* is "revealed" when some prince decides he loves the local wine and commissions an ode to it, or a merchant comes up with a compelling story about a patch of ground, or a journalist attempts to account for the flavors he's tasting, or a region comes together to conjure up a history of itself. *Terroir* is revealed through careful stewardship of the land, but it is also revealed through articulation, through speaking of it, through creating the narrative of its existence.

Which, of course, leaves the concept vulnerable to all manner of puffery and fakery. We might begin by reciting the monikers of countless industrial wines which invoke "place" and the land in their brand names, the Creeks and Glens and Hills and Ridges and Peaks and Rocks and Valleys, reinforcing an alleged closeness to the earth. Large wine companies have whole departments devoted to generating beguiling stories, with no fact-checkers in sight. If winery marketers are to be believed, nearly every wine on earth captures the essence of its special vineyards. (You'll *never* find one that proclaims, "Our vineyards are incidental, but our tanks are amazing!") In modern wine marketing, allusions to *terroir* are almost mandatory.

It is, in the end, a source of great skepticism, and as such we approach it that way here. We start with a selection on the relationship of commerce and *terroir,* a conceptual overview from Master of Wine Stephen Charters. The next section involves the marketing of regions, and examines how the reputations of Bordeaux and Burgundy were crafted, excerpting Benjamin Lewin on the famous 1855 classification, James Ulin on the invention of the Château mythology, and French sociologists Serge Wolikow and Olivier Jacquet on how growers in Burgundy banded together to establish quality standards.

How new *terroirs* stake their claims is explored with a selection from Sue Callahan, supplemented by Deborah Elliott-Fisk's account of the establishment of designated viticultural areas in Lodi, California. Finally, the rise of the Napa Valley is scrutinized through selections from historian James Lapsley, winery owner Jean Arnold, wine writer Gerald Boyd, and the Napa Valley Vintners Association, whose planning documents for the Destination Napa Valley publicity campaign employ notions of *terroir* liberally. To leaven those claims, we present a tasting of multiple California appellations as they compare with Napa, results assembled by wine historian Charles Sullivan.

FROM TASTE OF PLACE TO POINT OF SALE

While a good dose of skepticism is justified, the relationship between *terroir* and sales is based on more than sheer mendacity. Suspicions about *terroir* claims as mere marketing have circulated in the discourse on wine for a very long time. Since wine is, like it or not, a commodity, it is inextricably entwined with the logic and institutions of the marketplace. Wines that express—or claim to express—a particular *terroir* possess an attribute that can give them a competitive advantage against wines unable to make such claims, and it is no surprise that vintners, regions, even entire countries pursue that edge.

Stephen Charters, a Master of Wine teaching at the Burgundy School of Business (his job title is Professor of Champagne Management), clarifies the inescapable intersection of *terroir* and marketing in a 2010 presentation delivered to a wine business research meeting in New Zealand.

FROM STEPHEN CHARTERS MW, "MARKETING TERROIR: A CONCEPTUAL APPROACH"

For many producers the ability to mark their wine out as different from all others because of its origin can be invaluable. This is perhaps most apparent in Burgundy, where, for the best wines (the *grands crus* and *premiers crus*), the *appellation contrôlée* (A.C.) is coterminous with the vineyard boundary; the vineyard, in turn may be owned by just a few owners, or even a single person—and in the 1930s terroir was explicitly developed in Burgundy as a marketing tool to offer this differentiation. The need to distinguish site from site has spawned many books and regularly inspires tastings to compare wines of different vineyards. It has also led to the multiplication of appellations which represent different, often minute, terroirs. Crucially, environment alone may be insufficient to explain the reputation of a wine, so trade, scarcity and demand converge to enhance its fame. This use of terroir as an aid to marketing may not be relevant for most wines but, it has been suggested, wine marketers are aware that the extra value they offer some consumers is in conveying this sense of place. Moran, more controversially, has also suggested that terroir as territory gives legal and administrative power to a site or region, which is in turn justified by (unproven) environmental determinants, particularly soil. In this argument terroir becomes merely a political and marketing tool, with no substantial viticultural validity.

The result of this is that terroir is used across the world as a justification for and endorsement of the quality of wine. Another interpretation, however, may be that when a wine, or an entire region, becomes successful the profits of that success are put back into maintaining and improving quality. In practice this may mean improving the processes used for production (better clones, more new barrels, the most up-to-date equipment). That in turn enhances quality—but the enhanced quality is credited to the quality of the vineyard and/or the region rather than better technology.

This overall interpretation can be envisaged graphically, as shown in Figure 1 (*Figure 8.1 here—eds.*). This is a development of the model offered by Vaudour who discussed the interlinking aspects of vine, identity, marketing ("slogan") and territory. In this example terroir is conceived as a tension between the physical (or environmental), the mystical sense of place, and its role in marketing—with the use of terroir to make distinctions the unifying factor. Vaudour's idea of territory as space thus becomes the core, and "slogan" becomes more widely interpreted as a general marketing dimension. Distinctions reflect diverse viticultural environments, they are used as a means of promoting the wine as one which stands out from all others and they secure identity.

DEVELOPING TERROIR IN MARKETING

It may be that terroir offers a specific link to the authentic. The notion of authenticity in wine is complex, and it seems that different consumers may search for it in different aspects of a product. Nevertheless, it seems that contemporary consumers do seek authenticity as a key attribute of their consumption, and terroir in wine, the fact that the drink is inextricably linked to a place and cannot be precisely replicated elsewhere, may be crucial for some. Thus the British wine critic, Hugh Johnson, has said "with wine, unlike most products, where it comes from is the whole point."

In a saturated market, finding a point of differentiation is essential, and it has been noted that terroir may offer that possibility, with wines of similar organoleptic characteristics showing price differentials in California of up to 50%. It is certainly the case that region of origin (overlapping with—but not identical to—terroir) offers competitive advantage, though in a complex fashion. Many would argue then that terroir is essential for this differentiation, but equally it can be suggested that

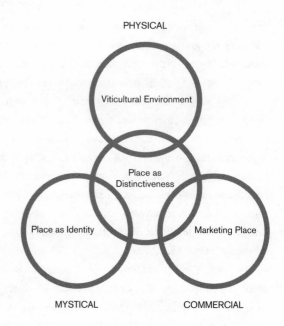

PHYSICAL

Viticultural Environment

Place as
Distinctiveness

Place as Identity

Marketing Place

MYSTICAL   COMMERCIAL

FIGURE 8.1
Interlinking aspects of vine,
identity, and marketing.

only high involvement consumers tend to have the interest and time to gain the kind of knowledge required to understand and separate terroirs.

Consequently, some in the wine industry consider that wines can be categorised into two types. In today's wine production a distinction should be made between "terroir wines" and "branded wines." Terroir wines are produced in a specified location. Branded wines are produced by blending wine or grapes from larger areas and from a variety of sources, which may vary from year to year. This reflects a wider wine industry view that somehow high quality wines are in some way not branded. As is evident from a number of critics, "branded" wines are somehow inferior to those displaying a particular terroir. The British wine writer Andrew Jefford has said brands "are for cereal and toilet paper, not wine. They underdeliver and rip the consumer off. They are more about distribution, marketing, and advertising than wine itself."

Ultimately these three aspects—differentiation, authenticity, and the contrast with "industrial" wine—may combine, as Maguire notes: "in general then, the terroir of a wine is a strategic device for adding and assessing value, as it brings into focus the tension between the large-scale, the industrialized and the artisanal, the inauthentic and the authentic."

HOW CLASSIC *TERROIRS* MADE THEIR NAMES

Many wine education classes and texts revolve around a region-by-region world *terroir* tour, dipping into Bordeaux and Burgundy, the Mosel and the Douro, Rioja and Chianti. In class after class, the usual regional suspects appear and re-appear. But how did they get to be on the list? What committee voted them into the wine Hall of Fame?

To achieve this sort of regional recognition, *terroir* requires consistency to thrive both as a geographical and marketable concept. Whether it's a wine region—a macro-*terroir*—or an individual estate, those that gain wide recognition among informed wine drinkers will produce a quality product, in good quantity, in a distinctive, recognizable style, year after year. But clearly there is more to it than that: some wine regions earn a reputation at the expense of another, and that reputation is based only partly on the quality of the region's wines. Some happen to be located near centers of transport and commerce, or near larger populations, which gives them a better shot at recognition than more remote areas. (Bordeaux's stunning success as a global brand certainly stems in part from its proximity to Atlantic ports—and to Great Britain.) Some wine-producing zones put enormous time, money, and effort into promoting their locales, while some just let their products speak for themselves. None of these factors directly affects how a wine tastes, but they certainly affect who tastes it, how often, and how those tastes are interpreted. The great *terroirs* of the world, in other words, provide more than just clues to flavor and texture: they provide *context,* a narrative framework for the flavors at hand. And behind every great *terroir* is a great deal of conscious, organized effort on its behalf, shaping its message while refining its geographical and philosophical scope.

THE MAKING OF BORDEAUX

The earliest official demarcation and ranking of vineyards came in the eighteenth century, with quality classifications developed for the Chianti region of Italy, the Tokay region of Hungary, and the Douro River Valley in Portugal. A hundred years later, the Bordeaux classification of 1855, identifying First-, Second-, Third-, Fourth-, and Fifth-Growths, laid down markers that still loom large in today's wine ratings and price structure.

The 1855 classification established a hierarchy in Bordeaux wines that is often mistaken for a ranking of vineyards, a sorting out of *terroir.* But in fact the classification had nothing to do with place, or vineyard locations, or *terroir* in the least; it was all about the money, as this selection from Master of Wine Benjamin Lewin explains in his incisive book on the subject, *What Price Bordeaux?*

FROM BENJAMIN LEWIN MW, *WHAT PRICE BORDEAUX?*
THE IMPORTANCE OF BEING CLASSIFIED

"INFIRM OF PURPOSE, GIVE ME THE DAGGER," said Lady Macbeth. Well Macbeth was positively resolute compared with the inconsistency of classification in Bordeaux. The lack of any significant hierarchy among appellations, together with a wide range of quality in most appellations, means that the appellation on the label conveys little direct information about quality. It's left to each appellation to decide how to classify its wines, and within the appellations, systems of classification vary widely.

At one extreme, Pomerol (the smallest appellation in Bordeaux) famously has no classification whatsoever, allowing Château Pétrus, selling at more than $1,000 per bottle, to have the same description as generic Pomerol from Moueix (the proprietor of Pétrus) selling at $25 per bottle. At the other extreme, the Médoc has five levels of Grand Cru Classés, three levels of Cru Bourgeois, and Cru Artisan. . . . With 9 different levels of classification there must be an incredibly detailed definition of land. Well, no, actually there is no classification of land at all involved in this system. It is all done on the price of the wine. And, in fact, the whole thing is based on an accident.

In 1855 the Emperor Napoleon III organized a Universal Exposition in Paris. This was intended to provide a showcase for French products (and not incidentally to outshine the Great Exhibition that had been held at the Crystal Palace in London in 1851). The Bordeaux Chamber of Commerce was invited to display the wines of the Départément of the Gironde as well as other regional products. The wines were to be submitted to a jury that would consider them for medals on the basis of tasting.

To accompany the display and to give some significance to the individual wines, the Chamber of Commerce commissioned a wine map of the Gironde. They asked the brokers who usually handled the wines in Bordeaux to provide a list of the leading châteaux, identified by class and commune, to accompany it.

Over the following two weeks, a committee of brokers drew up a list of 57 red wines (all from the Médoc except for Haut Brion) and a list of 21 Sauternes. The intention was not to classify the wines of the Médoc as such, but the quality of wine produced on the right bank was generally considered lower, and its producers were less well known to the brokers, who naturally enough concentrated on where their business came from: the left bank.

A great deal of guff is talked about the basis for the classification; exaggerated claims have been made about the procedures the brokers went through to assess each château. But the speed with which the classification was drawn up precludes any idea that extensive tastings were organized or visits arranged. The brokers cannot have had time to do more than consult their records before putting together a list representing received wisdom at the time. The main criterion for the classification lay strictly with the commercial basis of pricing in prior years, which the brokers would have been able to consult from their records. This lends little credence to fanciful comments suggesting that they made a detailed analysis of terroir or indeed took into account any factors other than price.

No one ever intended the classification to be more than a contemporary guide to display the wines at the Exposition. The brokers would have been enormously surprised at its subsequent longevity. But the system has stuck. So unlike other regions in France, classification of the top wines in Bordeaux is performed not in terms of the land but in terms of the producers.

■　■　■

THE MYTH OF THE CLASSIFICATION

The 1855 classification defies the usual reliance on terroir. The vineyards of a château today are not necessarily the same as those it held in 1855. Châteaux can (and often do) change their terroirs by trading land. Yet this does not affect their classification. The château is in effect a brand name, and the proprietor can change the terroir from which it is produced without affecting its classification. This is unique to the Médoc.

Many leading châteaux in Bordeaux now produce more than one wine from their vineyards, the Grand Vin being the château named in the classification, and the second wine being sold under a different name at a lower price. It's entirely within their discretion which vineyard plots are used for each wine, further reducing the connection between the terroir owned by the château and the wine it produces.

There is no particular reason to suppose that all châteaux are performing equally at any particular moment in time. No doubt it is true that the potential of each château is influenced by the quality of the vineyards that it holds at that moment, but some may be underperforming due to lack of interest or resources, while others possibly are over-performing due to larger investments or attention than the others. The snapshot of the châteaux' relative positions in the 1855 classification is no more than that: a freeze-frame of performance over the period leading up to the moment of classification. What is the basis for supposing that 1855 was a magical moment when the reputation of each château could be set in stone?

If great wines come only from great terroir, why should the classification remain valid where there have been changes in the quantity or quality of the vineyards? How can the persistence of a classification based on price in 1855 be reconciled with the view that terroir determines quality? Yet after 150 years the 1855 classification remains the only authorized hierarchy of the leading Médoc châteaux. Irrespective of its "official" or other status, it has undoubtedly proved to be one of the most effective marketing tools ever used for wine—or for that matter, anything else: how many marketing campaigns last 150 years? Never mind the quality, feel the width, as they used to say in the rag trade! ■

THE CHÂTEAU owners of Bordeaux didn't just produce a self-classification; they more or less invented a whole history for themselves. Anthropologist Robert Ulin has written extensively about French winegrowers; here, from a 1995 article in the journal *American Anthropologist,* he looks at how newly-enriched chateau owners raised their status—and their prices—by associating themselves and their properties with centuries of tradition, making their very human self-promotion seem like the inevitable outcome of excellent natural conditions for winegrowing.

FROM ROBERT ULIN, "INVENTION AND REPRESENTATION AS CULTURAL CAPITAL"

The wines produced in the immediate vicinity of Bordeaux in southwest France have long enjoyed a worldwide reputation for superior quality. This preeminent reputation is attributable not only to the special attention given to the vinification and aging of wines but most especially to a climate and soil regarded by most experts as ideal for winegrowing. However, if we look to the more distant past, the 15th century and earlier, we discover that Bordeaux wines were not always highly regarded and in fact were held in less esteem than those produced in the nearby interior. I argue that Bordeaux's ascendancy to its current paramount position follows conjointly from its political and economic history and from a more general process of "invention" that disguises what is social and cultural in "natural" attire. Bordeaux's particular winegrowing history thus illustrates the dialectical connection

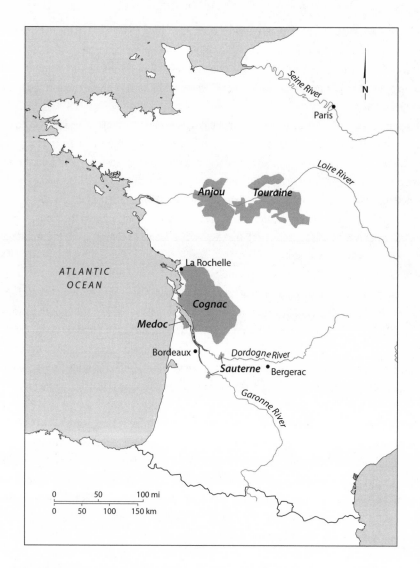

FIGURE 8.2
Map: Wine-growing regions of Southwest France.

between commodity production and invented culture or tradition. However, before we turn to the historical narrative of Bordeaux wines and winegrowers, it is imperative to address the theoretical and political implications of the invention theme, for I maintain that power differentials between classes of winegrowers significantly silenced all but the elite growers and merchants in the invention of a hierarchical and hegemonic winegrowing tradition.

The historical relation of Bordeaux wines to those produced in the nearby interior illustrates how socially produced differentiations can be constructed or invented as natural. This is especially the case since, with few exceptions, it is favorable natural conditions that are invoked retrospectively by winegrowing experts and elites to explain why interior wines have been repeatedly denied the prestige of the Bordeaux classification. For example, winegrowers of the Bergerac region, located 90 kilometers to the east of Bordeaux (see Figure 2 *[Figure 8.2 here—eds.]*), have sought several times in the last century to have their wines classified with those of Bordeaux, as they believed their wines to be equally good. However, in spite of the proximity of the two regions, the requests for a Bordeaux classification have always been denied to Bergerac winegrowers, ostensibly for reasons of quality attributable to a less favorable climate and soil.

THE CHÂTEAU AS INVENTION

While the development of French wines owes much to the early monasteries and estates of the Middle Ages, many aristocratic families lost their estates following the French Revolution as numerous aristocratic and church properties were seized by the new republican government, then divided and put up for sale. The powerful symbolic association of wine with the church and aristocracy nonetheless endured, as the production and consumption of wine was taken up by the nascent bourgeoisie and merchant families, many of whom purchased aristocratic titles.[1] Merchant families were the only ones with sufficient capital to afford the elite estates. Moreover, some of these estates were acquired inexpensively as they had fallen into ruin as a result of persistent economic crises. Many of these families built châteaus that were small-scale replicas of those constructed during the Middle Ages. The château was chosen as the architectural model because elite proprietors wished, as Roudié has noted, to distinguish themselves culturally from the masses by insisting on the ancient roots and quality of their wines. While the actual ties to aristocratic wines are dubious, given that the oldest plants rarely exceed 50 years, the nostalgia for an aristocratic past succeeded in carving out a symbolic place for the elite wines at the center of French civilization. This invented connection supplied elite growers with sufficient cultural capital to ensure their commercial success and to establish their role as leaders in the construction of French winegrowing discourse and knowledge

⁂   ⁂   ⁂

This is precisely what the elite winegrowers of the Médoc hoped to accomplish in building their homes as replicas of the celebrated châteaus of the Middle Ages. The expectation existed on the part of the elite growers that their wines would also become equated with the French national heritage, and thus recognized as a national treasure. That there were also commercial benefits to be reaped from this should be self-evident. However, by embodying the commercial benefits in a less strategic symbolic or cultural form, the message is much more subliminal and hence a good deal more seductive.

The association of elite wines with an aristocratic past builds upon and extends rhetorically, not to mention hegemonically, the discourse of wine as natural. That is, the value attributed to aged wines, older vine stocks, and aristocratic roots (metaphorically both age and depth) is represented in terms of a time that ostensibly passes naturally and therefore can be accounted for and measured objectively. Consequently, the positive value that is attributable to age is supported rhetorically through associating the superiority of age with nature or natural time. As is the case with unquestioned customs or habits, naturalized time as part of winegrowing discourse reinforces the privilege of elite wines and winegrowers through eclipsing the cultural mediation of time and hence the social constitution of the natural.

■   ■   ■

While it is true that elite estates invest more time and capital in the aging of wines, the association of quality with single domains or place is vastly overemphasized and serves principally to support the distinctiveness of elite wines and by extension elite French culture. This is especially apparent in elite regions such as the Médoc where it is not uncommon for small and medium growers, including those who are members of cooperatives, to have vineyards that border directly on those of the grands crus. This was confirmed by Madame Rambeau of the Paulliac cooperative, who explains why she would not sell her wine to the nearby renowned Château Latour, while expressing outrage over the difference in price between her wine and that of the chateau:

> There are two hectares of my property that come within 80 centimeters (two and a half feet) of château Latour. Chateau Latour has requested many times to buy this property. I've always said no. But what I ask those people, if some day I sell to château Latour, say they sell my wine. I sell my wine under Haut Pauillac. And overnight it will be sold under the label of château Latour, but not at the same price. When you see that it sells at 800 to 900 francs per bottle, then that my wine sells at an average of 35 francs. That is a large difference. Yes, the difference is the transformation of genre. Me, I do not understand how one can pay just for a label.

■   ■   ■

Although the proprietors of elite estates have sought to limit the proliferation of the château label through the judicial process, they have been largely unsuccessful with those producers who carefully follow the mandates of the law. Today, there are numerous smaller growers, merchants, and even cooperative members who have been able to claim a small part of the cultural capital associated with château wines. What this means, apart from the commercial interest, is that growers of all sorts have accepted the standards of quality and taste associated with the château wines. However, the acceptance is not resignation; these same growers are able through what Roger Keesing calls mimetic discourse to challenge and resist the very privilege to which they are subordinate.[2] Thus, in elite winegrowing regions such as the Médoc, virtually all the largest growers who are members of cooperatives commercialize their wines through the cooperatives but under château labels. This is accomplished legally by segregating their harvested grapes from the general lot of the cooperative and then pressing and vinifying the grapes separately. Moreover, cooperatives as a whole, even though they cannot commercialize the majority of their wines under the château label, have sought to emulate the standards of the elite wines by producing a greater volume of appellation contrôlée wine. The efforts from independents and cooperatives alike to emulate and thus reproduce the standards of the elite wines, combined with the not-to-be-underestimated influence of advertising, have done much to influence the consumption patterns of the general public.

THE STRUGGLE FOR THE SOUL OF BURGUNDY

The French AOC system (*Appellation d'Origine Controlée*), developed in the first decades of the twentieth century, brought together public officials and commercial winegrowers in an effort to define the boundaries, permitted grape varieties, and essential viticultural and winemaking practices of the important wine-producing regions of the country. The creation of dozens of AOCs was not just an exercise in applied geography; it had enormous consequences for the public face of French wine, its reputation, and its market potential.

Classifying the wines of Burgundy meant confronting a particularly acute tension between geographical boundaries and quality measures, played out as conflicts between small grower/winemakers, large estates, and even larger wine traders. Burgundy is a prime example of the high stakes and clashing vantage points involved in giving *terroir* formal definition. The battles between different interests and different conceptions stretching from the 1860s to the 1930s amounted to a struggle over the very meaning of *terroir,* a thoroughly human construct, according to French sociologists Serge Wolikow and Olivier Jacquet.

FROM SERGE WOLIKOW AND OLIVIER JACQUET, "A VICTORY OF THE UNIONS"

Using the example of Burgundy, and in particular vintage wines of the Côte d'Or, we plan to show in this article how terroir evolved from a literary concept into legal and commercial forms, before adopting its current meaning as a symbol of intrinsic quality connected to an area's soil. This crucial moment took place between the two World Wars, instigated by a handful of fine wine-makers in Burgundy who joined professional unions that were fighting to break away from the powerful merchants who at that time dominated the trade.

These wine-makers felt themselves to be far too dependent on the traders, and wanted to promote their own norms for the production and commercialisation of wine. In Burgundy, they drew up rules based on the notion of "Origin" and organised judicial hierarchies of production areas, confirmed in a 1935 government decree. From then on, any wine for sale had to be legally connected to a terroir that had been constructed economically, culturally, and socially, sometimes at great human cost.

The wine-makers' professional associations wanted to fight fraud (essentially fake denominations) and they wanted to use the law on Appellation d'Origine to establish clear geographical delimitations. The process of defining and protecting "good terroir" led to many angry confrontations between Burgundy's wine-makers, each estate owner wanting to impose his view of usage and to legitimise his particular definition of what constituted Burgundian terroir. They were fighting for different ideas of terroir—some legitimised by history, others by economic practice or based on each vineyard's internal professional reports.

Before the 1919 law establishing Appellation d'Origine, the certification granted to products from certain areas, Burgundy wine barrels were identified by the name of a village or area and by the trader's name. The business relied equally on the reputation of a handful of communes known to the buyers that were used as quality standards as on the trader's initials. Both the vinification process and commercialisation were in the hands of the trader, who staked his name and reputation on the wine he was selling. Yet some of these traders, taking advantage of the market deregulation prevalent at the time, would cut their wines or buy cheaper grapes from similar climates. Viticulture was in crisis and wine-makers were getting angry, engaging a number of lawsuits against merchants.

THE CONTROVERSIAL 1860 PLAN

To help the courts reach their decisions, the unions used some of the old literary, scientific and statistical definitions of vineyards, including an 1860 plan put together by Beaune's Agriculture and Viticulture Committee based on a study of the various communes. It classified production into first, second, and third vintages, and gave each parcel a commercial value. In the conflict over the delimitation of Corton that opposed the wine-makers of Ladoix and Pernand to the owners of Aloxe, only the latter evoked the plan. The head of the Ladoix-Corton union stated that the plan "has informative value, no more. It is often a useful reference, but not proof in itself." The small wine-makers of Ladoix referred to more recent commercial proof that traders had bought their wine under the name Corton. For the first time, terroir was acquiring commercial significance and was also expanding over larger areas.

The 1860 classification had been put together by doctors and educated men, and their view of the Côte was both cultural and geographic. The reference to this plan in the lawcourts and the fact that it was challenged showed that wine professionals were moving from a cultural representation of terroir to defining it as an object legitimising commercial practices.

The Beaune Committee's approach contradicted the more homogenous view held by some of the fine wine Côte de Nuits estate owners. The owners of the vineyards of Richebourg, la Tâche, la Romanée Saint-Vivant and Clos Vougeot did not want their "clos" to be divided into qualitative categories. The Beaune Committee, on the other hand, felt that wines should be assessed for their quality, regardless of the estate owners' views. This particular conflict was won by the influential estate owners, and the ruling allowed the fine wines they produced to maintain their qualitative homogeneity.

It is very complicated to define terroir, a notion based on a specific place, as well as cultural elements, a particular fine wine's reputation and the idea that an estate defines the cohesion of a top vintage.

TWO RADICALLY DIFFERENT POINTS OF VIEW

The 1919 law establishing Appellation d'Origine Contrôlée (AOC) created a number of problems. For reasons of economy and to increase the flavour of their wines, traders imported grapes from southern France and Algeria. The wine-growers saw this as unfair competition.

The conflict between trader Charles Bouchard of Maison Bouchard Aîné et Fils and the Marquis d'Angerville, owner of the fine wines at Volnay and president of the union for the Defence of Bourguignon Viticulture, is a perfect illustration of what divided those who defended a trade name from those who defended an appellation. Among many important social positions, Bouchard was president of Beaune's Chamber of Commerce.

D'Angerville was defending the idea of strict geographical delimitations, while Bouchard was fighting for trade names. During the 1932 trial, d'Angerville taxed Maison Bouchard of fraud over their Appellation d'Origine. The trial encapsulated the battle between two antagonistic reactions to the crisis in the wine trade, two ways of interpreting and demonstrating old usages, two views of the past. D'Angerville and the union for the defence of viticulture were to win. They had the better lawyers.

But there was no unity in this battle for the elaboration of norms, in this march towards appellations. Even the winemakers unions were in conflict. Each and every person had a view of terroir that was more or less restricted in the area it covered, and had different quality standards. The idea of creating delimitations for every climate was causing further division.

DISADVANTAGED VILLAGES FIGHT BACK

The 1860 plan could be read both vertically and horizontally. Vertically, viticultural zones could be organised into hierarchies according to climates strictly confined to villages. Read horizontally, the colours on a map indicating the quality of fine wines covered several communes and established large inter-communal areas of equivalent wine quality. A white wine from Monthélie, for instance, obtained from second vintage parcels, could be named after the more prestigious Meursault because the borders between villages were less important than the colour in the classification. This was called the practice of equivalencies.

Some wine-growing communes represented by their mayors and unions didn't want to lose that system's commercial added value. If their vineyards were little known, they wanted their wines to be named after the Côte "flag-carrier" communes. Union alliances developed between these so-called "disadvantaged" villages and other villages disappointed by the delimitation system.

The unions and the men running them were to grow increasingly aware of a network of national influences that would allow them to weigh in on parliamentary and ministerial decisions and become major economic players. Today, we are faced with an interesting paradox: how does one appropriate national judicial norms via local stakes? How does one define one's own collective delimitation rules, one's own conception of terroirs, when the practice of wine-making in France is so fragmented?

THE POWER OF COLLECTIVE ACTION

As an example of effective union action, let us focus on the Syndicat de Défense des Producteurs de Grands Vins Fins de la Côte d'Or founded in 1928 by the Marquis d'Angerville. This union made up of almost every single union and fine wine cooperative in the Côte d'Or played a decisive role in the elaboration of the Burgundy AOC. Despite its relatively light economic weight at national level, the union represented vineyards that had the same prestige outside France as the châteaux of Bordeaux.

D'Angerville represented the vineyards, backed by a large network of unions, politicians, wine-growing associations, and administrations. He also knew a lot of local politicians, corresponding with them across political divides. Opposite him stood the Confédération des Associations Viticoles de Bourgogne (CGAVB) who wanted a special status for Burgundy and counted among their members the "disadvantaged" villages of the Côte Dijonnaise and the Hautes Côtes. They too used political influence, and counted on the sheer number of their members across Burgundy. This complex network of commitments and power struggles was really the quest for cultural legitimacy that was then considered indispensable for the construction and normalisation of wine-growing areas.

THE PUBLIC FACE

The Marquis d'Angerville was well aware that the 20th century was the era of communication and he used this in his battle for the AOC in Burgundy. He made sure that the fraud trials instigated by his union made their way into the local and national papers. He and his friends wrote for the *Revue du Vin de France,* a magazine launched in 1925. But he also undertook to educate foreign buyers. For the business to pick up again in France and beyond, the consumer needed to be reassured about the quality of the wine on the market. As of 1934, d'Angerville concentrated his efforts on the United States where prohibition had just been abolished. The writings of US writers Tom Marvel and Julian Street explaining what the French winegrowers were doing to fight fraud were the result of an abundant correspondence with d'Angerville.

D'Angerville's trump card in his defence of AOC was his invention, with some 13 other wine-lovers, of the famous Académie des Vins de France. The undertaking was both cultural and commercial. They held their first meeting in 1934, bringing together winegrowers, doctors, journalists, and gastronomists. The Academy talked about wines and culture, health benefits, and the idea that the subject was worthy of articles and criticism.

In 1935, the State took over the norm, introducing new administrative and political legitimacies into the mixed system of AOC. During World War Two and for 30 years after, new ways of identifying the best zones for producing wine were introduced. These included rules about production methods from the culture of the vine to vinification. The connection tightened between the place of production and the nature of the product.

At the same time, the developing status of terroir reflected a general feeling of anxiety about food safety born in the days of an industrialised food industry. This "rediscovery" of terroir products, associated in people's minds with a control of production methods, also corresponded with a time when many people no longer wanted to drink table wine. The concept of terroir as guarantor of quality was a way of finding new consumers.

The last century has seen major technical changes, new modes of consumption, and evolutions in the market. Terroir was never a "natural" notion, always a social construction historically determined by such factors as the intervention of the State and professional bodies. Terroir is more closely connected to the history of laws than to nature.

CONCLUSION

The modern construction of terroir—judicial, economic, political, and social—is particularly interesting when looked at from the viewpoint of union action. The wine-makers' unions have clearly been central in building production and commercialisation norms for wine, and in the establishment of Appellations d'Origine. These unions were at their most active in times of economic crisis, stimulated by their confrontations with the other main actors in the trade—the wine-sellers.

A number of perceptions of terroir emerged—terroir associated with a brand, a broad terroir of equivalences, micro-delimited terroir. The latter has become law, and is defended by wine-makers involved in increasingly dense and efficient networks. In the Côte d'Or, the problem of delimitations involves the confrontation between the complexity of terroirs and the many views held by the professional organisations set up to defend wine-growing regions.

Among its many definitions, terroir is also about authenticity: If terroir wines are good, as wine-growing marketing tells us, it's because of the belief in a quasi-divine production area. When people approach it this way, they ignore such historical factors as the union activity of the inter-war years. Of course, folklore and national identity also affect our current views of terroir. But the legal and commercial notion of "origin" is fundamental to a definition of terroir because it is the way usages connected to the soil have been fixed.

Terroir is not a natural phenomenon that has been improved by unceasing human activity. Terroir is a historic construct, an object forever redefined by a tumultuous history. It is also the fruit of the unavoidable construction of norms without which no market can function.

GIVING DEFINITION TO NEW *TERROIRS*

The classic *terroirs* of Europe have been recognized for so many centuries that they seem entirely natural, predestined for greatness as much by tradition as by their inherent physical characteristics. In its most reactionary form, the notion that terroir is an Old World phenomenon runs something like this: "*Terroir* is inexplicable; we have it, you don't; therefore our wines are better." In fact, the great European *terroirs* and their reputations were constructed by human activity, the interplay of growers, winemakers, suppliers, merchants, regulators, consumers, and marketeers, and there is no rational reason the process could not be repeated elsewhere.

To capture the dynamics of this interplay, the late Sue Callahan of the University of Otago in New Zealand coined the term Interconnected Spatial System (ISS), which she defined as "blending physical, technical, and socially constructed elements in specific regions, generating place branding." It is how traditional *terroirs* earned their spurs, and Callahan was optimistic that New World regions can do the same.

FROM SUE CALLAHAN, "THE ILLUSIVE MATTER OF TERROIR: CAN IT BE DUPLICATED IN THE NEW WORLD?"

An incessant contention about terroir is the possibility of its duplication in the New World. An Old World viewpoint in the terroir controversy is that the New World wines have none and, in fact, have tarnished or even destroyed the very idea of "haut" wine. However, this is countered by New World producers in that they recognized and responded to changing consumer tastes. Through marketing research, producers in California, Africa, Chile, Australia, and New Zealand discovered that consumers wanted a full fruit, sweeter flavoured wine that was high quality, reasonably priced, and dependable (consistency in brand). Aylward indicates that the AOC standards, particularly restrictions on variety of grapes and viticulture methods, may be constrictive and display a producer's rather than a customer's viewpoint. The New World appears to employ a solid marketing orientation.

That leaves the dilemma of tradition and motivation. Can terroir be duplicated anywhere outside of France based on AOC standards and the metaphysical elements involving the people of a particular region? Several New World wine production areas have regulations attempting to replicate AOC standards, but fall quite short. Some New World regions claiming terroir apparently rely on the ISS concept to develop and maintain terroir. Laurence Bérard, a scientist of anthropology and ethnobiology at the Centre National de la Recherche Scientific in France, actually avoids natural factors in the definition of terroir and applies it to a variety of agricultural products, not just wine. He emphasizes that terroir products are derived from *shared knowledge and know-how* within the geographical area which may not be connected to nature at all. He does think that there will be a small percentage of the population who will continue to care about the origin of products. He hinted that perhaps other "social vectors" could re-create the idea of terroir, naming them "revival phenomena." Perhaps the *Interconnected Spatial System* with wineries striving for place branding is such a schema to accomplish the genesis of terroir.

Clearly the term *terroir* has many curves and facets to it. The physical process of viticulture and winemaking involving the soil, climate, and topography is well documented. Some proclaim terroir does not exist at all, others believe that it does, but that New World wine producers cannot possibly attain it. In reality, there are emotional and marketing stakes in the argument. If it can be duplicated in the New World, it dilutes the Old World's renowned position in premium wine categories while enhancing the prestige of New World wines. Marketing plays a weighty role in the terroir controversy. The link between terroir and *place* has been well-established. A particular region claiming terroir and using it in their external marketing communication could influence consumers or perhaps wine critics. However, terroir is defined ultimately by the customer, so marketing will have to continually monitor changing consumer tastes, as perhaps the Old World has not done. The controversy of terroir seemingly has no immediate end and is bound to have many altercations ahead.

This author postulates that one means of the New World conceiving terroir is through an Interconnected Spatial System of wine-producing competitors, buyers, suppliers, and other institutions working together in a geographical region. Long-standing research on such ISSs has determined that through frequent and informal interaction creating relationships, knowledge pertinent to the region is generated and exchanged. The recommendation for future research is to determine how interaction develops into networked relationships among constituents in a wine region. Secondly, how do constituents in a region work together to project a place brand? Additionally, there are the inquiries of how terroir is achieved and what is the link between terroir and place branding? This research will contribute to marketing by supplying a better understanding of inter-firm collaboration and interaction within networks of relationships, the process of knowledge transfer, the development of terroir, and place branding. Managers could also benefit by gaining insight into this seemingly paradoxical situation of competitors sharing information to promote place branding resulting in global marketing success. ■

MARKETING A regional macro-*terroir*—and by extension, promoting individual producers and vineyards within it—requires mapping the boundaries of the place and making a persuasive case for what's special about the grapes and wines from inside those lines. Starting with the French AOC system developed early in the twentieth century, the entire wine world, Old and New, has adopted procedures through which would-be appellations make their case to one form of regulatory body or another.

In the US, the system of American Viticultural Areas (AVAs) was instituted in 1980 by the Treasury Department to recognize winegrowing areas with common geographical features. The broad requirements for qualifying as an AVA are these:

1. Evidence that the name of the proposed new AVA is locally or nationally known as referring to the area;

2. Evidence, historical or current, that the boundaries are legitimate;

3. Evidence that growing conditions such as climate, soil, elevation, and physical features are distinctive. (Existing legal jurisdictions—states, counties—are also eligible for automatic AVA status.)

In contrast to most European appellation systems, the US AVA definitions do not include regulation on permitted grape varieties, yield targets, or specific viticultural or winemaking practices.

As we'll see below, some of the 250 or so American AVAs are reflections of the political muscle of important wineries, while some are carefully researched and scientifically-based attempts at capturing differences in physical environments. An example of the latter is the case made for creating seven sub-appellations within the larger Lodi AVA, at the northern end of California's Central Valley. The move to define sub-appellations came as part of a long-term campaign to raise the visibility and reputation of the Lodi area, long considered a source for value-priced grapes but not for high-end premium wines. The selection below comes from an account written by Deborah Elliott-Fisk, a geographer at the University of California at Davis, who served as lead scientist coordinating development of the successful AVA petition.

FROM DEBORAH ELLIOTT-FISK, "GEOGRAPHY AND THE AMERICAN VITICULTURAL PROCESS, INCLUDING A CASE STUDY OF LODI, CALIFORNIA"

AN EXAMPLE: LODI, CALIFORNIA AND SEVEN NEW AVAS WITHIN IT

In 2001, a small group of winegrowers and winemakers in Lodi, California, approached the author to gain her insights and expertise on establishing a new AVA within the large Lodi AVA around the town of Clements in the lower foothills to the east of Lodi. In meetings to review a series of maps and the process, then visiting various vineyards in the field, it became apparent that this group believed that there were several distinct areas within the region that could have appellation status. We undertook research across the region and decided to pursue an ambitious, but cost-effective, effort to petition TTB for seven new AVAs. A project like this had never been presented to TTB before, instead with "sub-AVAs" being submitted one by one, usually by different parties, across a decade or more.

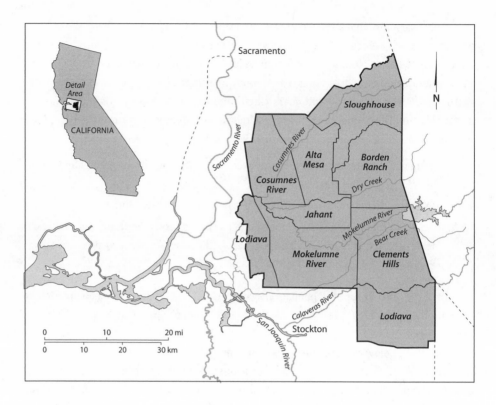

FIGURE 8.3
Map: Lodi Viticultural Areas.

    Thus, in 2002, after more than a year of discourse with the public in open meetings and much thought, the Lodi American Viticultural Areas Steering Committee, where the author was the lead scientist, submitted a petition, proposing the designation of seven regions as new American Viticultural Areas. They are fully within and inclusive of the entire established Lodi American Viticultural Area, with no overlapping boundaries or omissions. These AVAs are found on the map below, and fall within both Sacramento and San Joaquin Counties, California, as does the Lodi AVA.

■   ■   ■

Lodi is a region with a very rich agricultural history. Wine grapes have been grown, and wine produced here for well over 150 years. Many changes have occurred through time in the grape varietals planted, the style of wines made, and in both viticultural and winemaking practices. Through decades of farming, knowledge of site potential and the diversity of sites within the Lodi AVA have grown, especially since the Lodi AVA was established in 1986 and with the organization of the Lodi-Woodbridge Wine Grape Commission in 1991. Winemakers across California had recognized the Lodi appellation as an excellent source of high-quality fruit for wines carrying the "California" label, for Lodi AVA wines, and/or select vineyard designates. Definition of smaller viticultural areas with geographically distinctive land forms, soils, climates, and topographies contributing to *terroir* would provide more accurate information to consumers on grape origin, which we know has long been recognized in both the literature and in tasting as a key control of the characteristics of wine.

It was the LAVA Steering Committee's belief that the designation of smaller, geographically distinct viticultural areas within the large Lodi AVA would:

1. Further promote the grapes and wines of the Lodi viticultural area, the new and smaller viticultural areas within it, and the efforts of all groups.

2. Provide a better understanding of the viticultural geography of the Lodi appellation and the new AVAs, aiding winegrowers and winemakers in their selection of vineyard sites, grapes, viticultural practices, and winemaking techniques.

3. Provide the grape buyer, consumer, and media with more information as to the diversity and distinctiveness of the viticultural areas, their grapes, and wines.

4. Follow the successful history and approach of other viticultural regions, e.g., Sonoma Coast, Napa Valley, Sierra Foothills, in defining and promoting small viticultural areas that produce distinctive fruit and wines.

This vision was sold to the public, other winegrowers who were not part of the LAVA Steering Committee, and the Lodi-Woodbridge Wine Grape Commission, who were part of the discussions but not part of the LAVA Steering Committee, as the Commission represented a very large constituency of winegrowers.

Lodi is, from a physical geographic perspective, a fascinating part of California. With the low relief of the landscape in this valley floor setting, across distances exceeding 25 miles at the vista, the viticultural diversity of the region is best recognized by the winegrowers themselves who are cognizant of important differences in local climates, surface soils, and subsurface sediments. They recognize that the differences strongly influence the performance of rootstocks, clones, yields, and fruit characteristics.

Scale is important here. What appear to be very subtle environmental changes across long gradients to the casual observer are important, local, or site specific conditions for wine grapes. It is this diversity and distinctiveness that contributes unique character to Lodi wines across the region and the several unique viticultural regions which are now being showcased to increase the reputation and consumer recognition of the area.

Like the Napa Valley floor, the Lodi region is composed of a set of geologically old to young alluvial fans with the incursion of bay/delta deposits on its western fringe and the underlying bedrock of the Sierra Nevada to the northeast, with distinct climatic gradients across the region. Soils, as a function of geology, climate, and time, are a strong influence on vineyard site potential. The alluvial geomorphic surfaces in the Lodi region are much larger than those for the Napa Valley, Alexander Valley, or other valleys of the Coast Ranges of California due to the sheer size of the Sierra Nevada mountain range and its large watersheds from which the alluvial fans issue. The surfaces represent old Pleistocene, glacial age alluvial fans, and river terraces with the older soils on higher river terraces and fan sand buried at depth by more recent deposits, especially to the west.

Geographically, the Lodi area is characterized by a cooler, maritime climate with smaller daily and monthly temperature ranges than to the north (Sacramento), the south (Stockton), or east (Camp Pardee east of Clements Hills and Folsom Dam east of Sloughhouse). This has long been recognized by botanists and agronomists as relevant to both the native vegetation of the region and the regional potential for agriculture. With the passage of winds off the Pacific Ocean through the Golden Gate topographic gap which continues up the Carquinez Straits, these coastal winds encounter the landmass at Lodi—effectively, the toe of the Mokelumne River alluvial fan—and decrease in intensity (speed) and duration (time) toward the east due to friction with the ground as elevations increase. With low temperatures and elevations, precipitation in the Lodi AVA is the least in the lower Mokelumne River near the town of Lodi, e.g., in the southwest corner of the Lodi AVA and the proposed Mokelumne River and Jahant AVAs, with precipitation increasing upslope, in particular from southwest to northeast, associated with the increase in elevation, thereby creating a gradient in local climates. The climate of the entire Lodi AVA using the global-scale climate classification system of Koppen is Mediterranean warm summer (Csa), although the town of Lodi itself is almost a Mediterranean cool summer climate (Csb). [Note: Climate classification systems are really based on large geographical regions for the globe, similar to the classification of soils to orders for the globe.] Climates are not classified to the fine level, e.g., series, as are soils.

Distinct climatic gradients exist across the large Lodi AVA. The combination of climatological parameters, e.g., temperature, precipitation, wind, and evaporation, provides the basis for distinctive local or topographic climates that characterize each of our proposed seven AVAs. As shown by using the standard climate parameters of temperature, precipitation, growing degree days, actual evaporation, and wind, and as depicted graphically from the City of Lodi both inland to the east and north toward Sacramento, the annual precipitation increases; the mean annual temperature increases; the mean seasonal temperatures for winter, spring, summer, and fall all increase; the growing degree days over a base of 50°F, the standard used in viticulture, increase, yet evaporation and wind duration/speed decrease. What these data show is that the city of Lodi and the Mokelumne River AVA is cooler and windier than the areas to the north—Cosumnes River AVA, Alta Mesa AVA, Sloughhouse AVA, and Sacramento and Folsom—and to the east—Borden Ranch AVA, Clements Hills AVA, Camp Pardee, and the Sierra Foothills AVA. The climate of the Jahant AVA is similar to that of the Mokelumne River AVA where the city of Lodi is located.

The region is distinguished geologically by its tectonic history and complex of distinctive alluvial fans, with the current surface landforms spanning the last 1 Ma (million years). The region links the Sierra Nevada and the valley floor, with isostatic and tectonic processes leading to continued deposition by rivers to the south of the current river channels. The rivers themselves serve as good boundaries for viticultural areas. It is also evident that the subsurface environment for the vines is even more diverse than the surface climates, with this "terroir" significant to grape composition and wine flavors and aromas. Although some soils are very old as either surface or buried paleosols that formed in the past under different climates, such as the San Joaquin series, most soil profiles are largely immature, as the active erosion and alluviation (deposit) by streams has controlled soil profile development.

Diagnostic characteristics for soils in the region which attest to their various ages are (a) a dark mollic epipedon at the surface, with organic material accumulating and decomposing through time, (b) cambic and argillic B horizons where clays have formed as secondary minerals with pedogenesis and accumulated over time, eventually leading to the formation of a claypan, and (c) the formation of duripans, with soil horizons cemented by silica through weathering of granitic minerals and volcanic ash. Pedogenic accumulation of clays and reddening, both denoting leaching and time, are seen in the San Joaquin and Redding series where land surfaces are more stable.

Soil pH is near neutral for most soils, and there are no strong chemical imbalances in the soils except for the very oldest soils on the eastern margin. Viticultural challenges are mostly due to impervious subsoils and in places, to very fine, sandy, permeable surface horizons. The Storie Index, which is used to rate soil potential for intensive agricultural uses based solely on the soil characteristics (from 100 highly suitable to 0 unsuitable), shows the diversity of soil types for viticulture in the Lodi AVA, with the Tokay/Hanford soils in the proposed Mokelumne River AVA near the town of Lodi having the highest rating (80–95) and being the location where grapes were first grown. The San Joaquin series and Jahant clay loams and loams are next with ratings around 25–40—Cosumnes River, Alta Mesa, and parts of the Sloughhouse and Jahant AVAs—and then, the other more upland soils of the Sloughhouse, Borden Ranch, and Clements Hills AVA are rated as moderate to low suitability, 15–30, which often means they are excellent for contemporary viticulture where reduced yields often lead to enhanced quality.

From both geographical and viticultural viewpoints, it is thus apparent that there are large differences seen by winegrowers from north to south and west to east across the 458,000-acre Lodi viticultural area, from the deep, sandy loam soils around cool Lodi, to the rocky, clay rich soils in the sunny hills above Clements, to the moderate sites on clay loam soils north near Galt. This has resulted in new varietals being planted in the last several years by subregion, for example, more white varietals in the cooler, western areas and red varietals in the warmer, eastern areas, careful rootstock selection by site/soil, new canopy management, and trellising techniques, and all the other contemporary cultural practices that are dependent on the terroir-varietal interaction. It should also be noted here that the Lodi growers are well recognized in California and elsewhere for their Lodi Rules for Sustainable Winegrowing program, under the leadership of the Lodi-Woodbridge Wine Grape Commission (www.lodiwine.com).

The 458,000-acre Lodi American Viticultural Area is characterized by:

1. A relatively maritime climate for its inland location in the Central Valley.
2. By alluvial valley floor soils derived from Sierra Nevada glaciofluvial and river deposits.
3. By a long history of grape-growing and wine production.

This large region has considerable geographical and historical diversity, supporting the establishment of seven new American Viticultural Areas. These proposed new AVAs are the:

1. Cosumnes River
2. Alta Mesa

3. Sloughhouse

4. Borden Ranch

5. Jahant

6. Clements Hills

7. Mokelumne River

The names of these AVAs are derived from well-recognized, locally significant, geographic place names and historical family names as reflected in the communities, roads, and ranches of the region. The elevations, climate, as reflected by growing degree-days, temperature seasonality, annual precipitation, duration of fog, persistence of wind, and evapotranspiration demands, soil capabilities for viticulture, geomorphic surface type and age, soil types, and viticultural conditions are quite diverse. The framework of fully subdividing, with no exclusions or omissions, the Lodi AVA into these distinctive viticultural areas aids consumers, winemakers, and viticulturalists in their pursuit of crafting and enjoying fine wines from this historic grape-growing region of California.

THE RISE OF THE NAPA VALLEY—A CASE STUDY

Of all the wine regions of the New World, California's Napa Valley has staked the biggest and most successful claim to top-tier *terroir* status, worthy of mention in the same breath as Bordeaux, Burgundy, and the like. While the great regional *terroirs* of Europe developed their reputations and market shares over hundreds of years, Napa went from jug wine to cult status in little more than half a century, all of it after World War II.

Napa's astonishingly rapid rise was made possible by the easy availability of modern winegrowing and winemaking knowledge and technologies—a leg up those old Burgundian monks didn't have. But as everyone in the California wine industry knows, the identification of Napa with world-class wines was also fueled by world-class marketing efforts.

No one would deny that Napa producers make some great wine, and that some of it certainly does reflect its vineyard origins in striking ways. But it is also true that for decades, the vintners of Napa Valley have put an extraordinary amount of time, money, and creativity into promoting the Napa brand, sponsoring endless conferences and press tours, building up a huge tourist infrastructure, and using every communication channel available to link Napa wine with elegance, luxury, and the ultimate lifestyle. And almost from the beginning, the story told about Napa has centered not just on its terrific wine, but on the special characteristics of the place—in other words, on its *terroir.*

What follows is not a formal case study, but it does chronicle several episodes in Napa's meteoric ascent. We start with a selection from *Bottled Poetry,* an account of the development of the Napa wine industry by UC Davis wine historian James Lapsley. This selection comes from a chapter titled "Politics and Promotion," which focuses on the groundwork laid by its powerful marketing body, the Napa Valley Vintners' Association, in the 1950s and 1960s.

Following World War II, the Napa Valley Vintners' Association engaged in various forms of collective action to build awareness of Napa Valley wines, and its early efforts provided a model for larger-scale industry activity later in the decade. In the minds of some quality producers, the Wine Advisory Board's advertisements for "California" wine had backfired, since consumers had "been disappointed with some wines carrying the name 'California,' and had then 'erroneously assumed that all California wines are inferior.'" The goal of the NVVA was to distinguish the Napa Valley from other wine-growing areas by reiteration of the name "Napa Valley" and by promoting interaction with potential wine consumers. To that end, in 1949 the association erected a sign welcoming visitors to "this world famous wine growing region," and sponsored a cable car in San Francisco, which included giving away a bottle of Napa wine each day to "one lucky passenger," a practice that brought "considerable publicity . . . as far away as Fairbanks, Alaska, New Orleans, and Boston."[3]

The association also aggressively sought out influential groups visiting San Francisco and invited them to the Napa Valley. Most notable was its hosting of the Associated Harvard Clubs when they met in San Francisco in fall 1949. The association arranged transportation to the Napa Valley and catered lunch for six hundred Harvard graduates and their guests "at the cost of several thousand dollars." As *Wines and Vines* editorialized, it had done so in the belief that "close contact with the wineries" would "build future demand for the wines made there." The editorial concluded by commenting that eventually with such cooperative action *"where* the wine was made may achieve an equal importance with *who* made it."[4]

The association continued these personal-relationship-building activities throughout the 1950s and 1960s. Francis Gould, the editor of the Charles Krug winery's newsletter, *Bottles and Bins,* referred to the "large assemblages" consisting "of delegates to National and International conventions meeting in San Francisco" hosted by the NVVA, and *Wines and Vines* listed such groups as six hundred members of the American Bar Association and the thousand-strong Building Owners' and Managers' Association as visiting the Napa Valley as guests of the NVVA.[5] Although costly and time-consuming, such promotions helped to build a following for Napa premium wine.

The NVVA's pursuit of public relations extended to the media as well. Writers like Frank Schoonmaker were courted. In the case of Schoonmaker, this paid off when he published an article entitled "California's Vintage Vale" in *Holiday* magazine in 1952. "The Napa Valley is in a class by itself," the article began, and Schoonmaker wondered "why anyone who had ever seen [it] ever lived anyplace else." The piece was one of the first "lifestyle" articles written about Napa, concentrating as much on the gracious living possible there as it did on discussing Napa winemakers.[6]

Other examples of media relations were more prosaic, but helped lodge the name "Napa Valley" in the minds of national audiences. In 1955, the NVVA hosted Leonard Wibberly, author of a popular *Saturday Evening Post* serial set in the mythical European duchy of Grand Fenwick. In one segment, the duchy declared war on the United States over a Napa Valley wine labeled "Pinot Grand Enwick." An exchange of letters to the editor and an invitation to the author kept the appellation "Napa Valley" before the *Post's* readers as long as possible.[7] Another subtle example of the NVVA's public relations was the sponsorship in 1958 by individual members of the association of *This Earth Is Mine,* a motion picture set in the Napa Valley. Starring Rock Hudson and Jean Simmons, and loosely adapted from Alice Tisdale's novel, *The Cup and Sword,* based on the life of Georges de Latour, the movie clearly identified the Napa Valley as California's leader in premium wine production.[8]

Such creative uses of public relations represented a new and indirect approach to image-building, in which potential consumers were "surrounded," either physically by visiting the Napa Valley or psychologically through strategically placed articles or pictures calculated to influence perception and behavior. These techniques contrasted with the more static advertisements of the Wine Advisory Board and later became the basis for a public relations campaign sponsored by the premium wineries of California.[9] ∎

WITHIN THE common cause of promoting the Valley as a whole, individual Napa wineries found ways to tell their own stories, as did nearby regions like Sonoma. Increasingly, those stories contained references to this or that element of *terroir,* even if the word wasn't always used. The essay below, from a collection on the global wine business, describes campaigns by two California wineries, Chalk Hill in Sonoma and Hanzell in Napa, to raise the profiles of their brands by creating *terroir*-like associations with their brands. (Napa is hardly the only California or New World wine region to pursue this tack; Napa has simply been more successful than most.) The authors of the essay include Jean Arnold, owner of Hanzell, along with wine marketing educators Linda Nowak and Paul Wagner.

FROM LINDA NOWAK, PAUL WAGNER, AND JEAN ARNOLD, "MARKETING AND BRANDING WINE"

In 1996, Chalk Hill was a little known grocery store brand barely making enough revenues to keep the winery going. By 1998, they had repositioned themselves to be one of the top US Chardonnay producers, with a waitlist of customers wanting to buy their wine at over US$40 per bottle. How was this achieved? How was Chalk Hill Winery able to change the "brand image" of their wine in such a short time?

The answer lies in their decision to make the commitment and investment to realize a vision for a high-end Chardonnay. Working with a team of wine-marketing consultants, the owners and winemaker went through a process of identifying what was truly unique about their wine. What was the story behind the wine? How was it different from any other Chardonnay in the world? The answer came from the land—the special chalky soil in which their Chardonnay grapes grew was like no other soil in the world. It was a unique soil, from a very special appellation, Chalk Hill, in Sonoma County, California. Furthermore, the vineyards were situated in rolling hills, with 300 acres planted. The combination of all of these factors, plus a small production of cases, grew into the brand image of Chalk Hill.

Once this special, authentic image was realized, then the brand-building process began. A second label sold by the winery, which was eclipsing the Chalk Hill label, was pulled from the market. The Chalk Hill wine label was then updated to reflect the new brand position and story. Wine marketers and a public relations team then put together a communication strategy to describe the "story" to high-end distributors, retailers, and wine writers. The story of the special chalky soil and hillside vineyards was so appealing that high-end restaurants around the nation began to purchase the wine. Sommeliers and wine waiters told the story again and again to customers. Customers then repeated the story to friends, and soon Chalk Hill Chardonnay became a top-selling brand with a waitlist of customers trying to purchase it.

A second successful brand story is the one behind Hanzell Vineyards. Once a little known family winery in the foothills of the Mayacamas, today Hanzell is a prestigious, luxury wine that is sold only by allotment to high-end wine shops, fine wine restaurants, and an elite group of wine club members for US$65 per bottle and higher. Again, this was achieved by an in-depth focus and research into what made Hanzell Vineyards unique. What was the story that set them apart from all other vineyards and wineries in the world?

Similar to Chalk Hill, Hanzell made the investment in developing a world-class wine brand by hiring experts to help them "articulate their story." Fortunately, they didn't have to look far, because the quality of the wine was already there—and had been for more than 50 years. The issue, at the time, was that the winery was not that well known.

Hanzell's unique story is the "grand cru" farming standards that they use. Focusing only on the Burgundian varietals of Chardonnay and Pinot Noir, they use sustainable farming techniques, which are environmentally friendly to the special mountain vineyards of their Mayacamas, Sonoma Valley estate. With only 42 acres of their 200-acre estate planted, they use the same farming techniques as the top 2% of Grand Cru Burgundy. This includes stressed growing conditions and rigorous pruning to intensify the flavors of the wine, which results in very low yields—less than 3 tons per acre—and only 3000 cases produced per year. This creates a very rare, scarce, and exclusive wine that is sought after by wine connoisseurs around the world.

Wine marketing methods to support both brands include using only the finest materials for their embossed and engraved labels; expensive parchment letterhead and brochures; and bottles hand-polished and wrapped in tissue. Every detail of the grape-growing, winemaking, and wine-marketing process receives the utmost care and attention. Likewise, Hanzell Vineyards targets only the top 20% of wine writers, distributors, and retailers, who are invited to special educational events at the winery and other locations. During these meetings, the special story of Hanzell Vineyards is communicated in a relaxing and intimate environment to a very elite and influential group of buyers who will continue to "tell the story" to their customers. Thus, the brand grows successfully, with grace and exclusivity.

These two examples of wine brands illustrate the importance of understanding the uniqueness of the wine's story and then attempting to develop a complete brand strategy to communicate consistently the image and position of the wine. This takes much time, talent, dedication, and vision, but in the end it can pay off handsomely in customer loyalty and positive cash flow. ∎

FOR SEVEN decades, the Napa Valley Vintner's Association has represented Napa producers. Starting in 2001, the NVVA supported three major research and analysis efforts aimed at documenting the physical characteristics—soil and climate—of the region: *The*

Foundations of Wine in the Napa Valley: Geology, Landscape and Climate of the Napa Valley AVA, by EarthVision in 2002; and *Soils and Wine Grapes in the Napa Valley* and *Weather and Wine Grapes in the Napa Valley,* both by Terra Spase in 2003. The EarthVision study was conducted by Jonathan Swinchatt and provided much of the research that went into a book he co-authored with David Howell, *The Winemaker's Dance* (excerpted in chapter 3). The Terra Spase studies were done by Paul Skinner: Skinner's and Swinchatt's work are both cited in the selection below.

The main themes of the research reports were synthesized in a more popular account by wine writer Gerald Boyd and published as a pamphlet by the NVVA, *The Science Behind the Napa Valley Appellation,* in 2004. It has received wide free distribution ever since.

Several things should be noted about this selection and the pamphlet as a whole. First, the starting premise is that Napa makes great wine, so the research is focused on demonstrating how that came to be, not whether it's true. Second, what is highlighted as special about Napa soils is that they are so varied (the number 33—as in "33 types"— keeps appearing), in distinct contrast to the single soil types claimed as the foundation of most world-class *terroirs*—slate in the Mosel, limestone in Burgundy. The logic implies that the reason all this diversity is good for grapegrowing and winemaking is because they are in the Napa Valley. Finally, like so many soil-centric explorations of *terroir,* it makes no attempt to tie any particular soil to any particular wine characteristic (or grape variety), but simply presents the fact, as if its links to wine quality are obvious.

FROM GERALD BOYD, *THE SCIENCE BEHIND THE NAPA VALLEY APPELLATION*

NAPA VALLEY is a name that conjures up many images, thoughtful reflections, and names of legends and leaders that are emblematic of the contemporary meaning of the valley as a major wine region. In 1966, with the opening of his eponymous winery in Oakville, Robert Mondavi, Napa's elder statesman, recognized the unique qualities of Napa Valley soils and the wines they produced. "We knew then that we had the climate, the soil, and the varieties that made our own distinct style of wine that could be the equal of the great wines of the world, but it did require the winegrowing and the wisdom to know how to present it to the world."

■ ■ ■

Within the delimited space known as the Napa Valley AVA, there are no fewer than 33 different soil series, a variety of microclimates, steep mountains and sloping foothills on the eastern and western flanks, and a network of pocket valleys that meander eastward off the valley floor. Nowhere in the United States is there such a compact and varied spot for growing premium wine grapes.

Historians are better trained and informed about the history of the valley and its environs. What we are concerned with here is the evolution of the Napa Valley AVA as it relates to geology and soils, climate and weather, and landscape, the factors that when combined form the foundations of wine in the Napa Valley. Having an in-depth source of knowledge about what makes the Napa Valley tick is of value to the established and neophyte grower alike, as well as anyone considering growing grapes and making wine within the Napa Valley AVA. Because in today's high-profile, high-cost wine market, nurturing quality wine grapes is an expensive undertaking that requires a thorough knowledge of the total vineyard environment that may help defray some of the costs in time, effort, and money.

■　■　■

Geologist Jonathan Swinchatt, of EarthVision, summarized the value of first understanding the geology of a vineyard in his "Foundations" study for NVV. "Our charge is to summarize the scientifically characterizable attributes of the (Napa Valley) AVA . . . the topography, the bedrock and surficial geology, the distributor of surface sediments and soils, and regional variation in temperature and precipitation . . . and how they make the Napa Valley an ideal place to grow a wide variety of wine grapes."

■　■　■

The learned words of scientists like Skinner and Swinchatt persuasively further the argument that regional flavor and character do exist in wine. This is good news since some critics maintain that regional flavor sometimes suffers at the hands of winemaking, despite the fact that wines with purer fruit character, reflective of the vineyard terroir, are once again crowding wine shop shelves.

■　■　■

If it were possible to view a cross section of the Napa Valley, divided on a north-south axis, showing the mountains, the valley floor, the surface soil underlying stratification, the dense layered scene would resemble a complex, enigmatic painting by the 16th century Flemish artist, Hieronymus Bosch. Texture, form, color, and other components blend and whirl, separate and diverge, presenting a matrix far too complex to be digested in one take. Within the Napa Valley AVA there are 33 separate soil series that are alike in all aspects except the texture of the surface layer.

Skinner believes that everything is interconnected and dependent. "Each of the five soil forming factors (time, climate, biota, topography, and parent material) plays an important role in defining the vigor of soil series with respect to grape growing in the Napa AVA." In turn, he notes, the potential of soil vigor influences such vineyard practices as irrigation, fertilization, canopy management, and disease control. At the surface, the soil series has an effect on the selection of rootstocks, irrigation systems, vine and row spacing, and trellis types. An understanding of each of these factors and how they work together, places the focus on Skinner's belief that the diversity of the wide range of soils in the Napa Valley influences the quality and complexity of all of the premium wine grapes grown there.

■　■　■

Skinner maintains that the Napa Valley AVA is unique in that precipitation patterns combined with the distribution of soil types allows growers to use different combinations of irrigation practices, grape varieties (or the same grape variety), rootstocks, and spacing and trellis configurations to achieve different quality and complexity levels in the grapes harvested.

■　■　■

THE MERCEDES EFFECT

Another contributing factor that works with weather is the combination of summer heat and the marine fog layer, which Skinner calls "The Mercedes Effect." Hot dry summer conditions coupled with the development of a cooling marine fog layer slows grapevine growth while shifting to fruit, ripening "as smoothly as a Mercedes changing gears." Pausing between gears, Skinner drives home his point about the Mercedes Effect. "In my opinion, the smoothness of this transition and the frequency with which it accelerates at an optimum time in the vine growth cycle, may be one of the most important but overlooked climatic characteristics that set the Napa Valley AVA apart from other grape growing regions of the world." It would appear that Skinner's auto-analogy provides an explanation and, perhaps, justification for the popular practice by many winemakers today toward achieving physiological ripeness . . .

We wine drinkers have been told for years, first by the French who hammered away at the concept and then, more recently, by California growers and winemakers who climbed on board the terroir bandwagon, that without understanding terroir, you don't understand wine. Having thoroughly read the studies by Skinner and Swinchatt, I can say my new knowledge of soils, bedrock and climate, the essential components of terroir, give me a new appreciation and understanding of the Napa Valley AVA. And understanding the character of a region and how it translates to the wines produced there is vital in today's wine market, where the most often heard buzz phrase is about regional influence, the relationship of local terroir and wine character and taste. ∎

THE TITLE of James Lapsley's *Bottled Poetry* comes from a novella-length story by Robert Louis Stevenson, *The Silverado Squatters*. The passage seems to be a paean to Napa wines, with Stevenson unabashedly extolling their virtues. But when you read the passage in its entirety you realize that the quote has, somewhat egregiously, been taken out of context, as co-editor of this volume, John Buechsenstein, explains.

As a younger wine aficionado driving up to Napa Valley, my first glimpse of the iconic billboard always heightened my excitement. That landmark sign greeted visitors in both directions entering the valley. In an earlier incarnation, it provided a list of pioneering valley wineries, a "who's who" of early Napa.

At some point the winery names were removed, replaced by the now famous quote:

". . . and the wine is bottled poetry . . ."

The quote came from the great Robert Louis Stevenson, author of numerous works in the American and young adult canon, most notably *Treasure Island*. I was a staunch fan of Robert Louis Stevenson, having spent many a boyhood day kidnapped or digging for buried treasure—and I was always thrilled to see one of my literary heroes quoted in this way; it made the journey a little more special.

Stevenson wrote these words in a short work of prose called *Silverado Squatters,* published in 1883; he was recuperating from illness and honeymooning in Napa Valley when he wrote this trave-

FIGURE 8.4
Napa Valley welcome sign.

logue about his adventures in the Valley with his new bride, with accounts of visits to some of the first winemakers of the region. In several passages he expressed the romantic notion that vines brought forth flavors from the precious geology of the earth, assisted by the cosmos, with "poetic" results—a concept that shares a great deal of fidelity with the ideas of *terroir* traditionalists at the time.

But when you reinsert that ellipsis to its original passage, it's plain to see that Stevenson's assertion that the Napa wines are "bottled poetry" is altogether more speculative:

> Wine in California is still in the experimental stage; and when you taste a vintage, grave economical questions are involved. The beginning of vine-planting is like the beginning of mining for the precious metals: the wine-grower also "prospects." One corner of land after another is tried with one kind of grape after another. This is a failure; that is better; a third best. So, bit by bit, they grope about for their Clos Vougeot and Lafite.

Stevenson is plainly "excusing" Napa for being years away from its peak expression, conceding that it has yet to discover the vineyard land that will produce its Clos Vougeot and its Lafite. He then adds a cogent observation about terroir—poetry in a bottle—that Napa producers can aspire to:

> Those lodes and pockets of earth, more precious than the precious ores, that yield inimitable fragrance and soft fire; those virtuous Bonanzas, where the soil has sublimated under sun and stars to something finer, *and the wine is bottled poetry*: these still lie undiscovered; chaparral conceals,

thicket embowers them; the miner chips the rock and wanders farther, and the grizzly muses undisturbed. But there they bide their hour, awaiting their Columbus; and nature nurses and prepares them. The smack of Californian earth shall linger on the palate of your grandson.

Stevenson was describing a future reality. In the very next paragraph, he makes this painfully clear:

Meanwhile the wine is merely a good wine; the best that I have tasted better than a Beaujolais, and not unlike. But the trade is poor; it lives from hand to mouth, putting its all into experiments, and forced to sell its vintages. To find one properly matured, and bearing its own name, is to be fortune's favourite.

None of this, of course, concerned the Vintners. Instead, they appropriated a phrase that was plainly aspirational in intent, claiming it as a kind of gospel truth. Even more remarkably, they drew from a passage set within an accurate, if rudimentary, description of soil terroir. Such is marketing, I guess, where the message is so important that even the meaning of literary heroes is besmirched. Tourists driving by will glance at this sublime rendering, where a sense of place is transformed into a magical liquid, and never know that its true context describes the potential of terroir as well as any American author had to that point. ∎

BY THE 1990s, Napa had clearly established itself as the premier wine region in California: the most expensive wines, the most press coverage, the most tourist buses, even though the region generates only a sliver of California's total wine production.

In 2009, a group of influential Napa Valley leaders came together as the Napa Valley Destination Council and mapped out a multi-year, multi-million-dollar campaign to market Napa as a prime destination for high-end tourists. By the time the program went into operation in 2010–11, Visit Napa Valley (visitnapavalley.com) was in play, and the budget had roughly doubled. This planning/working document from 2011 gives a sense of the resources and determination Napa's adherents have put into play, and also conveys the region's high opinion of itself.

By the time the Destination plan was being developed, Napa had changed its official logo to include the word "legendary," a status usually conferred onto a region by other people.

FROM THE NAPA VALLEY DESTINATION COUNCIL DESTINATION MARKETING PLAN
THE NAPA VALLEY EXPERIENCE

The Napa Valley is home to the founders of America's fine wine industry—legendary entrepreneurs who showed the world that the high art of winemaking had crossed continents.

The Valley itself is one of the most rare and precious agricultural preserves on earth—a place that moves in perfect synchrony with the seasons. Nature is our bounty.

Here, we enjoy virtually perfect soil and climate conditions for the cultivation of the wine grape. Lakes, rivers, wetlands, and geothermal hot springs surround us. Wildlife flourishes.

The ambiance of the region is warm, gracious, and culturally vibrant. Scenic rolling hills dotted with vineyards inspire the eye.

Authentic towns and villages invite guests to explore and engage with artisans, restaurateurs, and local business owners. Arts and culture thrive.

Boutique hotels, resorts, and spas invite relaxation. And our natural environment challenges those with active lifestyles to outdoor fitness and wellness pursuits.

For the discerning traveler, the Napa Valley strikes an ideal balance between an authentic rural lifestyle and informal elegance—from the tables we set, to wine and foods prepared with great care, to an impeccable service standard that greets every guest.

The full enjoyment of life at an unhurried pace is a way of life in the Napa Valley. It is a pinnacle experience that defines fine food and wine destinations the world over.

THE NAPA VALLEY EXPERIENCE PROMISE

Here, in the birthplace of America's fine wine industry, you will enjoy the charm of rural life, the pleasure of the outdoors, and the art of living well.

You will feel at ease and inspired in a setting that is intimate and inviting.

Every condition for the pinnacle wine experience will be met. Authentic Towns and Villages welcome you. Rolling hills and scenic vineyards evoke a calming beauty; while nature works its magic—for the Napa Valley is inherently earth's perfect wine country.

We invite guests to experience our genuine nature, time and time again. ∎

CHARLES SULLIVAN, California's pre-eminent wine historian, devoted an entire book to tracing the development of the Napa Valley. At the end of the final chapter of his *Napa Wine: A History,* he takes up the question of whether Napa wines really are of superior quality, inventorying critics' ratings and several years' worth of results from structured comparative tastings by wine professionals, who gathered over the course of a decade in San Francisco at the Vintners Club. The results are decidedly mixed.

FROM CHARLES SULLIVAN, *NAPA WINE: A HISTORY*

The Vintners Club usually held six Cabernet tastings each year, twelve wines in each, with the top twelve meeting in the Tasteoff. From 1977 to 1988 the Club evaluated about 720 Cabs, of which 66 percent were produced from Napa grapes. About 20 percent were Sonoma wines. At the eleven tasteoffs, 80 percent of the wines making the cutoff were from Napa. In the preliminary tastings Napa wines had an average place in twelve of 5.41. (If pure chance had determined the outcome, the expected score would have been 6.50.) Sonoma wines averaged 6.72. This is a significant difference in quality perception, although certainly not overwhelming. (I was a member of the Club from 1975 to 2003 and sat at most of these tastings. My averages were Napa 5.91 and Sonoma 6.79.)

Now let us look at some evaluations by wine writers and wine publications to see whether this objectively calculated perception carries through in the published media.

Wine Spectator has made systematic and comprehensive evaluations of California wines since 1988. The results derive from blind tastings and are expressed on a 100-point scale, which has become increasingly popular, since the numbers remind us of school days when everyone understood the difference between an 87 (not quite a B+) and a 93 (no minus on that A). From 1988 to 1990 the magazine evaluated almost 400 Napa Cabernets and almost 200 from Sonoma. Each year the average Napa wine received a slightly higher average rating than that of Sonoma. For the three years combined the average Napa score was 87.68, Sonoma 84.85. If we expand our analysis of the Wine Spectator evaluations, to include vintages since 1975, there is no significant change in averages (Napa = 87.78, Sonoma = 84.23). Close, like the Vintners Club results, but a clear Napa win.

Robert Parker publishes a respected newsletter that regularly evaluates California wines. He tastes his wines blind, unless otherwise noted, and employs the 100-point rating scale. From 1986 to 1988 Parker gave Napa Cabernets a close nod, 84.70–83.39.

Charles Olken, the publisher of Connoisseurs' Guide to California Wine has been rating wine much longer than the other publications. Until recently he has not used the 100-point scale. The newsletter has depended on written evaluations, but singles out superior wines by giving them one, two, or three puffs for: "fine examples," "highly distinctive examples," and "exceptional examples." I sat on the publication's red wine panels for ten years and can attest to the blindness of their tastings. For the California vintages 1977–1986 they gave stars to sixty Napa and twenty-six Sonoma producers of Cabernet Sauvignon. The average number of stars granted Napa wines was 1.46, to Sonoma wines 1.27. Close again, but Napa clearly on top.

James Laube in 1989 brought out his 450-page evaluation of California's post-Repeal Cabernets. Seventy-four of the ninety-five "great" producers listed are from Napa, seventeen from Sonoma, and four from the Central Coast region. For vintages 1980–1986 his average scores for the best wine evaluated by producer from each vintage gave Napa a slight edge: Napa's 87.73 to Sonoma's 86.08. For the great vintages 1984, 1985, and 1986 the spread was similar: Napa 90.50, and Sonoma 89.47. Close, but Napa wins again.

I find these evaluations puzzling, not for the closeness but for the consistency. It is true that many parts of Sonoma's winegrowing areas have trouble ripening Cabernet, but there is not much Cab planted in these areas. There is also a lot more Napa Cabernet planted on soil derived from volcanic sources than in Sonoma. We wait in vain for any scientific explanation for such a relationship.

We must ask, of course, whether grape prices confirm these data. The answer is a definite YES. In 1990 Napa Cabernet Sauvignon sold for an average of 26 percent more than that from Sonoma. In 2003 the Napa lead was 58 percent. Now let us look at Chardonnay, the variety usually paired in wine lovers' minds with Cabernet Sauvignon. By reversing my data path, but following the same line of logic, it would seem that I could look at Chardonnay prices in 1990 and guess correctly the perceived quality comparison between Napa and Sonoma wines in this category. Napa Chardonnay grapes sold for an average 17.4 percent higher than those from Sonoma, and 28.7 percent higher than those of the Central Coast region. In 2003 Napa's lead over Sonoma was 37 percent, and 65 percent over Central Coast.

Does the perception of quality support this price relationship? Since 1977 the Vintners Club evaluated well over 900 Chardonnays, about 750 of them from Napa, Sonoma, and the Central Coast. In these blind tastings Napa wines averaged a ranking of 5.86, Sonoma 5.73 and Central Coast 5.57.

(Remember, a lower number is a plus here.) All three areas had very good scores, indicating that rarely did a wine from any other region place high. The scores are very close, but Napa placed third.

The Wine Spectator's evaluations yield almost identical results for ratings published between 1988 and spring 1991. Napa scored 83.66, Sonoma 83.79, and Central Coast 84.08. Laube's evaluations of "great" Chardonnays appear in a 370-page volume published in 1990. Again, Napa had the lion's share of producers, a whopping 64 percent. His average scores for Vintages 1984–1988 closely follow the pattern already indicated for Chardonnay. Sonoma averaged 88.87, Central Coast 88.50 and Napa 87.76. Very close, but Napa again third. I thought that the Connoisseurs' Guide's star ranking would show a deviation from this pattern, but I was wrong. For California vintages 1985–1988 the Central Coast Chardonnays singled out for special recognition had the highest average number of stars, 1.39. Then came Sonoma with 1.33 and Napa with 1.25. Again very close, but it's another third place showing for Napa.

The data for Cabernets partially confirm the Napa image as the home of the Western Hemisphere's greatest wines from red Bordeaux varieties. But the numbers are too close to those of other regions to explain the overwhelming nature of this perception. I believe the power of this image derives from the historical status given such wines as BV's Private Reserve, Inglenook's Cask wines, and Heitz's Martha's Vineyard. The introduction of several highly priced, internationally publicized bottlings, such as Opus One, Dominus, and Rubicon have reinforced this reputation. The superstructure of this powerful image was constructed in the last quarter century by the consistently world-class performance by producers such as Caymus, Dunn, Diamond Creek, Phelps, Robert Mondavi, Stag's Leap, and Chateau Montelena. (It is difficult to stop adding to this illustrative and illustrious list, when so many other Napa producers who started work in the sixties and seventies so clearly deserve mention. And the number continued to grow in the eighties and nineties.)

The Chardonnay data are more difficult to interpret. It appears that Napa's image boosts its grape prices overall, even when there is no clearly perceived or measurable quality differential in Napa's favor. We have seen that the quality perception scales tilt slightly in favor of Sonoma and the Central Coast for this varietal, even though Napa Chardonnay grape prices are considerably higher than those of the other regions.

It would be interesting and instructive to examine the ratings for other varietal wines, but I think we learn enough from the numbers for Cabernet and Chardonnay. There is, however, one other variety that no California wine lover can overlook in this comparative analysis. We have to look at Zinfandel, the grape that made Napa "claret country" between 1880 and 1895. For all major grape varieties Napa prices in 1990 exceeded those of Sonoma by 27 percent overall. Napa's red wine grapes were 35.4 percent more costly. But hidden in these statistics is a tantalizing pair of numbers. They show that for their Zinfandel grapes Sonoma growers received $44.00 per ton more than their fellows to the east. Napa received top dollar for all other important varieties. By 2003 the Sonoma lead was $445.

For the longer term Sonoma Zinfandels have run up a very impressive record at the Vintners Club, with an average place of 5.14 to Napa's 6.04. The Sierra Foothills are a close third at 6.13. But in the annual Zinfandel Tasteoffs, Sonoma Zinfandels have clearly triumphed, doubling the number of Napa wines making the finals, with a place differential of 5.80 to 6.88.

I should be happy to be able to draw weighty conclusions from these data. But I wouldn't think of it. I hope that their significance and my methodology provoke some arguments. I delight in the fact that wine, like music, has a powerful intellectual appeal, which heightens the sensual nature of both these pleasures.

POSTSCRIPT

The thing I've been asked most often about the statistics in this book's previous edition is whether they hold up in later years. Since I am most impressed by the results of blind tastings, I looked at the Vintners Club's evaluations in tastings for the years 1991–1997 for the three varietals.

The results showed no change in ranking. Napa was again supreme for Cabernet. Chardonnays from the Central Coast again took first, ahead of Sonoma and Napa. And Sonoma was still the master of Zinfandel. But there were some obvious differences in the magnitude of each area's victory. Napa's lead over Sonoma in the Cabernet category was cut by 62.6 percent. In fact, for the tastings from 1995 through 1997 Sonoma had a tiny advantage.

The Central Coast's Chardonnay lead over Sonoma and Napa increased significantly, as did Sonoma's lead over Napa. All three areas' place average increased, mostly because of better results for Lake and Mendocino counties. For Zinfandel, Sonoma increased its lead over Napa, but both areas' place average increased, in this case from better showing by the Sierra Foothills, the Paso Robles area and Mendocino County.

ON THE OTHER HAND . . .

The selection above from Deborah Elliott-Fisk defining the seven sub-appellations within the broad Lodi AVA is a fine example of characterizing a wine-growing region carefully and objectively. That care and objectivity is not always so evident.

One of California's most ill-defined AVAs is the Sonoma Coast; a quick glance at this region on a map shows that much of it is not on the coast at all, but takes up huge tracts of inland territory. Nonetheless, through the efforts of the Sonoma-Cutrer winery— whose flagship Chardonnay routinely topped the national list of restaurant by-the-glass wines—the boundaries for this AVA were drawn to include its far-flung vineyards, located miles from the true coast.

The following piece by *San Francisco Chronicle* wine editor Jon Bonné describes some of the frustration experienced by growers and winemakers caught in this fuzzy appellation and their efforts at guerrilla marketing on behalf of the "true" Sonoma Coast.

FROM JON BONNÉ, "DRAWING NEW LINES FOR WINE ON THE SONOMA COAST"

When is the coast not necessarily the coast? When it's marked on a bottle of Sonoma wine, apparently.

In wine terms, the Sonoma Coast appellation has been a mess since its approval in 1987. It stretches over 750 square miles, from the eastern end of San Pablo Bay, on Napa's edge, to the far northern reaches of Sonoma's actual coast, near the hamlets of Annapolis and Gualala. Along the

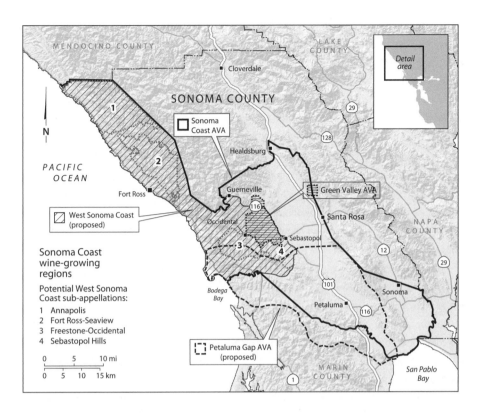

FIGURE 8.5

Map: Wine regions of the Sonoma Coast.

way it engulfs most of Russian River Valley, Green Valley, the Sonoma portion of Carneros, and nearly half of Sonoma County.

The original motives of the appellation are draped in the sort of politics that attend so many American Viticultural Areas, or AVAs. The original intent, as always, was to delineate a growing region with unique character. But the map for this particular sprawl of an appellation was largely sketched to include the vineyards of Sonoma-Cutrer, which needed an area that encompassed its winery and farther-flung vineyards for the "coastal" Chardonnay it promoted in the 1980s. "It's big," says Brice Jones, Sonoma-Cutrer's founder, of the Sonoma Coast, "but it's true to the purpose of having appellations."

In subsequent years, every logical twist has been applied to explain this draw-outside-the-lines appellation: the predominance of certain soils, the impact of ocean wind through the gap in the coastal range near Petaluma (the Petaluma Gap is now pushing for its own appellation, of course), the presumptive cool climate.

"We felt the Sonoma Coast appellation was beginning to be used on wines for marketing purposes, and wasn't being used on wines that showed the natural style of the area," says Carroll Kemp of Red Car, which has vineyards in remote Fort Ross. "In that sense, it is deceptive."

So there's what has come to be called the "true" Sonoma coast—vineyards within a few miles of the Pacific coastline that, as it happens, are defining spots for some of the country's top Pinot Noir and Chardonnay. This includes names like Hirsch, Peay, Flowers, and Marcassin, along with wineries like Littorai, Williams Selyem, Kistler, and Freestone—all together a litany of California outperformers.

At long last, many have had it with the funny geography. Last month they unveiled a new organization, the West Sonoma Coast Vintners (see more at westsonomacoast.com), with two dozen members and its own festival to be held in August in Occidental.

This is hardly the first attempt to bring order to the gerrymandered blob of the Sonoma Coast. Around 2002, several of the group's founding members submitted a proposal for a Fort Ross-Seaview appellation that included coastal ridges from Cazadero to Annapolis. That effort was shot down after concerns were raised by Jones, some of whose vineyards were excluded, along with Fort Ross Vineyards, who saw its brand name imperiled, and Hartford Family Winery, who similarly had vineyards outside the dotted line.

There are hopes that Fort Ross-Seaview will rise again. Efforts for Freestone-Occidental, an appellation south of the Russian River, are coming along, as are plans for an Annapolis appellation to the north, a Sebastopol Hills area south of the Bodega Highway and the Petaluma Gap.

But for now, the new group's creators have moved to other tasks. They would rather promote the common culture of West County than start marking up maps once again. So they have devised a large, and unofficial, "West Sonoma Coast" area, with Highway 116 as a dividing line that cuts northwest through Sebastopol and Forestville.

Their hope is that rather than mire themselves in more bureaucracy, they can take their case directly to their customers.

"Looking at it from my point of view and some of my neighbors, we don't really care anymore," says David Hirsch of Hirsch Vineyards, who spearheaded the original Fort Ross-Seaview effort. "There's been a learning curve around the real Sonoma Coast versus the generic, so I'm not sure how much interest there'll be in putting these AVAs on the label."

NOTES

1. In actuality, prior to the English occupation, wines coming from the interior by way of the Dordogne would pass north of Bordeaux through Liboume. They would then go to La Rochelle where they were shipped to distant markets. After the fall of La Rochelle and the ascendancy of Bordeaux, these wines from the interior had no choice but to pass through Bordeaux on their way to foreign markets.

2. Keesing argues that subaltern populations often make use of the very forms and discourse through which they are dominated to resist their oppressors. There are two important points represented here. First, those who are dominated should not be regarded as passive or mere victims of their oppression. Second, hegemony contains within itself the possibility of its own negation.

3. "See-Through Poster Welcomes Napa Valley Visitors," *Wines and Vines,* October 1950, 32; "Vintners Sponsor Cable Car," ibid., November 1949, 32; "Give-Away Too Good," ibid., April 1950, 86. The title of the last article derives from the fact that the cable car company

curtailed the practice, fearing that it would make it a "freight carrier" in the eyes of the Public Utility Commission.

4. "Cooperative Promotion" (editorial), *Wines and Vines*, October 1 949, 3–4. Italics in original.

5. Francis Gould, "Napa Valley Vintners," *Bottles and Bins*, October 1962, 2; "Lawyers Guests of Calif. Vintners," *Wines and Vines*, August 1962, 7; and "Napa Valley Vintners Play Host," ibid., June 1962, 14.

6. Frank Schoonmaker, "California's Vintage Vale," *Holiday*, August 1952, 103–7. Quotations are from p. 103.

7. See "Wise and Otherwise," *Wines and Vines*, February 1955, 16, and "Napa Vintners Honor Author," ibid., April 1955, 9. The duchy earned wider notoriety in the 1959 Peter Sellers comedy *The Mouse That Roared*, but in the movie war was declared, not to defend national honor, but to win U.S. aid as a vanquished foe.

8. "Vintners, Movie Stars Meet," *Wines and Vines*, October 1958, 9; "Wine Movie Opens to Considerable Fanfare," *ibid.*, July 1959, 7.

9. Philip Lesly of the Philip Lesly Company, which was hired to promote premium California wines in 1959, spoke of the need to "surround the public with an atmosphere favorable to the idea we want to get across."

9

THE FUTURE OF *TERROIR*

After several millennia of grapegrowing and winemaking on nearly every continent, how certain can we be that all the great *terroirs* have been discovered? How confident can we be that they will continue to produce distinctive wines as long as there are still wine drinkers around? And what assurances do we have that the world wine industry is united in the desire to nurture and celebrate these special wines?

The truth is, we have no such assurances. *Terroir* does not stay put, nor is it universally valued. Just as wine itself is a living thing, forever changing and developing even long after it's bottled, so *terroir* is an evolving environment of rocks and weather, aromas and flavors. There's no way to tell if we've discovered all of the world's great *terroirs*, but the chances are slim at best that we have. We only know the great ones that have been planted and marketed. The slowest, subtlest, and most inexorable of all the changes applied to distinctive vineyards is that of climate, and climate, of course, is changing. If grapes are shaped by the climate they grow in, how can the *terroirs* of the last century survive the next? Setting aside the natural world, *terroir* wines are threatened by dynamics within the wine industry itself, both the ocean of industrial wines that come from nowhere in particular and the stream of flashy, international-style blockbusters that could come from anywhere.

These powerful factors guarantee that the world *terroir* map will look vastly different in fifty years, and in a couple centuries may be unrecognizable. *Terroir* has a future, and not all of its prospects are good.

This chapter is broken into three sections: the discovery of new *terroir*, the struggle to preserve wines of distinction in a market that offers advantages to homogeneity, and the challenge of climate change. For new *terroir,* we look at the sudden appearance in the past twenty years of delicious and attention-grabbing wines from Spain's Priorat, New Zealand's Marlborough, and the Champagne district of the Aube, all areas considered wine backwaters for decades. The trends in modern global wine styles are addressed by Jamie Goode and Sam Harrop MW, and writer Dan Berger looks into the loss of a sense of place in Napa Cabernet. Finally, the daunting issue of climate change gets an overview from Greg Jones and Hans Schultz, two leading scholars, and a close-up look by Robert Pincus at the changing environment for the delicate white wines of Germany, Austria and Alsace. A minority dissent on climate change and viticulture by John Gladstones concludes the chapter.

DISCOVERING NEW *TERROIR*

When we think about *terroir,* we usually focus on the natural endowments of the vineyard and the sensory payoff in the glass. But without conscientious winemaking, the expression of place gets lost, or never gets found in the first place.

Spain has been growing grapes since before the Romans arrived and today leads the world's nations by a good margin in the number of planted acres. But its wines have lived in the shadow of their French, Italian, and German counterparts for centuries. A wave of new investment in vineyard renovation and modernized facilities has changed all that: Spain has become the source of some of the most exciting and distinctive new offerings in the market—Albariño from Rias Baixas, Bierzo from Galicia, and the stunning red wines from the Priorat.

As this 2007 account by Stephen Brook from *Decanter* magazine describes, it took a relative handful of place-conscious winemaker-visionaries to transform a perennial wine backwater into a compelling regional *terroir.*

FROM STEPHEN BROOK, "PRIORAT"

From the valley floor it is a tortuous drive up to the vineyards that Sallust Alvarez, manager of the Vall Llach winery, wants me to see. The hairpin bends are bad enough, but his four-wheel drive is so elongated that it takes a number of manoeuvres to negotiate each one. If he miscalculates it will be a long slither down the steep slatey slopes to the valley floor.

It is worth the trip. Up at 500 meters, the views are stupendous, but it is the vineyards that I have come to see. All around me the slate glints in the autumn sunlight, and here and there a few gnarled vines are somehow dragging nutrients out of the impoverished soil. Alvarez shows me a neighbouring vineyard too: "This is owned by a woman in her 60s, who walks up here whenever she has to tend the vines. She doesn't own a car."

Twenty years ago, he explains, the Priorat region was on the verge of extinction, with grape prices so low that this was one of Spain's most impoverished regions. The revival of Priorat in the 1990s has rewarded the growers, many of whom are now too old and frail to enjoy their relative prosperity.

The story of Priorat's revival is well known, but it bears re-telling. In the 19th century there were about 15,000 hectares (ha) under vine in this mountainous region, but by the late 1980s there were a mere 900 ha, mostly old vineyards high on the slopes, hard to get to and punishing to work. The decline seemed irreversible, until a few visionaries realised Priorat's potential. These included René Barbier, now the owner of Clos Mogador and Manyetes; Alvaro Palacios, owner of Finca Dofí and L'Ermita; Josep Lluís Pérez of Mas Martinet; and Dafne Glorian of Clos Erasmus. Barbier had sensed the potential in the early 1980s, when Priorat was dominated by the Scala Dei winery and various cooperatives that mostly produced bulk wine, but each year vineyards were being grubbed up and replaced by olive and almond groves.

The revivalists worked together, buying small parcels of mostly venerable vineyards, restoring stone terraces and access roads, and producing a single wine together at Clos Mogador. In 1992 they went their separate ways, all dedicated to quality but pursuing their own hunches. It took until 1994 for the wine press and international merchants to realise that Priorat had been reborn. Outside investors such as Torres and Freixenet began acquiring land, and the number of producers has risen from 10 in 1989 to about 80 today; the surface under vine now stands at around 1,700 ha. It is unlikely to increase much further.

What drew the pioneers to Priorat was the abundance of old vines and the remarkable terroir. The soil, except on the fringes of the region, is pure slate or schist, known locally as llicorella, with virtually no topsoil. Yet there is no shortage of subterranean water, so the vines, though their parched roots must descend many metres, can always find sufficient moisture to keep them alive. Certain vineyards are up to 100 years old; most of these veterans are Cariñena, but there is a good deal of Garnacha too—it varies from village to village.

Other grape varieties planted include Cabernet Sauvignon, Merlot, and Syrah. René Barbier Jr. notes, and my own experience confirms: "In Priorat, terroir reigns supreme over variety. Sometimes in a blind tasting it is difficult to identify the varietal composition of a wine." Another feature of the best wines is their freshness. Alcohol levels can be high—up to 15.5%, especially for Garnacha-based wines—yet they are rarely heavy or overbearing, thanks to their enlivening acidity.

Josep Lluís Pérez's daughter, Sara, the winemaker at Mas Martinet and the wife of René Barbier Jr., tells me: "On average the wines I make have about 14.2% alcohol. But at the same time the pH is around 3.4 or less, compared with 3.9 or 4.0 for a Napa Cabernet for example. Our llicorella soils retain freshness and disguise the alcohol. When my father and father-in-law were starting out here in 1989, they didn't know any of this, but it soon became apparent. Here in Priorat the soil speaks."

In the winery, Priorat is a hotbed of innovation, with many variations in practices such as retention of stems, maceration times, and use of new oak. There is some over-extraction and excessive alcohol—defects shared by numerous wines worldwide—but the overall standard of winemaking seems remarkably high, with little dilution or astringency. What is striking and unusual about the wines of Priorat is the ease with which Garnacha in particular takes to new oak. In the southern Rhône or Châteauneuf-du-Pape, ageing Grenache in new oak has to be done with great care, because of the ease with which the variety oxidises. Yet in Priorat this doesn't seem to be a problem.

Joan Asens, winemaker for Alvaro Palacios, has an explanation: "Yields in Priorat are lower than in Châteauneuf, for example, and the berries are small. That means the fruit has greater concentration. Our vineyards also have less direct exposure to sunlight than those of southern France. So the fruit has a different structure, with better acidity and lower pH, and it's far less prone to oxidation."

The source of old-vine Garnacha and, especially, Cariñena, will gradually shrink as vines die off, so it is understandable that growers are planting international varieties too. Unlike Merlot or Syrah, Cariñena needs to be at least 15 years old before it has a chance of becoming expressive and interesting; consequently, growers are reluctant to plant it, and those who do, such as Sara Pérez, are wary of productive clonal selections and prefer the more complex massal selections from old vineyards. Significantly, the proportion of Cabernet Sauvignon in the vineyards is probably diminishing. Clos Mogador used to have as much as 30%, but today there is far less, and at Mas Martinet, Sara Pérez is courageously grafting over both Cabernet Sauvignon and Merlot to Garnacha. She worries that the gradual disappearance of old-vine Cariñena will alter the typicity of Priorat for good.

Yet llicorella stamps its personality on the wines, whatever their composition. The profile of Priorat may change somewhat in the 21st century, as, for example, Syrah increases in importance, yet it's safe to predict that the grandeur of the wine, its minerality and staying power, should remain unaltered.

THE NEW CHAMPAGNE

In the Champagne region of France, the district of the Aube, a growing region that's not contiguous with Champagne's central growing and production areas, has labored for decades under a reputation for second-rate grapes and undistinguished wines. But, as this piece by Eric Asimov of the *New York Times* reveals, all it took was some determined winemakers to show that the Aube has a personality well worth capturing. Asimov details the differences in approach that make the Aube an island of *terroir*-driven wines in a sea of standardized opulence. This is how Asimov gets into his subject:

NEW *TERROIR* IN CHAMPAGNE

Hectare for hectare, France's Champagne region leads the world in turning vineyard space into money. Within this high-profile territory, the poor relation has always been the district of the Aube, a bit to the south and physically separate from the grand estates of the north. The northerners tried to keep the Aube out of the Champagne appellation a hundred years ago, leading to riots, and when it was officially included in 1927, none of its vineyards received grand or premier cru designation.

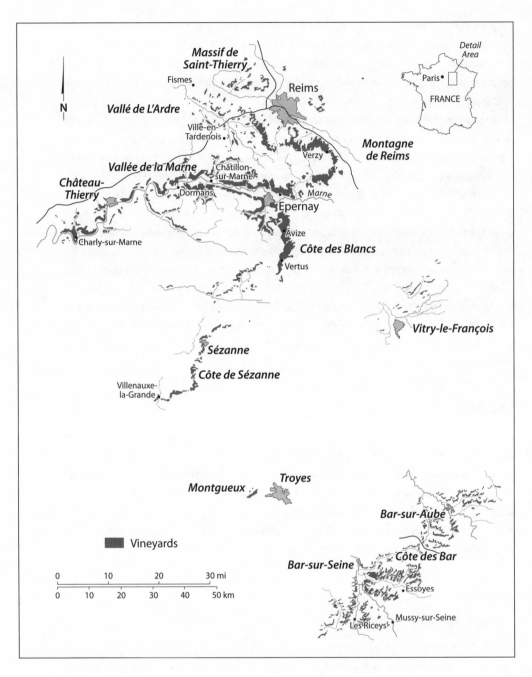

FIGURE 9.1

Map: Wine regions of Champagne.

Troyes, France

Unlike Reims and Épernay, the Marne cities to the north that are rivaled only by caviar in their close association with Champagne, this pleasant medieval city in the Aube, with its cobblestone streets and timbered architecture, is rarely considered the hub of a thriving Champagne region.

Perhaps that's because for years the Aube has served anonymously as the workaday supplier of grapes to the production areas to the north, a sort of scullery in the elegant house of bubbly, essential to the smooth operation of Champagne, but best ignored.

Yet today, the spotlight is unexpectedly shining on the Aube, and its primary growing area, the Côte des Bar. Now, the region is coming to be known for its independent vignerons, whose distinctive, highly sought wines have caught the attention of Champagne lovers the world over.

The grandes marques of the Marne made Champagne one of the world's leading luxury brands by marketing it as an urbane beverage for special occasions. They emphasized the art of blending, in which the distinctions of terroir, grape, and vintage are absorbed into a house style.

By contrast, many Aube producers are taking their cues instead from Burgundy, with its emphasis on farming and on being able to trace terroir through the wines. Rather than the hushed pop of the cork and the silken rush of bubbles, these Champagnes suggest soil on the boots and dirt under the fingernails.

■　■　■

The focus on terroir in the Aube reflects a larger discussion throughout the entire region, in which small producers making distinctive, terroir-specific Champagnes from grapes they farm themselves have seized initiative from the big houses. These small grower-producers account for barely an eyedropper's worth of the Champagne that flows from the region, but they now lay claim to an outsize portion of the fascination among Champagne lovers.

"Before, it was Champagne, singular," said Michel Drappier of Drappier, the largest and best known producer in the Aube, which was founded in 1808 but didn't begin to bottle its own wines until the early 20th century. "Now it is Champagnes, plural, as sophisticated and complex as Burgundy, with as many villages, winemakers, and styles as any place."

Mr. Dosnon studied viticulture and enology in Beaune, the heart of Burgundy, and he brings a Burgundian passion for the land to his work. Strolling through a hillside vineyard in the hamlet of Avirey-Lingey, about 25 miles southeast of Troyes, one parcel among 17 acres or so that they farm, I noticed another similarity to Burgundy, tiny fossilized seashells in the earth, like those often seen in the vineyards of Chablis.

Indeed, the Côte des Bar is closer to Chablis than to Épernay, and its limestone and clay soils are more like those of Chablis than the chalky soils to the north. Yet, despite the geological resemblance to Chablis, which makes the most distinctive chardonnay wines in the world, the vast majority of the grapes in the Côte des Bar are pinot noir.

"The soil is also interesting for pinot noir," Mr. Dosnon said. "There's a lot of volume and complexity."

The Dosnon & Lepage Champagnes are superb, especially the 100 percent pinot noir Récolte Noire, powerful yet graceful, wonderfully fresh and aromatic, and a blanc de blancs, Récolte Blanche, a wine of finesse and nuance, with savory, focused floral and mineral flavors.

If the evolution of the Aube seems a bit of a Cinderella story, it's with good reason. A century ago, in 1911, riots tore through Champagne as, among other issues, the big houses in the Marne tried to exclude the Aube from the Champagne appellation. Eventually, a compromise was reached in which the Aube was granted second-class Champagne status. Even after the Marne finally, if gingerly, embraced the Aube as a full part of Champagne in 1927, none of its vineyards were designated grand cru or even premier cru, marks of quality reserved only for the Marne.

And so the Aube served primarily as a faceless source of grapes. While a small amount of Champagne has always been made here, the grapes mostly traveled 80 miles or so north, through the flat farmland that separates the Côte des Bar from the production areas of the Marne.

■　■　■

The vineyards of Montgueux, largely on an imposing south-facing hillside, are distinct from the Côte des Bar, and are one of the few places in the Aube that emphasize chardonnay. In Montgueux, achieving sufficient ripeness is rarely a problem. Indeed, the exotic, tropical fruit flavors of Montgueux chardonnay are highly unusual for Champagne. Mr. Lassaigne's aim is to capture the aromas and flavors of this singular terroir.

"My job is to say, 'Montgueux is good,'" he said. "It's not better, but it's absolutely not worse."

His non-vintage blanc de blancs Les Vignes de Montgueux is very much its own Champagne, with light aromas of tropical fruit and flowers. It feels broad yet is dry and refreshing. His vintage blanc de blancs are a step up in elegance, with more mineral flavors yet still with the distinctive Montgueux fruit.

Foremost, perhaps, among the region's new stars is Cédric Bouchard, whose single-vineyard Champagnes are exquisitely delicate and subtle, gently expressive of their terroir. His dark, tussled hair and piercing olive green eyes give him the brooding look of a young philosopher. Indeed, his uncompromising winemaking might be called highly philosophical.

"I'm only interested in the wine, the grape, the parcel and the terroir," he said. "It's got to have emotion to it; otherwise, it's going to the négociants."

Mr. Bouchard's father grew grapes and made a small amount of his own Champagne, but as a young man Mr. Bouchard left for Paris, where he worked in a wine shop. There, he said, he discovered the wines of vignerons he described as working naturally, and decided that he, too, wanted to make wine. He returned to the Aube only because his father offered him land.

Right away, he proved himself independent. "Whatever my father did, I did the opposite," he said. "Spiritually, I'm the first generation because it's my own style and philosophy. I think my father is proud of the wines, but he would never admit it directly."

Mr. Bouchard tries to be as natural in his approach as possible, even rejecting the use of horses in his vineyards, which he now plows by hand. In that sense, he said, he is lucky to have only small parcels.

Another rising star in the Côte des Bar, Bertrand Gautherot, named his label Vouette & Sorbée, after the two vineyards he farms biodynamically. His family grew grains and grapes and raised animals around the town of Buxières-sur-Arce. As a young man he left, to design lipsticks, but the call of agriculture was great, and he soon returned.

"We were not in the business of Champagne," he said. "We were more farmers than winemakers."

Mr. Gautherot, too, focused on farming, selling off all his grapes to cooperatives or the big houses. Among his good friends were superb grower-producers from the north, like Anselme Selosse and Jérôme Prévost, who he said urged him to begin making his own wines.

"But I understood I had to learn the terroir of my village," he said. "A big problem in Champagne is that wines are easy to make by recipe. It's much harder to learn the taste of your vineyards. That's why it's called Vouette & Sorbée rather than Bertrand Gautherot." ■

WINEMAKERS AREN'T the only important actors in unleashing the power of *terroir*. The abrupt entry of the wines of New Zealand in the 1990s—Sauvignon Blanc from Marlborough, Pinot Noir from Central Otago, Merlot from Hawke's Bay—reflect an effort built on two centuries of trial and error.

New Zealand had the good fortune to lie in an hospitable climate zone for winegrowing, and British missionary settlers in the early nineteenth century started planting grapes. The burgeoning industry was fortified by a wave of Dalmatian immigrants who brought their winemaking skills (and thirst) with them; even today, several prominent New Zealand wine brands bear Dalmatian/Croatian names. Croatian-born, Italian-trained Romeo Bragato took the lead in discovering new areas for viticulture and introducing new varieties around the turn of the twentieth century. That's where economist Mike Veseth picks up the story in this excerpt from his book *Wine Wars*.

FROM MIKE VESETH, *WINE WARS: THE CURSE OF THE BLUE NUN, THE MIRACLE OF TWO BUCK CHUCK, AND THE REVENGE OF THE TERROIRISTS*

THE THIRD WAVE

It would be nice to be able to say that New Zealand's comparative advantage in fine wine was quickly developed once Bragato and others identified it, but that wasn't the case. While some good wine was made, the bulk of production lacked distinction apart from sugar and alcohol. When Kiwis drank wine, they went for the same sweet, fortified production that their Australian neighbors were gulping down.

It took one more wave of globalization to put this particular wine in your glass. Seagram's investment in Montana in the 1970s gave the firm the capital it needed to expand production. On the recommendation of another consultant—this one, I'm told, from California—Montana planted the first Sauvignon Blanc vineyards in Marlborough on the South Island at the Brancott Estate, virgin wine territory at the time. The wines were immediately recognized for their distinctive quality. It is tempting to say that the rest is history, but in fact just the opposite was the case.[1]

The birth of Marlborough wine had the bad luck to coincide with an era of spectacularly misguided economic policy in New Zealand. Like many less developed countries, New Zealand adopted the policies that political economists call Import-Substituting Industrialization. In an attempt to emulate

Japan's postwar success, New Zealand raised import barriers, subsidized domestic production, picked "winning" sectors, and generally tried to grow a modern economy from the inside out. This is never an easy task and more so given New Zealand's limited population and resources.

The result in the wine industry was a vast increase in plantings of high-quantity, low-quality grapes as everyone aimed for the captive domestic bulk wine market—the market for Dally plonk (*the Kiwi term for junk wine—eds.*). This led, predictably, to a glut of bad wine, which could not be sold at home or abroad, falling prices, and a general collapse of the industry. Having failed to develop by keeping imports out, New Zealand changed course and liberalized its economic policies, lowering trade barriers, eliminating agricultural subsidies, and embracing competition and global markets. The wine in your glass owes its existence—dare I say it?—to these neoliberal market reforms!

NEW ZEALAND WINE TRANSFORMED

A transformation of the New Zealand wine industry followed. Cheap bulk wines from Australia and then also Chile flooded in, capturing the bottom end of the market. Domestic Dally plonk was replaced with imported Aussie wine. The government paid winegrowers to rip out their old vines (a process called grubbing up) and replant with high-quality classic varieties like Pinot Noir and Sauvignon Blanc. Winegrowers began to focus on quality both because quantity no longer paid and because, with good imported wines now readily available, quality was the only way to compete. The crisis of the 1980s finally created the foundation for the wine industry that Romeo Bragato envisioned eight decades before.

That's how these great New Zealand wines came to be produced, but how did they get from way over there to way over here, in your glass? How did New Zealand supply connect with you and me, the demand side market? Globalization is the answer again.

In 1985 a Marlborough winemaker named Ernie Hunter entered his Sauvignon Blanc in the *Sunday Times* of London's annual wine festival competition, where it was unexpectedly voted the top wine. His Chardonnay scored big the next year, showing that Hunter's wines weren't flukes. This public success opened up the world's most important wine market, British supermarkets like Tesco and Sainsbury's, to Hunter's wines and soon to Marlborough wines and New Zealand wines generally.

At the same time Cloudy Bay, a Marlborough wine produced by the quality West Australian firm Cape Mentelle (now owned by the French luxury goods conglomerate LVMH—Moët Hennessy Louis Vuitton), also hit the British market through the parent company's distribution network and achieved spectacular success. A firm foundation for New Zealand wine exports was established.

Today the New Zealand wine industry is remarkably globalized, with more than two-thirds of its production earmarked for export. Britain remains the largest foreign market followed by Australia (where Kiwi white wines routinely outsell domestic products), the United States, and Canada. Foreign ownership dominates the wine sector. Pernod Ricard of France (Montana and Corban's, plus other brands) is the largest producer followed by the U.S. conglomerate Constellation Brands (Nobilo, Selaks, Monkey Bay, Kim Crawford, and several others). Other important brands including Cloudy Bay, Craggy Range, Clos Henri, and Whitehaven have international owners, too. New Zealand is both a destination and a home base for dozens of "flying winemakers."

British missionaries, Dalmatian gumdiggers, Canadian investors, California consultants, British supermarkets, French multinationals—and three waves of globalization—that's the unlikely story of how Marlborough Sauvignon Blanc came to be poured into your wineglass.

STYLE OVER SUBSTANCE

Climate change will likely alter traditional notions of *terroir* (and traditional *terroirs*) irrevocably, but a far more insidious threat comes from another man-made development: the homogenization of wine styles, on a global scale. Industrial production methods and the internationalization of the wine marketplace yield a vast supply of beverage-grade wine, designed not to be distinctive or challenging but to go down easily and reliably, year after year. For the vast majority of the world's wine drinkers, recent advances in viticulture and winemaking make today's "ordinary" wine light years better than the comparable wines of even thirty years ago. But even though industrial wine serves a purpose, that purpose is not the expression of *terroir;* that serviceable bag-in-box Chardonnay you've gotten fond of could be a blend from tankers full of wine from Argentina, Australia, Italy and California's Central Valley. These wines taste more like the wineries where they were blended than any place of origin.

At the high end, controversy has erupted in the past couple of decades over the dominance of the so-called "international style" winemaking focused on high extract, robust alcohol, and lavish oak treatment, on varieties that aren't usually treated this way, in regions not accustomed to such practices. For consumers who like these wines and the critics who quantify their merits, there are plenty of them around, from nearly every major wine region. Those not enamored of the trend argue that these wines do not reflect distinct regional and vineyard origins—they could come from anywhere. As with industrial wines, high-end, high-alcohol, fruit-forward, well-oaked reds tend toward the monochromatic, with all the rough edges and quirks once prized as badges of *terroir* obscured, smoothed out, or eliminated altogether.

In their book *Authentic Wine*, British wine writer Jamie Goode and New Zealand Master of Wine Sam Harrop see real promise in the effort to preserve the place of place in wine. But they also draw a powerful picture of the forces that endanger *terroir*. This excerpt also includes their assessment of the role of critic Robert Parker, responsible more than any other individual for the trends and controversies in contemporary fine wine.

FROM JAMIE GOODE AND SAM HARROP MW, *AUTHENTIC WINE*

In the past, fine wine was an aesthetic system, based on benchmarking and learning. Students of wine explored the classic styles and learned to discern what constitutes a great wine as opposed to an ordinary one. It's worth remembering here that wine appreciation operates on two distinct but intersecting levels. First, we have the hedonic level. You sip a wine and then report how much you enjoyed the experience: how nice did it taste? Second, there is the learned component: people have traditionally learned what constitutes a great example of Bordeaux, white Burgundy, or Champagne. Of course, the two methods of appreciation overlap, and most people use both in tandem. However, a novice is really capable of only the first level of appreciation.

It follows, then, that when experts assess a wine, they do so from inside this tradition of an aesthetic system of fine wine. They don't evaluate a wine just on the basis of what is in the glass, and how much pleasure it brings them. Some knowledge of context is necessary in the process of wine appreciation. To be a good critic, a novice must first seek to gain some knowledge of the wine style he or she is evaluating. When we are tasting blind, there is a limit to what we can say about the wine that is in front of us. The wine trade has traditionally acted as a custodian of a fine wine tradition, and those entering it have undergone a sort of apprenticeship in wine.

Of course, this aesthetic system of fine wine is terribly Eurocentric and rather elitist, which puts it firmly in the searchlights of those outside the system, who are gunning for a target. The system has also been shaken to its core over the past couple of decades by the emergence of the world's most powerful wine critic, Robert M. Parker, Jr. Probably just a minority of readers of this chapter will be unaware of Parker and his work, so we'll introduce him only briefly.

Robert Parker was a lawyer from Maryland with a passion for wine. In 1978 he began publishing the *Wine Advocate,* a simple magazine, but one that was to revolutionize the fine wine market. Parker's approach was to position himself as a consumers' advocate, and his aim was to give people the sort of impartial guidance that would help them make informed wine-buying decisions. His stroke of genius was to score wines on an easily understandable 100-point scale, where any wine scoring under 80 wasn't much and anything over 90 was pretty special. Parker was rating fine wine, but it could be suggested that he was operating outside the established, British-dominated aesthetic system of fine wine, perhaps unintentionally. Firmly on the side of the consumer, he was positioning himself outside the wine trade with a view to maintaining an independent voice.

Suddenly consumers were empowered. While tasting notes are an important part of the *Wine Advocate* and his book *(Parker's Wine Buyer's Guide)*, it is the scores that make relative performance transparent. "Parker points" offer a way into wine for those daunted by its complexity. They introduced an element of competition in the world of fine wine and enabled overperforming new producers to rub shoulders with the classics: rather than building a reputation over generations, all that was now needed for entry into the wine world's elite was a string of scores in the high 90s. They also allowed wine to be traded by those with no specialist knowledge, because prices track scores, and young wines with the highest Parker scores have tended to increase in value spectacularly after release.

Consumers have found Parker points to be a shortcut mechanism for making buying choices, but wine merchants have found them equally useful as a sales tool. Most wine lists are punctuated by the likes of "RP 92" or even "RP 100" (for something very special). The ratings, by making fine wine easier to understand and buy, have also been a significant factor in opening up new markets for fine wine, most specifically in Asia. "From the perspective of wine buyers and merchants in Asia, Parker's influence can hardly be overstated," says Nicholas Pegna, managing director of Berry Bros. and Rudd in Hong Kong. "His views are very, very significant. People do not need to understand the wines, nor even his notes; they can simply look at the scores. This is vital where people might not be confident in what they like drinking."

The problem with the Parker phenomenon is that by operating outside the aesthetic system of fine wine, he has changed the way in which quality is defined. The old rule book has been discarded. A Bordeaux wine is assessed not as a Bordeaux wine but as a red wine and suddenly is being compared directly, by means of its score, with a Napa Valley Cabernet or a leading Hermitage from the northern Rhône. Is a Parker score a valid judgment of wine quality? Consumers and merchants think so, and in some fine wine markets a good score is a prerequisite for a high price tag.

This discussion is in no way intended to be read as a criticism of Parker, who is an able, hardworking, and consistent taster, and who deserves his success. But the way his ratings have been understood and used by many, coupled with his huge impact on the fine wine market, hasn't been completely and unreservedly good for wine diversity. Because all wines are now comparable by means of a score, ratings are the effective definition of wine quality. Consequently, some commentators have suggested that there has been movement towards an "international" style of red wine that lacks a real sense of place, because sense of place is not rewarded with points in this new definition of wine quality. A wine that might otherwise be regarded as a good, typical example of a particular appellation may see its score and desirability improve if the grower reduces yields, harvests later, and uses interventionist cellar techniques, such as ageing the wine in barrels made exclusively of new oak, which imparts a distinctive flavour to the wine. So what if it no longer represents a typical reflection of what this patch of ground is capable of? For white wines, there has been less of a trend towards internationalization, perhaps with the exception of Chardonnay. This is because the main focus of critics such as Parker has been on red wines, and for many whites, there has not been a move towards picking later and using lots of new oak.

Some may argue that by making these international-style wines, producers are simply respond-ing to the demands of the market. Well-heeled collectors enjoy these sorts of wines; they are grati-fying and, with their dense, sweet, powerful fruit, are quite easy to appreciate. The impression you gain is that many of these collectors don't really care whether a particular wine fits into the aesthetic system of fine wine; they are happy as long as they have a cellar of suitably highly rated wines. In this new synthesis of fine wine, typicity (the way a wine displays characteristics shared among wines from this particular location) and diversity are unwelcome complications. They are irrelevant. Indeed, notions of terroir are seen as undemocratic tools used by members of the wine establishment to maintain their privileged position. But others contend that the new, simplified world of fine wine, with its easily understood measure of quality, is impoverished and less interesting because it fails to celebrate diversity and individuality in their own right.

Is the whole fine wine venture heading for the disaster of dull uniformity? Probably not. While too many wines are being made in the international style, a quick comparison of the wine list of a good merchant today with one from thirty years ago will show that we've never been so lucky. It's fair to say that the real situation is complex, and the critical viewpoint outlined above is somewhat simplis-tic. There's room for a diversity of wine styles, even within a particular region, and it would be wrong to impose from above a single taste standard on all the producers within an appellation. Besides, who would be the arbiter of this standard? Additionally, the world of wine can't be expected to stand still. Innovation, experimentation, and progress are inevitable whenever winegrowers are intelligently curi-ous. However, one person's idea of progress is another's idea of iconoclastic trashing of valued tradi-tions, and the positive gloss of the modernizers, who argue that there is now more fine wine being made than ever before, fails to mask entirely the disturbing trend towards darker, richer, more alcoholic red wines. An interesting aside in this context is that the whole discussion of the influence of critics and the changing nature of fine wine styles concerns to a large extent red wines. There hasn't been a similar fight over top whites—notably white Burgundy, German Riesling, Alsace whites, Champagne, and Sauternes. These are wine regions for which Parker points are less important in driving sales.

Thus the first great pressure on wine diversity is a new, simplistic definition of wine quality that treats all wines the same and fails to recognize sense of place as a valid criterion of assessment. This definition has led to a shift in wine style as producers who make high-scoring wines reap substantial rewards.

Another great, rather more ordinary pressure on wine diversity stems from changes in the way wine is sold. The key issue for winemakers worldwide is access to the market. Wine producers used to make wine and then try to sell it. Many still do, with varying degrees of success. Increasingly, however, this approach doesn't work. People in many countries now largely buy wine as part of their supermarket shopping. As a result, the wine trade is consolidating into two rather different markets. At one pole we have the fine wine niche, dominated by the classics, which are sold by specialist merchants to a discerning market of wine geeks and wealthy restaurant diners. At the other we have wine for the masses: wine as a commodity, dominated by branded wines and sold through supermarkets and convenience stores. The middle ground is disappearing fast, which is a minitragedy, because this is where much of the diversity lies.

Modern retailing doesn't suit interesting wine well. Generally speaking, wine is best made by families because of the continuity they offer; making good wine is a long-term investment that usually requires tying up large sums of capital, notably in vineyard ownership, for many years. Families also are usually better able to manage small to medium-sized properties of tens of hectares rather than hundreds. Modern retailing needs volume, demands high margins, and slashes prices to the bone. In this environment it is difficult for smaller producers who aren't able to climb to the relative safety of the fine wine niche to find an outlet for their wines, other than direct sales to consumers and restaurants. Modern retailing much prefers branded wines, made to a style, scalable, and with the inconsistency of vintage variation ironed out, to more idiosyncratic, variable, and honest products made in smaller quantities. Only the big brands have the marketing budget necessary to get the attention of increasingly distracted consumers. Hence the sameness apparent in many supermarket wine offerings.

Another pressure on diversity is the societal shift in attention from the local to the global that has occurred over the past few decades, paralleled by the rise in celebrity culture. With our tight schedules and limited attention spans, we all end up with the same sports heroes, listen to the same music, and chat about the love lives of the same "celebs." The rewards are disproportionately spread: the gulf between the earnings of the tennis players or golfers ranked 1 to 10 and those ranked, say, 50 to 60 is much larger than any differences in ability. Wine faces a similar situation. Frequently, when affluent but time-poor consumers develop an interest in wine, they don't want to fuss with lesser wines; they want the best. A large portion of available wine spending is therefore focused on relatively few wines, with the rewards going to the top producers from the top regions. Prices of the very best wines are inflated massively above prices of wines from producers and regions in the next perceived tier: most collectors can cope only with so many names, and the few top-ranking producers reap a disproportionate share of the spoils.

Together, these pressures have reduced wine diversity. It is worrisome that at both ends of the market, wines are beginning to bunch into a limited number of styles. Go into your local supermarket, and there'll be a selection of Sauvignon Blancs, Chardonnays, Cabernet Sauvignons, Merlots, and Shiraz; if you are lucky, you'll also find some Rieslings and Pinot Noirs, but not much else. In your local specialist wine shop the wine selection may seem more diverse, but many of the top reds will taste rather similar, with lots of fruit and some noticeable oak: ripe, sweet flavours offset by some spice and tannin. True, in both outlets consumers will be faced with what looks like a bewildering array of wines, but this impression of diversity is increasingly illusory. Ironically, factors that have led to the democratization of wine, notably, clear labeling of wines by grape variety and bright, accessible, fruity flavours from modern winemaking techniques, could threaten the future of wine diversity if consumers never progress to appreciating and purchasing more individual, interesting wines. ∎

NO PLACE reflects the debate over international style wines more than the Napa Valley and its prized Cabernet Sauvignons. Napa Cabernets (and some of its Chardonnays) were the stars of the famous Judgment of Paris tasting in 1976, at which a panel of French wine experts preferred the upstart wines of the New World to their own domestic entries.

And thirty years later, Napa Cabernets, weighing in at 15% alcohol, sweetened with "200% new oak," and priced at hundreds of dollars a bottle, are but a faint resemblance of their former selves.

Wine writer Dan Berger, longtime columnist for the *Los Angeles Times* and now publisher of a widely-read independent wine newsletter, penned a two-part series in 2007 for the online wine magazine Appellation America about the Cabernets of the Rutherford American Viticultural Area within Napa, home to many legendary labels. In the first part of the series, Berger bemoaned the similarity of the wines to those of other Napa regions, and in the second part, reprinted here, he looks into why that might be. Berger does not like what he finds, and doesn't mince words laying it out.

FROM DAN BERGER, "REGIONAL TRANSPARENCY—WHEN MARKETING TRUMPS WINEMAKING"

When it became clear to me that a range of Rutherford AVA Cabernet Sauvignons from the 2004 vintage lacked much distinctiveness of regional character, I decided it was time to look at how this might have occurred.

Some might argue that the loss of varietal and regional characteristics has occurred as a result of factors totally out of the hands of man's control (such as "global warming" or "super yeasts" or some other nonsense). It is clear to me that what has happened in Napa and other areas in California is a *conscious* effort on the part of wine makers to deal with a number of difficult factors. Winemakers are trying to make wines that both justify a high price and appeal to some of the more recognizable wine critics who prefer wines that are powerful and flavored rather atypically to what was previously rated as great. (Such as the wines that vanquished the French in 1976.)

This was not the case in the 1970s and 1980s when wines were made from grapevines grown far differently from the way they are today, and when wines weren't being compared with some ethereal paradigm that, frankly, does not represent very good wine. Back then, the top Napa Valley Cabernets were seen as collectibles mainly because they were balanced, improved in the bottle, and delivered a charming level of fruit and complexity over a decade or two. Or more.

Today's more heavy-handed styles of wines don't appear to be aging as long or delivering as much complexity. Indeed, the lack of regional definition has left many of the wines with a hollow mid-palate that is filled mainly with alcohol and oak. In fact, the instant hedonism that some ascribe to some of these wines is more likely a result of aging in expensive French oak barrels. Caramelized flavors are, after all, appealing in a circus atmosphere.

GOODBYE PHYLLOXERA, HELLO BIG VINES

The key factors in how we got to this point—where alcohol is up, regional character is out, and flavor definition is compromised—probably began two decades ago when we began to hear reports of phylloxera hitting Napa Valley vineyards (and elsewhere). A lot of the vines were planted on the AxR1 rootstock, which was well known to be susceptible to that particular root louse. Soon after it was developed, AxR1 was prized for its ability to produce significantly larger crops than did other rootstocks.

AxR1's vigorous root system allowed vines to produce as much as 16 percent more fruit per year (according to one scientific report), and of a quality good enough to make a wine as near-perfect as Heitz's Martha's Vineyard Cabernet—and with consistency. And many other great Cabs came from AxR1, whose abundance was due to the fact that the A in the name of the root system came from Aramon, a huge producer of red wine fruit of rather inconsequential character from France. (It has since been banned by presidential decree from French vineyards!)

When phylloxera hit, most wineries began to tear out the infected vines and many in the North Coast decided to use devigorating rootstocks, a complete reversal from the vigorous AxR1 that had existed before.

Moreover, the top wood (called scion wood) varietal material they chose to graft on top was virus free. Prior to phylloxera, much of the vine material in the North Coast was infected with viruses. Leaf roll virus in particular was actually beneficial, say many growers, because it helped to retard sugar development and allowed the grapes to ripen in terms of flavor maturity longer into the season without high sugars threatening the ultimate balance of the wine with excessive alcohols. It was the root system's vigor that kept the vine in balance.

This was the situation for California's North Coast for some 20 to 30 years or more, and growers got to know how to deal with their vines under this regime: vigor below ground, retarded development above ground. Result: even ripening while sugar developed slowly. Almost all of the new vineyards were thus inverted from what they had been.

Then came another innovation: a new trellising system called vertical shoot positioning (VSP). Older trellising systems had larger canopies of leaves, which also slowed sugar accumulation by not allowing as much sunlight onto the grapes. The VSP method, which was chosen partially because it was easy to manage, allows the vine's fruit-bearing arms to be lifted well above the level at which they once grew, making the leaf canopy thinner and accordingly, exposing the fruit more directly to sunlight. Among other things, this insured what some wine makers said was complete ripeness of the fruit.

MAKE MINE ULTRA-RIPE PLEASE

Many wine makers saw this as a crucial factor in pleasing some wine critics who liked the more ultra-ripe flavors and seemed to dislike the more varietal/regional elements found in Cabernet Sauvignon from more shaded vines. Such wines are slightly greener and more herbal. The key ingredient in Cabernet Sauvignon as well as Merlot and other varieties was methoxypyrazine (or simply pyrazine for short). It was soon the target of near-universal derision and scorn.

The California Sprawl trellising systems of the past seemed to encourage the grapes to retain traces of this naturally occurring component, which is related to both varietal as well as regional character. It was regularly seen in Bordeaux of vintages prior to 1982. By switching to VSP trellises, California growers made life a bit easier in terms of canopy management, and they were able to get more direct sunlight onto the grapes, which generally had the effect of diminishing pyrazine character.

As a result, three major changes had occurred in the vineyards (new roots, new virus-free scion wood, new trellising). So those growers and wine makers who had learned what was happening in their vineyards for decades under varying weather conditions now had to discard what they knew and had to re-learn how to deal with their vines. Moreover, the (new) demand to make a wine with no pyrazine elements encouraged growers to harvest later and later.

Alas, ultra-late harvested fruit can be trickier to ferment to dryness, and wineries don't like stuck fermentations. Nor do they like 16 percent alcohol in their Cabernets, which could well have resulted from some of the must-weights they were getting from later picking. And with high sugars, stuck ferments were a distinct possibility as many yeasts sputter in the presence of high alcohol levels.

So wineries quietly went to the state of California a few years ago and asked for the right to ameliorate with greater amounts of water (usually at the crusher) "to facilitate the completion of the fermentation," according to the propaganda that was fashionable at the time. The State, which knows nothing of the grape growing or wine making process, complied and allowed greater amelioration. It was clearly unaware that one of the pernicious aspects of this ruling was that it basically flew in the face of growers. Most of them were aware that they had virtually ideal ripeness when the grapes averaged 24° to 25° Brix and that 27° or higher would produce what they began to call "dry port."

WATERED-DOWN WINE?

However, under the newly liberalized water-addition regulation, wineries could then tell growers to wait until their grapes got to 27° or even 29° Brix. They didn't fear an alcohol content of 16 percent or even more because the new regulations allowed them to "water back" the fruit at the crusher and make a lower-alcohol wine. Many growers became irate since later harvesting caused their grapes to dehydrate. The loss of water meant that a vineyard that once would deliver four tons of fruit now would deliver only three tons. Most growers are paid on a per-ton basis.

After paying a lot less for ultra-ripe fruit, wineries could now put back the lost ton of water by adding water from a hose. Net result: the growers realized they were being shafted out of revenues and the wineries were still getting all the "tonnage" they desired. It was estimated by one major grower that by waiting as late as the wineries desired, he was losing more than 30 percent of his revenues.

At the same time, it was getting worse for the consumer seeking a balanced wine. The VSP trellising system wasn't working out well for that older, more food-friendly, age-worthy style of wine, since many canopies under the VSP system are so thin, with far less leaf surface area than under other systems. So much so that many red wine grape varieties were getting too much direct sun, which is known to cause sunburn, creating a more raisiny aroma, as well increasing certain harsh tannins in the resulting wine.

But many wine makers didn't have to deal with strong tannin levels because they already were making the wines with elements that mitigated that astringent effect:

Alcohol. Even though they were using later-picked grapes and reducing the alcohol with water, few wineries reduced alcohols to the 13.8 percent of the 1970s or 1980s. In fact, the majority kept the alcohols in the 14.5 percent to 15 percent range, which adds a certain "sweet" feel to the wine and masks the tannins. Alcohol delivers a "sweetish" taste.

The pH. Wine scientists suggest strongly that a red wine pH of 3.6 or less is appropriate for a wine to have proper balance, and to have the potential to age, and to have the structure to work well with food. But the pH levels of most North Coast red wines in the last decade have risen to almost ridiculously high levels. Most were 3.75 or so, some exceeded 4.0, and at that level the wine is considered to be unstable. But the immediate impact was to compromise the acidity. So again the tannins were masked a bit more by lower acids and high pH levels. This was done regularly by some high-end producers by simply adding calcium carbonate.

Oak. The use of new well-toasted French oak can also mask tannins because of the added texture from the barrel. So can use of oak chips. And then we began to hear about "double-oaked" wines that were aged for a year in new French oak, then racked into new barrels for the second year!

IT ALL COMES BACK TO THE ORIGINAL PARADIGM:
WINE IS MADE IN THE VINEYARD.

Those wine makers who claim that so-called "super yeasts" are really to blame for higher alcohols may be hoping the consumer will accept the notion that the basic rules of chemistry no longer apply. Yeasts convert sugar in wine to carbon dioxide and alcohol (ethanol) in a precise, unchanging formula. About 60 percent of the sugar converts to alcohol, so grapes harvested at 24° Brix will have an alcohol level of 14.4 percent if no water is added. With a must weight of 25° Brix, a dry red wine will have 15.0 percent alcohol; at 26° Brix, the alcohol would come out at 15.6 percent. Yeasts, which are catalytic agents, do not create sugar; they only convert what sugar is in a liquid to alcohol on a formulated basis. (The above figures are for wines in which no ameliorating agent or alcohol-removal technique is used.)

As for global climate change, that subject was extensively discussed at Australia's 13th Wine Industry Technical Conference, and most experts said that, thus far, the increases in temperature we have seen in the last few years don't account much, if at all, for higher sugar levels—and most certainly it is not as great a factor as are the demands of wine makers seeking to pick later and later.

In the last decade, a patented reverse osmosis system developed by Vinovation of Sebastopol, Calif., has been widely used by literally hundreds of wineries to remove alcohol from wines. Clark Smith, the developer of the process, estimates that 45 percent of the wines in California are alcohol-reduced by either his technique or the use of the Australian-invented Spinning Cone. (Another patented concept, developed by Memstar in Australia, also allows the use of reverse osmosis to reduce alcohol levels.)

Think about it: if wineries were not completely aware that excessive alcohol posed a serious problem, would they spend the money and effort to reduce the alcohols in their wines? But here's the curious part: once having made the decision to "de-alc" their wine, why have so few dropped the final alcohol below 14 percent?

Reducing wine below 14 percent is a logical process since wine at 13.9 percent is taxed at a lower rate than wine above 14 percent. One would think that, once the de-alc-ing was under way, wineries would drop their alcohols to a more cost-effective, tax-saving level. Yet the vast majority of wines made in California today have alcohols of 14.5 percent or so.

Thus do I suspect that wineries are fearful that their wines could possibly have distinctive varietal or regional character at lower alcohols, which could cause their scores in the glossy magazines to drop. As a result, I believe there is a desire to have high-alcohol to insure that wines have a certain weight, density, impact, and concentration that appeal most to the handful of critics who seem to disregard varietal and/or regional distinctiveness as a factor in their wine evaluations.

In sum, we have started a new vineyard system that is 180 degrees topsy-turvy from what existed just 20 years ago; we have begun to employ new winemaking tactics that discard the teachings of the past that made for classically structured wines. And to justify it all, we have developed a series of buzzwords for today's wines that seem perfectly logical to the newer generation of buyers who have been brain-washed into thinking that today's monster wines are the best use of some of the greatest vineyards in California history.

And the purists be damned.

CLIMATE CHANGE

We know that if we take vines from one patch of earth and plant them in the soils of somewhere else, the resulting wine will be different. But what if that place, that patch of earth that's been occupied by vines for a thousand years, what if that place changes?

Climate change is already a reality in many of the world's vineyards. Places that have relied on natural rainfall for centuries have resorted to driplines in the face of higher temperatures and dwindling precipitation. Vineyard pests that used to be confined to the sunny Mediterranean have made their way to the Mosel. More and more, extreme and erratic weather is the norm, from the heat wave that cooked European grapes in 2003 to the relentless, abnormal fall rains that spread rot all up and down the West Coast of the US in 2011. Record temperatures in Australia feed disastrous bush fires, making smoke taint an issue in certain vintages. In some parts of the world conditions are actually improving for grape growing, worlds are opening up in the wake of more amenable climates. In other parts, it's a world of pain, with options shrinking, or altered irrevocably. In the span of a generation, thousands of years of tradition and cultural practice are becoming unviable.

Farmers, of course, are an adaptable lot. If drought years loom, irrigation systems could be installed, though such efforts wouldn't be much help if the next season is the wettest on record. The trellising system could be tweaked to provide more or less sunlight, but chances are that would also alter grape composition. If a new disease threatens, the vineyards could be ripped out and replaced, but then the wines would be made from vines that were five years old, not fifty. All of these scenarios could keep farmers in business, but all of them could threaten the expression of *terroir*, or at least alter it substantially.

Refer back to the chart from Greg Jones in chapter 4, which displays the temperature ranges in which different grape varieties perform best. Those ranges, you may recall, are relatively narrow. If a small change in annual temperatures could make a variety that has thrived for hundreds of years taste flat and boring, planting another variety might be economically sensible. If it gets too hot in a particular region for Riesling, the locals could always replant with Muscat instead, or maybe switch to reds. But what happens in Burgundy, for example, if the region becomes too warm for Pinot Noir? What happens to the region's reputation when out of climatological necessity the first Syrah vines are planted in the Cotes du Beaune, when they creep into Premier Cru vineyards? And what does it mean to be Grand Cru or Premier Cru if the climate is too warm in these sites to make great, distinguished wine?

Grapes are a sensitive, reactive crop, responsive to the slightest change in their environment, and thus serve as a kind of early warning system for the whole of agriculture. As a result, researchers have given considerable attention to the challenge posed by climate change in numerous conferences, journals, and books. The following is an excerpt from an article by Hans Schultz, viticulturalist at Geisenheim University in Germany, and Greg Jones, climatologist at Southern Oregon University, two of the leading researchers of wine and weather. This selection offers an overview of the climatic trends already in motion and requiring adaptation.

FROM HANS SCHULTZ AND GREG JONES, "CLIMATE INDUCED HISTORIC AND FUTURE CHANGES IN VITICULTURE"
CLIMATE AND VITICULTURE TODAY AND IN THE FUTURE

Traditionally the main grape growing and wine producing areas have been situated between 30° to 50°N and 30° to 40°S, although grapes have been grown outside these limits in the tropics for a long time. These latitudes approximate the 10°C and 20°C annual isotherms, which due to the incorporation of winter temperatures do not accurately reflect the real geographical distribution. The current extension of viticulture is best represented by the 12–13°C (lower threshold) to 22–24°C (upper threshold) isotherm limits of the average growing season temperature. The 12°C lower limit would be situated just south of London, England (1961–1990 average), whereas the 24°C upper limit would include some tropical areas. For many viticultural regions a warming trend has been observed over the past 50–60 years and has been particularly strong over the past 20 years. An analysis of 27 wine regions worldwide showed that the average winter and summer temperatures have increased by 1.3 and 1.4°C, respectively between 1950 and 2000 with a greater increase for regions in the northern hemisphere. Of these regions, 18 showed an increase in temperature variability, which confirms the projections of increased variability by the IPCC. There is evidence that in many regions night temperatures have increased stronger than day temperatures. The current observations in the trends corroborate the projections of the first UN climate report of 1990.

One of the early analyses of the impacts of climate change on viticulture, shortly after the release of the first UN state of the climate report in 1990, were conducted by Kenny and Harrison. The work indicated potential shifts and/or expansions in the geography of viticulture regions with parts of Southern Europe predicted to become too hot to produce high quality wines and northern regions becoming viable once again. These results have largely been proven correct. Other recent research

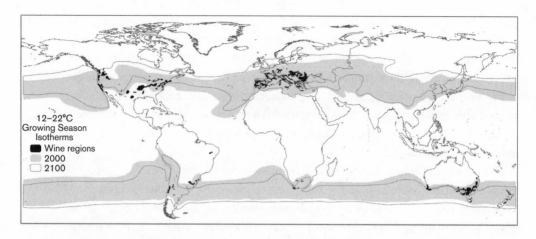

FIGURE 9.2
Map of growing season average temperatures (Northern Hemisphere April–October; Southern Hemisphere October–April).

by Jones et al., examining growing season climates in 27 of the world's top wine producing regions reveals significant changes in each wine region with trends ranging from 0.2°C to 0.6°C per decade for 2000–2049. Overall trends by 2049 average 2°C across all regions.

Depending on the underlying scenario, climate models predict an increase in global temperature of 1.5°C to > 5.0°C by the end of this century. An increase in temperature of this magnitude would substantially alter the geography of wine regions with the potential for relatively large latitudinal shifts in viable viticulture zones (Figure 2 [Figure 9.2 here—eds.]). Some increases in suitable area might be seen on the poleward fringe in the Northern Hemisphere (NH), but decreases in area are likely in the Southern Hemisphere (SH) due to the lack of land mass. Within wine regions, spatial shifts are projected to be toward the coast, up in elevation, and to the north (NH) or south (SH). Furthermore, climate variability analyses have shown evidence of increased frequency of extreme events in many regions, while climate models predict a continued increase in variability globally.

IMPLICATIONS FOR VINE GROWTH AND WINE COMPOSITION

Although many individual climate factors play a role in yield formation, grape development and grape composition (i.e. solar radiation, temperature and temperature extremes, precipitation amount and distribution, wind, humidity, etc.), temperature and water supply are among the most important. These are also the two factors most frequently addressed in reflections on the possible effects of climate change on viticulture.

In a comprehensive worldwide analysis of long-term time trends in vintage ratings for major wine-producing regions, Jones et al. showed a sustained increase in vintage quality, possibly resulting from complex combinations of improved winemaking and crop management technologies, and warming trends. Whereas in this analysis a quadratic effect of temperature on wine quality was assumed, with an implicit expectation of quality improvement with warming in cool regions, and a decrease in quality in hotter regions, trends were positive in 25 out of 30 regions, with only one case of a negative, but statistically non-significant trend in quality. However, Jones *et al.* showed that many of the wine regions may be at or near their optimum growing season temperatures for high quality wine production and further increases will likely place some regions outside their theoretical optimum growing season climate.

In addition, several studies have shown that grapevine phenology has significantly advanced in many wine growing regions in the past and will continue to shift forward in time with the main ripening period occurring at much higher temperatures. This is likely to affect grape composition, as evidenced by long-term increases in temperature in the past being implicated in altered fruit composition in Europe, North America, and Australia. While many studies have used temperature indices to predict shifts in the varietal spectrum, these approaches do not incorporate possible mitigation strategies through cultivation methods. Several of these strategies aim at retarding ripening in order to shift the maturity timing back to periods with cooler conditions that are more optimum for phenolic ripeness and flavour development.

It is also very likely that different varieties will respond differently to warming. For example, an increase in temperature from 20 to 30°C increased the weight of bunch primordia four-fold in Riesling but Shiraz was unaffected. Shiraz also showed very little response in basic yield components in a 2 to 4°C warming experiment. In principle, red varieties appear to tolerate warm conditions better than white varieties. In an in-depth analysis of the relationship between vintage quality and the long-term daily mean temperature during the month prior to harvest, Sadras et al. found contrasting responses for red and white varieties across 24 Australian wine regions. There was a positive correlation of quality ratings and daily mean regional temperature for red but not for white wines, whereas the apparent influence of temperature on vintage variability was strong for white wines but irrelevant for red wines. However, when wine score data were correlated with the average growing season temperature (October to April), there was a negative trend for red and white wines in some of the analyzed regions.

Under hot climate conditions red fruit varieties including Cabernet Sauvignon usually achieve sugar concentrations suitable for quality wine making, but often fail to colour appropriately. The relative stability of sugar in comparison to the plasticity of anthocyanins is partly related to the relative ranges of temperature for optimum activity of sugar (18 to 33°C) and pigment producing enzymes (17 to 26°C). Specifically high temperatures (>30°C) directly after véraison can inhibit anthocyanin formation. These apparent relationships are valuable information, but we only have a small data base of grape composition data, which seems insufficient to extrapolate to past or future climates in terms of wine quality.

Despite the low plasticity of sugar accumulation, Jones and Davis found a continuous, albeit not significant, increase in sugar concentration at harvest between 1970 and 2000 for selected vineyards in Bordeaux. Acidity, however, may be a better proxy for changes in grape composition, since both the pre-véraison formation of malic acid and its degradation after véraison are highly temperature sensitive, thus malic acid shows much more plasticity than sugar. We therefore exploited a relatively unique data set from Schloß Johannisberg in the Rheingau region, Germany, with yield, sugar

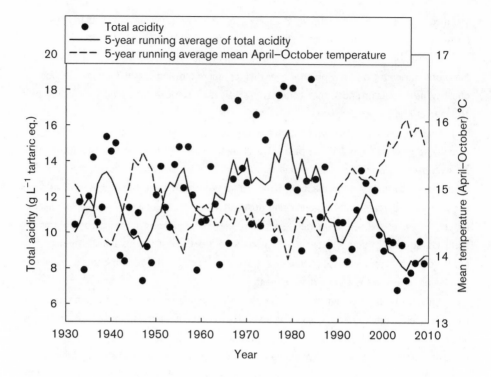

FIGURE 9.3

Total acidity data at harvest for Schloß Johannisberg since 1932.

concentration, and total acidity (malic and tartaric acid) data going back to at least 1932. Schloß Johannisberg has exclusively cultivated the variety Riesling since 1720. The data shown in Figure 3 (*Figure 9.3 here—eds.*) are averages across all individual vineyards of any particular year at time of harvest (between a minimum of eight to a maximum of 125 values per year). There is a clear negative correlation between total acidity and seasonal average temperature (Figure 3). In particular, the strong decrease in acidity over the past 20 years or so was concomitant to a strong increase in growing season temperature. This correlation was better than with any other temperature data of individual months or parts of the growing season (data not shown). Since tartaric acid is largely unaffected by temperature, the variation in acidity concentration is related to changes in malic acid. These data and their correlation with temperature confirm that total acidity is a good proxy for seasonal temperature and that it may be possible to model its development using various climate scenarios.

CONCLUSIONS

Climate change has the potential to greatly impact nearly every form of agriculture. However, history has shown that the narrow climatic zones for growing winegrapes are especially prone to variations in climate and long term climate change. While the observed warming over the last 50 years appears to have mostly benefited the quality of wine grown worldwide, projections of future warming at the global, continental and wine-region scale will likely have both beneficial and detrimental impacts through opening new areas to viticulture and increasing viability, or severely challenging the ability to adequately grow grapes and produce wine.

The spectrum of vineyard sites, climatic conditions, soil types and varieties across the world's grape growing areas is so large, that a general type of adaptive strategy is not possible. In many cases, the balance between vineyard site, climate, soil, variety and the applied cultivation measures, which has developed sometimes over centuries, will be perturbed or even completely altered.

The simplest deduction would be that the varietal spectrum would change substantially since the suitability for the cultivation of a given cultivar is largely temperature driven. However, in many cases only the lower threshold of varietal suitability has been well defined, while the upper suitability threshold of many varieties remains uncertain. Since very often the image of a wine growing region is determined by one or a limited number of cultivars (e.g. Riesling for the Mosel and Rheingau regions in Germany; Pinot noir and Chardonnay for Burgundy, Cabernet Franc, Cabernet Sauvignon, Merlot for Bordeaux, France; Nebbiolo for the Piedmont region of Italy) changing these cultivars may not be the immediate appropriate solution. Additionally, within Europe, the use of specific varieties is often regulated by law, which may become problematic in light of future climatic developments. ∎

IN A 2003 article in the journal *Gastronomica,* food and wine writer Robert Pincus (a former cloud physicist) covers much the same ground as Schultz and Jones, above: the mounting reality of climate change and the perils it poses for traditional expression of *terroir.* In the second half of the article, he delves into the details of the prospects faced in regions that produce some of the world's most exquisite and finely-honed white wines—the elegant whites of Alsace, the Mosel, and Austria's Wachau—and talks with veteran winemakers in each endangered place.

FROM ROBERT PINCUS, "WINE, PLACE, AND IDENTITY IN A CHANGING CLIMATE"

The following accounts are only the beginning of the story. The geography of wine regions is intensely interesting to both vintners and aficionados because the exact location, slope, and exposure of a particular parcel can dramatically affect the way grapes grow and ripen. Given the coarse resolution of the climate simulations (about 280 km, or 175 miles), the available predictions are only a rough guide to the changes that may actually occur. Furthermore, neighboring winegrowers can hold diametrically opposing views on how to grow their grapes, not to mention how to adapt to climate change.

FRANCE'S ALSACE

The vineyards of Alsace nestle up against the eastern slope of the Vosges Mountains, running parallel to the Rhine River in the northeast corner of France.[2] The mountains create a rain shadow, and the resulting sunny days and mild temperatures allow grapes to ripen at this extreme latitude (of the French wine producing regions, only Champagne is slightly further north). The current climate is just warm enough for wine to be produced in most years, and chaptalization (the addition of extra sugar to increase alcohol levels) is not uncommon. The varieties permitted in the *Appellation Contrôlée* are those that favor cool weather, many of which are grown across the river in Germany but not elsewhere in France—Riesling, Gewürztraminer, Muscat, and a little Pinot Noir, the only permissible red variety.

The climate projections for Alsace are as follows:

- Rain and snowfall may decrease slightly, though they will be more uniform from year to year. Springtime will tend to be drier. Summers will vary more dramatically from year to year than they do today, and falls will generally be more consistent.

- Sunshine over the growing season will increase by about 1.5 percent, with most of this change occurring in the summer months. The variations from year to year will be larger, and some years will be much sunnier than they currently are. Fall sunshine will be more variable but spring sunshine more consistent.

- Frosts in late spring may become more common, with the last frost occurring one to three weeks later than it now does. The first frost of fall, however, will be delayed by about a month.

- There will be about three hundred more degree-days during the growing season than there are today. This is a very small percentage increase, corresponding to an average daily temperature change of about 1.1°C. Most of the change will occur in summer and fall.

Jean Hugel of Hugel & Fils in Riquewihr has worked fifty-five vintages and remembers seventy-seven. He takes the long view. "Normally we have blossoming around the 10th of June, in early years at the end of May. In 1307 the blossoms emerged at the end of April, and the crop came in at the end of July," he says, quoting historical records.[3] Then again, the vines froze on May 22, 1146.[4] Vintners have always contended with the vagaries of weather and climate, and in Hugel's view the current climatic changes are not extraordinary. "If grapes ripened early in more years, I'd be delighted, but not convinced."

Alsatian wines are remarkable for their delicacy, floral aromas, and relatively low alcohol levels. Hugel feels that even in the presence of overall warming these traits can be maintained by making careful changes in vineyard and cellar practices. He foresees earlier harvests to limit sugar concentrations and, to a limited extent, changes in crop levels. "We pick Pinot Gris at 13.5 degrees potential alcohol," he asserts. "At 14.5 degrees it won't finish fermentation." He does worry a little about the impact of earlier harvests: "In my experience, a longer ripening period leads to better microchemistry and a better nose." He can imagine, perhaps, greater planting of relatively late-harvested varieties, like Riesling, at the expense of more reliable but less unusual selections like Pinot Blanc or Pinot Gris. But, he adds, "I am opposed to the planting of more Pinot Noir. We are a white wine region."

GERMANY'S MOSEL

It's hard to believe that anyone would seriously try to cultivate anything along the banks of the Mosel. The river's course through western Germany is so tortuous that one might start facing east, walk following the flow for a kilometer or two, and wind up facing west. The banks are vertiginous slopes of loose sheets of rock, climbing several hundred meters to wooded hilltops at angles of forty-five degrees or more, terrain so convoluted that it gives rise to entire taxonomies of microclimates. But in the warm oxbows, where the slopes face the low sun and the light and warmth are reflected from the river onto the dark slate, some of the finest Riesling in the world is grown. Grapes in the Mosel typically spend long, mild falls on the vine before harvest and so retain enough acidity to balance the apple and citrus fruit flavors and to age magnificently. In exceptional years, some growers may also take the risk of leaving grapes hanging in the vineyards to develop the fungus *Botrytis cineria,* the "noble rot" that dehydrates the grape, concentrating the juice inside the berries and adding a haunting flavor of honey. *Botrytis*-affected grapes become the sweet *Beerenauslese* and rare *Trockenbeerenauslese* wines. A few vintners make the further gamble of leaving fruit on the vines until December or later, when the hard frosts arrive and the water in the berries begins to freeze, concentrating the sugars within the grape. The fruit (sometimes picked at night) is pressed frozen to become intensely sweet *Eiswein* (ice wine).

The climate changes projected for the Mosel are:

- Precipitation will decrease by about 5 percent on average. Drier falls will become more common, although the wettest winters will be much wetter than today's.

- Sunshine over the growing season will increase by about 1.5 percent, though this will be more variable from year to year than it is now. Summers, in particular, are likely to be sunnier than today. Extremely sunny or cloudy years will become somewhat more frequent, though still rare.

- There will be about three hundred more degree-days during the growing season. Spring and summer will be uniformly warmer, and fall somewhat warmer but more variable than today.

- Frosts of—7 0°C (hard enough to make Eiswein) will arrive about a week later than they do now, but they will arrive in about 25 percent fewer years.

To Johannes Selbach of Selbach-Oster and J.H. Selbach in Zeltingen these predictions sound as much like history as a possible future. "It was the rule that in a decade you would have a couple of bad vintages [when grapes didn't ripen sufficiently] and a couple of great ones. Since 1988 every year has been good to outstanding," he explains.[5] Warmer weather in the past has been a boon for the Mosel, which is "on the brink of viticulture," and further warming in coming decades will allow Selbach to produce richer, riper wines. German law mandates ripeness levels for various designations of wine; warmer weather would allow Selbach to "put more meat on the bones of every category," as he has begun to do in recent years.

The greatest threat from the changing climate has nothing to do with grapes. "If the Mosel floods," Selbach points out, "we sit in the water." The valley is so narrow that villages lie right next to the river, and the mountains come down steeply to the shore. Like many of his colleagues, Selbach's cellar is in the basement of his home. "If all your money sits under water, it's a bad feeling." Flooding has become more frequent in recent years, a trend that is likely to continue as the winters become wetter. Selbach and others of his generation think it inevitable that they'll have to move from their homes, some of which are hundreds of years old, to higher ground.

Selbach also worries that there will be a vinicultural (*winemaking—eds.*) downside to climate warming in the Mosel. "Hot summers and early ripening would be detrimental to the cause of Mosel Riesling," he asserts, since grapes will lose the acidity that sets Mosel wines apart and lets them age so well. The climate projections don't indicate much danger of this in the next few decades, but they don't bode well for *Eiswein.* Even in the current climate, leaving perfectly good grapes on the vine in the hope of a hard frost is a big risk, because if the freeze does not arrive by January, the grapes will have been damaged by wind, sleet, and hail, and may be wasted. If the frequency of hard frosts in early winter diminishes, growers may become much more reluctant to make the wager. "Not to be able to make *Eiswein,* and not to be able to drink it, would be a real loss," says Selbach. "It's an exciting wine. I'll still try to make it, but I'll make a smaller gamble."

AUSTRIA'S KAMPTAL

To read an Austrian wine label is to get a crash course in a whole new vocabulary of varieties grown nowhere else in the world—Blaufränkisch, Zweigelt, Sankt-Laurent. Wine is produced throughout eastern Austria, near the borders with Slovenia, Hungary, Slovakia, and the Czech Republic. Some of the loveliest Austrian wine is produced in the regions along the Danube—the Wachau, Kamptal, and Kremstal. In Austria, which is substantially further south than Alsace or Germany, it is much easier to ripen grapes fully, so the wines tend to be richer and fuller than their northern counterparts. Austria's vinous heart is Grüner Veltliner, a white variety with structure, mineral content, and complexity, which pairs almost flawlessly with food, especially legumes and vegetables. Grüner Veltliner is a picky grape, requiring fastidious geology and just the right combination of warm days to mature and cool nights to stay fresh, and it thrives in Austria as in no other place on earth.

The climate projections for Austria are that:

- Rain and snowfall may increase by about 10 percent on average, with most of this precipitation occurring in spring. Summer rainfall may tend to cluster, so that a given summer will be either rainy or dry, but not enormously different than today's summers in regard to precipitation.

- Sunshine over the growing season will increase by about 2 percent, though this will vary from year to year. Summers, in particular, are likely to be sunnier than they are today. Extremely sunny or cloudy years will become somewhat more frequent, though they will still be rare.

- Growing season degree-days will increase by about two hundred and fifty (i.e., the average daily temperature will increase by about 1°C). Fall, in particular, will be warmer, as will summer, though to a lesser extent. Summers will also be more variable: years are more likely to be either cool (and cloudy) or hot (and sunny.) but not as moderate as today.

Like Johannes Selbach, Willi Bründlmayer of Weingut Bründlmayer in Langenlois is not surprised by the predictions. "We appreciate by now that a climatic change has already taken place. Our wines are getting more body and less acidity [and growers are] beginning to plant red wines."[6] But because grapes in the Kamptal already receive enough sun and warmth to ripen, Bründlmayer is concerned that more insolation and higher temperatures will push some of the vines too far. "We will drop out of optimum position for Grüner Veltliner," he says bluntly. And although Chardonnay is a more adaptable variety than Grüner Veltliner and is already grown with great success in Austria, Bründlmayer wants to keep producing something more distinctive. "The advantage of Zweigelt or Sankt-Laurent is that they're something new in the world," he explains. These are, in other words, the varieties that distinguish Austrian wine from all others, and Bründlmayer is devoted to the special qualities of what he grows. "If this forecast were 99 percent sure," he says, "I would pursue with vigor new sites for Grüner Veltliner. I would find new vineyards for the new situation." But the decision is not his alone to make. European Union regulations don't permit a net increase in any country's vineyard area, so new, exploratory plantings would have to come at the expense of existing sites. Adjustments at this scale would involve the whole Kamptal, or perhaps all of Austria.

For Bründlmayer, climate change "puts the complete system of *Appellation Controlée* into question." Ten years ago, in an action steeped in tradition, he helped found a small group[7] devoted to finding the best matches among soil, rootstock, and varieties in the Kamptal and neighboring Kremstal appellations. More recently, though, he has begun to believe that "if the future climate [proves] to be fluctuating and uncertain, a long-term fixed relation of vineyard site, rootstock, and variety may not be sensible, [and] a more dynamic approach might be reasonable." The group had already begun discussing these ideas when Europe was unexpectedly drenched by torrential rains in the summer of 2002, which caused one-hundred-year floods along the Danube and inundated several winery cellars in the Kamptal and nearby Wachau. They were forced to turn in a moment from patient investigation to the very short-term, pragmatic question of how to help the growers devastated by the waters. Although the summer's dramatic weather can't definitively be attributed to human activities or climate change, extreme events (like droughts and floods) that were previously rare have become more frequent almost everywhere on the planet.

One could say that we experience weather, as we feel the heat of the day or find that the winter has been especially wet. But we do not experience climate, which occurs on a time scale longer than we can really perceive. Yet climate inexorably affects the way that we live. It's almost certain that the climate will change over the next several decades, and wine producers, like everyone else, will have to adapt. Some will try to use their skill and experience to continue making wines in the traditional style of their region. In colder regions, where ripe grapes and great vintages have historically been rare, their jobs may become a little less stressful. Some adaptations will involve choices made by individuals, such as changes in vineyard management or the replacement of one grape variety for another; others will require more far-reaching changes, including cooperation with governing bodies and appellation boards. But where the culture of wine production is tightly coupled to the current climate, something will have to give.

What is certain to change as a result of adaptation are the links between particular wines and their traditional home. Ice wine, though rare even in Germany, is one of the purest expressions of that chilly climate, a vinicultural turning of lemons into lemonade. Grüner Veltliner is Austria's beloved treasure. Yet both of these wines will likely be much more difficult to produce in the coming decades than they currently are. Even where the impact of climatic change is less dramatic, decades, even centuries, of vinicultural experience will be rendered irrelevant.

ON THE OTHER HAND . . .

Most people involved in the global wine industry acknowledge the reality of climate change, because they believe they can already see its effects in their vineyards. Most viticultural researchers, like most scientists generally, see the evidence as overwhelming and voice a critical need to adapt to an evolving environment.

But as in the broader scientific community, a few notable viticultural researchers have their doubts. Prominent among them is Australian John Gladstones, author of numerous publications on climate and viticulture, including *Wine, Terroir and Climate Change*, from which we excerpted a section on climate and topography in chapter 4. Much of Gladstones' book is devoted to a conventional—and excellent—overview of the functioning of climate in the vineyard, but the later sections on climate change dissent radically from mainstream opinion. The following is a summary section from the final chapter of the book.

FROM JOHN GLADSTONES, *WINE, TERROIR AND CLIMATE CHANGE*
14.2 THE CLIMATE

Chapter 12 examined in some detail the evidence on natural and anthropogenic (man-caused) climate change, and Chapter 13 that of climate in recent decades and its relations to viticulture.

The main conclusion is that warming by anthropogenic greenhouse gases has been much overestimated. The widely publicized claims of the Intergovernmental Panel on Climate Change (IPCC) and other greenhouse proponents have depended too much on computer models unable to encompass the complexity of real climates; on uncertain data, dubious assumptions, and in some key cases biased statistical procedures; and particularly on ignoring the historical record of past climate warmth. Much of the thermometer record of warming over the last 100–150 years, which the IPCC ascribes more or less exclusively to greenhouse gases, has more likely other causes.

Of the estimated rise in recorded temperatures over the 20th century of 0.6°C, about half is almost certainly spurious, caused by urban warming around thermometers and an increasing proportion of topographically warmer recording sites. The methods claimed to have removed these biases from the record have been seriously inadequate.

The 20th century's true warming, as recorded in sea surface temperatures, is at least largely accounted for by natural climate fluctuations, for which the most credible cause on decadal to centennial timescales is fluctuation in solar output and magnetic field.

That rising anthropogenic greenhouse gases should make some contribution is theoretically to be expected. The question is how much, with great uncertainties both as to initial effects and to negative feedbacks that, as is now clear, must greatly moderate them. According to mainstream modelling, anthropogenic greenhouse gases should have increased global temperature over the 20th century by 1°C or more. Patently that has not happened. Properly taking into account biases in the land thermometer record and the impact of natural temperature fluctuations, the best conclusion is that greenhouse gases can have caused no more than 0.2°C of warming, which equates to only 0.4–0.5°C temperature rise for each successive doubling of atmospheric CO_2, or its combined greenhouse equivalent.

The semi-regular temperature fluctuations of the last two centuries suggest further that the natural warming of the late 20th century is at or close to ending, with a likelihood of natural cooling in the near future that will offset any greenhouse warming for some decades to come.

Rising atmospheric CO_2 concentration will itself probably increase the optimum minimum and mean temperatures for vines. This might already have been happening. Thus the possibility cannot be ruled out that the best terroirs will continue their historical shift to warmer locations.

I conclude that the widely held expectation of a viticultural flight to existing cold areas is misplaced. Optimum locations for particular wine styles will probably change little over the coming half century. Any minor shifts will be into areas with higher actual or effective night temperatures, e.g., closer to coasts, to warmer soils or to slopes with superior night air drainage. Favoured sites in warm to hot climates will have higher afternoon relative humidities, most desirably from afternoon sea or lake breezes.

NOTES

1. *Montana* is Croatian for mountain—a good name for wines but confusing here in the United States, where they are marketed under the Brancott Estate brand.

2. A useful source on the geography of viticultural regions worldwide is *Clarke's Wine Atlas* (London: Websters International Publishers, 1995).

3. Telephone interviews with Jean Hugel on 21 November and 27 November 2001.

4. M. Boesch, *800 ans de viticulture en Haute-Alsace* (Guebiller: Privately published, printed at L'imprimerie Art'real, 1983).

5. Telephone interview with Johannes Selbach, November 2001, and ongoing correspondence.

6. Telephone interview with Willi Bründlmayer on 26 November 2001 and ongoing correspondence.

7. The group Traditionsweingüter maintains a Web site (in German) at http://www.traditionsweingüter.at.

10

POSTSCRIPTUM

There's a certain irony in trying to establish a concluding chapter in a book about *terroir*. If we've proven anything in this lengthy tour, it's that *terroir*, the concept, the definition, the proof, is inconclusive, hard to grasp, and it eludes us no matter how hard we try to pin it down. After hundreds of thousands of words and hundreds of points of view, after spilling opinions onto the page like so many splatters of red wine, it seems as if we're no closer to a high-definition picture of this concept than we were when we began.

And yet no one who loves wine wants to deny its existence, or concede that it's not real, detectable, within our reach. We can without much difficulty discern similarities between wines made from the same mesoclimate, in the same soils, or differences between wines made in two different mescoclimates. Even the naysayers concede that site matters, that something accounts for the difference between one site or another, that the same fruit grown in granite, sand, and limestone soils, processed in a similar fashion, often (though not invariably) results in different wines. And no one would argue that the same grapes grown in a warm climate as opposed to a cool climate taste different, feel different in the mouth. Pinot Noir grown in Volnay doesn't resemble Pinot Noir grown in the Russian River Valley, the Santa Rita Hills, or the Eola Hills, and *terroir*, in the broadest sense of the term, accounts for these differences on a conceptual level.

Yet the absurd lengths to which regions go to protect, promote, and exploit such differences are maddening, as are claims of naturalism, of the inherent integrity of organic practices, or native indigenous yeasts, or varying degrees of laissez faire winemaking.

One can be sure that when *terroir* is invoked in the sale of wines, it is invariably distorted, trumped up, its trace elements viewed as certainties.

So we thought that as a concluding gesture it would be worthwhile to revisit, briefly, some of the tenets we set forth in our introductory chapter, and see whether we can still hold them to be truthful and self-evident.

We believe the effects of terroir *are real and undeniable: wines made from grapes grown in different places smell and taste different, vintage after vintage, for multiple, experienced, trustworthy tasters.*

This holds true. The thrill of experiencing *terroir* is derived from acquiring enough sensitivity to capture similarities and distinctions between wines, which unquestionably adds to wine's allure and our enjoyment of it. But every taster is different, every impression is subjective; every time you draw a conclusion about the *terroir* in a wine, you predispose yourself for the next time you taste it. In other words, for every frisson of *terroir* detection, there are distortion, self-deception, and self-fulfilling prophecy at play.

We believe that many if not most of the standard depictions of this phenomenon, however, merit skepticism—at least from the standpoint of modern science.

We remain flummoxed by the fact that grapevines do not communicate, refusing to reveal their inner workings in a private language we can understand. It is frustrating that so little is known about how they take up nutrients, how they interpolate light, how they create the flavors that are the markers of *terroir*. We are perplexed that modern science has been largely denied a gateway to understanding *terroir* better, that most attempts to "prove" its existence come up short. Nevertheless, skepticism is called for, no matter how much we want to believe.

We believe that for regional variations (macro-terroirs), climate and grape variety are both more important than soil; for vineyard variation, soil can be critical.

We can say with some certainty that this tenet has been upheld, that for much of the world the differences bestowed by climate and in grape variety are powerful vectors which distinguish one place from another. And when these are measurably similar, it's probably true that soil accounts the last shred of difference, more than other factors.

We believe that despite the obvious allure of that famous marketing trope, "Great wines are made in the vineyard," we wish to point out that great wine is in fact never made in the vineyard: it's made in the winery.

Our sixth chapter adheres to the notion that the human factor is an irreplaceable cog in the grand wheel of *terroir*, and that without a winemaker's intervention, stewardship, management, sensitivity, perceptual abilities, observational skills, and tasting ability, *terroir* would be undetectable and therefore indefinable.

More conceptually, dirt and place, climate and aspect, these are important elements of *terroir* but they are also fallback positions, appearing concrete, employed as foundations, but as data, rarely as solid as they seem. That is because *terroir* is ultimately about perception, and perception is always a subjective business. When tasters claim they can sense the limestone in the wine or the coolness of the breezes that ripened the fruit, that claim is inevitably a metaphorical construction. For this reason, more than any other wine concept, *terroir*'s subjectivity and mythological component will always be exploitable, and marketing departments will be more than happy to draw from its ephemeral tendrils in its efforts.

Furthermore, it is worth remembering, and the documentation here supports this, that the vast majority of wines in the world are only marginally expressive of place; the trend in wines at present is moving away from *vins de terroir;* industrial, mass-produced beverage wine, self-caricaturing fruit-and-oak bombs, are not and will never be wines of place, whatever their back labels claim.

Terroir, in the end, remains engaging at least in part for the uncertainties that arise in its expression. It is ambiguous, open to interpretation; it calls on a set of perceptual gifts, sensory skills that must be cultivated. And when it is revealed, its revelation is frequently an act of discovery, a breakthrough that is often emotionally transcendent and thrilling. And the satisfaction you feel on discovering it for yourself in a wine you love is nothing short of a Eureka! moment—you'll never look at that wine, or the experience of drinking that wine, or any wine, the same way again.

In this sense, the pursuit of *terroir* and its understanding is, you might say, a leap of faith. It requires belief. Belief in complex biochemical mechanisms outside of our understanding, belief in the power of place to convey a powerful message without language, belief in our own skills to tease out the tiniest sensory details, belief that all these moving parts will fall in line with something like unwavering consistency—knowing full well that variability goes to the very heart of what makes wine great.

Perhaps, in the final analysis, the lure of *terroir* is in the pursuit. Always hovering just beyond our grasp, *terroir* is that ephemeral *je ne sais quoi,* that element of magic that can never be fully defined or described, that lies at the heart of a truly transcendent experience. Maybe, after all is said and done, all we really have to do is close our eyes and allow the magic of the grapes to come to us.

ACKNOWLEDGMENTS

Nancy G. Freeman

Bringing a book to fruition after one of the co-authors has passed away is an act of love and devotion. As Tim's widow, I know he would want to thank everyone, family, friends, and colleagues, who encouraged his dogged pursuit of the many aspects of *terroir* and left us such an immense amount of completed work. Thanks to co-author John Buechsenstein, for shouldering the remainder of the project and carrying on. Patrick Comiskey played an essential role, weaving together new material and editing the text on hand. The editorial staff at UC Press was supportive throughout, particularly Kate Marshall, Bradley Depew, Francisco Reinking, and Blake Edgar. Thanks also to copy editor and indexer Meridith Murray, cartographer Bill Nelson, and to Sheri Gilbert and Teddi Black, who obtained the many necessary permissions.

Not to be left out are Tim's editors and close colleagues over the years who urged him to pursue his examination of *terroir*, among them Jim Gordon, Tina Caputo, and Alder Yarrow. Many other wine professionals stoked his curiosity and hardened his determination to write this book. Some of you will remember a long chat or hard-headed argument as you leaf through these pages. Thanks to you as well.

Most of all, my deepest gratitude goes to my cousin Aaron Belkin. In the wake of Tim's death, I was stunned, barely able to put one foot in front of the other. At that moment, Aaron stepped forward and offered to carry this project across the finish line. It was a precious gift. May this book, in turn, become the gift to the wine world that Tim and John always intended it to be.

Nancy G. Freeman
February 27, 2017

ACKNOWLEDGMENTS

John Buechsenstein

Intriguing ideas are often kindled with stemware in hand. Thus it went with this one—a hotel bar at an ASEV conference a bunch of years ago, first over glasses of wine, but martinis may have followed. So my earliest acknowledgment is to Tim Patterson and Nancy Freeman, who heard me out and pounced on the whole concept. I will leave it to Nancy to enumerate all of the wonderful editors who helped us navigate the final stages of our manuscript. But I do want to offer special thanks to Blake Edgar and UC Press for accepting our proposal and lending support.

A thousand thanks are due to Professor Robert Mayberry, who has been my special *terroir* mentor and friend since the beginning of my career, and to Rory Callahan in New York who talked the subject to death with me many times.

Our cited authors are too numerous to mention here, as you will see. But I thank them all for the inspiration of their individual points of view. Many friends and colleagues helped with my literature quest and served as sounding boards along the way: Bo Simons at the Sonoma County Library; sensory-enology professor Hildegarde Heymann at UC Davis; Scott Frost, my scientific private investigator; Professor Greg Jones at Southern Oregon University; and Glenn McGourty, the Mendocino Cooperative Extension grape expert.

Our book would not have completed its final furlong without the talented writing and editing assistance of Patrick Comiskey and the diligent, caring project management of Aaron Belkin.

I will be forever grateful to my coauthor and favorite home winemaker Tim Patterson, who kept me going with his strong coffee, savory scones, masterful turns of phrase, and disciplined writing schedule.

John Buechsenstein
February 27, 2017

BIBLIOGRAPHY

Amerine, Maynard and Roessler, Edward. *Wines: Their Sensory Evaluation*. San Francisco: W.H. Freeman and Company, 1976.

Asher, Gerald. "Rìas Baixas—Albariño: A Fragrant Wine of the Sea," from *A Vineyard in My Glass*. Berkeley: University of California Press, 2011.

Asimov, Eric. "Champagne's Servants Join the Masters." *New York Times*, July 12, 2011.

Ballester, J., Mihnea, M., Peyron, D., and Valentin, D. "Exploring minerality of Burgundy Chardonnay wines: a sensory approach with wine experts and trained panelists." *Australian Journal of Grape and Wine Research* 19, no. 2 (June 2013): 140–52.

Barcenas, P., Pérez Elortondo, F.J., and Albisu, M. "Projective mapping in sensory analysis of ewe's milk cheeses: A study on consumers and trained panel performance." *Food Research International* 37, no. 7 (August 2004).

Berger, Dan. "Regional Transparency—When Marketing Trumps Winemaking." http://www .appellationamerica.com (accessed September 4, 2007).

Berkowitz, Natalie. *The Winemaker's Hand*. New York: Columbia University Press, 2014.

Bird, David. *Understanding Wine Technology*. 3rd ed. San Francisco: Board and Bench Publishing, 2010.

Bisson, Linda F. "Geographic Origin and Diversity of Wine Strains of Saccharomyces." *American Journal of Enology and Viticulture* 63, no. 2 (June 2012): 164–76.

Bokulich, Nicholas A., *et al.* "Microbial biogeography of wine grapes is conditioned by cultivar, vintage, and climate." *Proceedings of the National Academy of Sciences* (October 2013).

Bonné, Jon. "Drawing new lines for wine on the Sonoma Coast." *San Francisco Chronicle*, April 1, 2011.

297

Boulton, Roger B., Singleton, Vernon L., Bisson, Linda F., and Kunkee, Ralph E. *Principles and Practices of Winemaking*. New York: Chapman & Hall, 1996.

Bourguignon, Claude and Bourguignon, Lydia. "Soil Searching." *Tong*, no. 2 (2010): 12–18.

Boyd, Gerald M. *The Science Behind the Napa Valley Appellation*. Napa: Napa Valley Vintners Association, 2011.

Brook, Stephen. "Priorat." *Decanter*, no. 5 (May 30, 2007).

Busby, James. *Journal of a Tour Through some of the Vineyards of Spain and France*. Sydney: Stevens and Stokes, 1833.

Cadiau, Paul. *Lexivin (Français-Anglais)*. Paris: Paul Cadiau, 1988.

Callahan, Sue. "The Illusive Matter of Terroir: Can it be duplicated in the New World?" *Proceedings of the Australian and New Zealand Marketing Academy (ANZMAC) Conference* (2007).

Carey, Victoria, Archer, Eben, and Saayman, Dawid. "Natural terroir units: What are they? How can they help the wine farmer?" *WineLand*, February 2002.

Casamayor, Pierre. *How to Taste Wine*. London: Cassell Illustrated, 2002.

Cassell's French Dictionary. New York: Macmillan, 1962.

Charters, Stephen. "Marketing terroir: A conceptual approach." 5th International Academy of Wine Business Research Conference, Auckland, New Zealand, February, 2010.

Clarke, Oz with Rand, Margaret. *Grapes & Wine*. New York: Sterling Epicure, 2007.

Colette. *Tendrils of the Vine*, Second Movement. 1908.

Comiskey, Patrick. "Back to Nature." *Bon Appetit*, February 2009, 58, 61.

Coombe, Bryan G. "Research on development and ripening of the grape berry." *American Journal of Enology and Viticulture* 43 (January 1992): 101–10.

Dagueneau, Didier. "Man and Terroir," from *Terroir and the Winegrower*. Edited by Jacky Rigaux. Translated by Catherine du Toit and Naòmi Morgan. Dijon: ICO, 2008.

de Blij, Harm Jan. *Wine, A Geographic Appreciation*. Totowa, New Jersey: Rowman and Allanheld, 1983.

Debuigne, Gerard. *Dictionnaire des vins*. Paris: Librairie Larousse, 1992.

Dehlholm, C., Brockhoff, P.B., Meinert, L., Aaslyng, M.D., and Bredie, W.L.P. "Rapid descriptive sensory methods—Comparison of free multiple sorting, partial napping, napping, flash profiling and conventional profiling." *Food Quality and Preference* 26, no. 2 (December 2012).

Derenoncourt, Stephan. "Terroir," from *Terroir and the Winegrower*. Edited by Jacky Rigaux. Translated by Catherine du Toit and Naòmi Morgan. Dijon: ICO, 2008.

Elliott-Fisk, Deborah. "Geography and the American Viticultural Areas Process, Including a Case Study of Lodi, California," from *The Geography of Wine*. Edited by Percy H. Dougherty. Heidelberg: Springer, 2012.

Estreicher, Stefan K. "Dark Ages, Light Wines," from *Wine from Neolithic Times to the 21st Century*. Algora Publishing, New York, 2006.

Fanet, Jacques. *Great Wine Terroirs*. Berkeley: University of California Press, 2004.

Fischer, Ulrich, Roth, Dirk, and Christmann, Monika. "The impact of geographic origin, vintage and wine estate on sensory properties of *Vitis vinifera* cv. Riesling wines." *Food Quality and Preference* 10, nos. 4–5 (July 1999): 281–288.

Foulkes, Christopher, ed. *Larousse Encyclopedia of Wine*. Paris: Larousse, 1994.

Gayton, Don. *Okanagan Odyssey: Journeys through Terrain, Terroir & Culture*. Victoria, B.C.: Rocky Mountain Books, 2010.

Geiger, Rudolf. *The Climate Near the Ground.* Translated by Scripta Technica. Cambridge, MA: Harvard University Press, 1965.

Gladstones, John. *Viticulture and Environment.* Adelaide: Winetitles, 1992.

Gladstones, John. *Wine, Terroir and Climate Change.* Kent Town, South Australia: Wakefield Press, 2011.

Goode, Jamie and Harrop, Sam. *Authentic Wine: Toward Natural and Sustainable Winemaking.* Berkeley: University of California Press, 2011.

Goode, Jamie. *The Science of Wine: From Vine to Glass.* 2nd ed. Berkeley: University of California Press.

Goodwin, Ian. "Managing Water Stress in Grape Vines in Greater Victoria." Website of the State Government of Victoria, Australia, Department of Environment and Primary Industries, November, 2002. http://www.agriculture.vic.gov.au/agriculture/horticulture/wine-and-grapes/managing-water-stress-in-grape-vines-in-greater-victoria.

Grahm, Randall. "A Meditation on Terroir: The Return." *World of Fine Wine,* edition 21 (September 2008).

Grahm, Randall. "The Search for a Great Growth in the New World," from *Been Doon So Long, A Randall Grahm Vinthology.* Berkeley: University of California Press, 2009. Originally published in installments in *World of Fine Wine* in 2006 and 2007.

Gray, W. Blake. "Demoting Terroir." Wine Review Online. October 19, 2010. http://www.winereviewonline.com.

Greene, Joshua. "Rock's Face." Editor's note. *Wines & Spirits* (Fall 2012).

Harrington, Robert J. *Food and Wine Pairing: A Sensory Experience.* Hoboken: John Wiley & Sons, 2008.

Hellman, Edward. *Oregon Viticulture.* Corvallis: Oregon State University Press, 2003.

Heymann, Hildegarde, Hopfer, Helene, and Bershaw, Dwayne. "An Exploration of the Perception of Minerality in White Wines by Projective Mapping and Descriptive Analysis." *Journal of Sensory Studies* 29, no. 1 (February, 2014).

Holt, Helen, Pearson, Wes, and Francis, Leigh. "Napping—A Rapid Method for Sensory Analysis of Wines." *Australian Wine Research Institute Technical Review* 208 (2014): 10–14.

Hopfer, Helene, and Heymann, Hildegarde. "A summary of projective mapping observations—The effect of replicates and shape, and individual performance measurements." *Food Quality and Preference* 28, no. 1 (April, 2013).

Iland, Patrick, Gago, Peter, Caillard, Andrew, and Dry, Peter. *A Taste of the World of Wine.* Adelaide: Patrick Iland Wine Promotions, 2009.

Jackson, David, and Cherry, Neil J. "Prediction of a District's Grape-Ripening Capacity Using a Latitude Temperature Index (LTI)." *American Journal of Enology and Viticulture* 39, no. 1 (1988): 19–26.

Jackson, David, and Schuster, Danny. *The Production of Grapes and Wine in Cool Climates.* Christchurch, New Zealand: Gypsum Press, 1994.

Jackson, David. *Climate: Monographs in Cool Climate Viticulture—2.* Wellington, New Zealand: Daphne Brasell Associates, 2001.

Jackson, Ronald S. *Wine Tasting: A Professional Handbook.* 2nd ed. Burlington, MA: Academic Press, 2008.

Jayer, Henri. "I Have a Dream . . . " from *Terroir and the Winegrower*. Edited by Jacky Rigaux. Translated by Catherine du Toit and Naòmi Morgan. Dijon: ICO, 2008.

Johnson, Hugh, and Halliday, James. *The Vintner's Art*. New York: Simon & Schuster, 1992.

Jones, Gregory. "Grapevine Climate/Maturity Groupings." Chart as of 2012, courtesy of the author.

Keller, Markus. *The Science of Grapevines: Anatomy and Physiology*. Burlington, MA: Academic Press, 2010.

Kennedy, Jessica, and Heymann, Hildegarde. "Projective mapping and descriptive analysis of milk and dark chocolates." *Journal of Sensory Studies* 24, no. 2 (April 2009).

Kliewer William, and Dokoozlian, Nick K. "Leaf Area/Crop Weight Ratios of Grapevines: Influence on Fruit Composition and Wine Quality." *American Journal of Enology and Viticulture* 56, no. 2 (June 2005): 170–81.

Kramer, Matt. "The Notion of Terroir," from *Making Sense of Burgundy*. New York: William Morrow and Company, 1990.

Kreydenweiss, Marc. "My philosophy as a man of the vines," from *Terroir and the Winegrower*. Edited by Jacky Rigaux. Translated by Catherine du Toit and Naòmi Morgan. Dijon: ICO, 2008.

Lapsley, James. *Bottled Poetry*. Berkeley: University of California Press, 1996.

Laville, Pierre. "Natural terroir units and terroir. A necessary distinction to give more coherence to the system of appellation of origin." Département Cartes et Synthèses Géologiques. B.P. 6009, 45060 Orléans Cedex, France: Bulletin de L'O.I.V., 1993.

Lewin, Benjamin. *What Price Bordeaux?* Dover: Vendange Press, 2009.

Librairie Larousse. *Wines and Vineyards of France*. New York and Paris: Arcade Publishing, 1991.

Lukacs, Paul. *Inventing Wine, A New History of One of the World's Most Ancient Pleasures*. New York: W. W. Norton & Company, 2012.

Lund, Cynthia M., Thompson, Michelle K., Benkwitz, Frank, Wohler, Mark W., Triggs, Chris M., Gardner, Richard, Heymann, Hildegarde, and Nicolau, Laura. "New Zealand Sauvignon blanc Distinct Flavor Characteristics: Sensory, Chemical, and Consumer Aspects." *American Journal of Enology and Viticulture* 60, no. 1 (January 2009).

Lynch, Kermit. *Adventures on the Wine Route: A Wine Buyer's Tour of France*. New York: The Noonday Press, 1990.

Maltman, Alex. "Minerality in wine: a geological perspective." *Journal of Wine Research* 24, no. 3 (May 2013).

Martini, Alessandro. "Origin and Domestication of the Wine Yeast Saccharomyces cerevisiae." *Journal of Wine Research* 1993 4, no. 3: 165–75.

Matthews, Mark. *Terroir and Other Myths of Winegrowing*. Berkeley: University of California Press, 2016.

Mayberry, Robert W. *Wines of the Rhône Valley: A Guide to Origins*. Totowa, NJ: Rowman & Littlefield, 1987.

McGee, Harold and Patterson, Daniel. "Talk Dirt to Me." *New York Times Magazine*, May 6, 2007.

McGovern, Patrick E. *Uncorking the Past: The Quest for Wine, Beer, and Other Alcoholic Beverages*. Berkeley: University of California Press, 2009.

Michelsen, Clive. *Tasting & Grading Wine*. Malmo: JAC International AB, 2005.

Moran, Warren. "You said Terroir? Approaches, sciences, and explanations." Keynote address at *Terroir 2006*, a Symposium held at the University of California, Davis, March, 2006.

Moussaoui, Karima A. and Varela, Paula. "Exploring consumer product profiling techniques and their linkage to a quantitative descriptive analysis." *Food Quality and Preference* 21, no. 8 (December 2010).

Mullins, Michael G., Bouquet, Alain, and Williams, Larry E. *Biology of the Grapevine*. Cambridge: Cambridge University Press, 1992.

Napa Valley Destination Council, Preliminary Draft, FY2001–2015, TBID Destination Marketing Plan, V. 1.9, January 2010.

Noble, Ann C., Williams, Anthony A., and Langron, Stephen P. "Descriptive Analysis and Quality Ratings of 1976 Wines from Four Bordeaux Communes." *Journal of the Science of Food and Agriculture* 35, no. 1 (January 1984): 88–98.

Nowak, Linda, Wagner, Paul, and Arnold, Jean. "Marketing and Branding Wine," in *Wine: A Global Business*. Edited by Liz Thach and Tim Matz. Elmsford, New York: Miranda Press, 2008.

Office International de la Vigne et du Vin. *Lexique de la Vigne et du Vin*. Réimpression en facsimilé de l'édition de 1963. Paris: O.I.V., 1963.

Pagès, Jérôme. "Collection and analysis of perceived product inter-distances using multiple factor analysis: Application to the study of 10 white wines from the Loire Valley." *Food Quality and Preference* 16, no. 7 (October, 2005).

Parker, Thomas. *Tasting French Terroir: The History of an Idea*. Berkeley: University of California Press, 2015.

Parr, Wendy V., Ballester, J., Valentin, D., Peyron, D., Sherlock, R., Robinson, B., Breitmeyer, J., Darriet, P., and Grose, C. "The nature of perceived minerality in white wine: preliminary sensory data." *New Zealand Winegrowers* 78 (Feb/March 2013): 71–75.

Patterson, Tim. "Rocks in my Head." *Wine Enthusiast* 20, no. 11 (November 1, 2007): 136.

Peynaud, Emile. *The Taste of Wine*. Translated by Michael Schuster. San Francisco: Wine Appreciation Guild, 1987.

Pincus, Robert. "Wine, Place, and Identity in a Changing Climate." *Gastronomica* 3, no. 2 (Spring 2003).

Pinney, Thomas. *A History of Wine in America: from Prohibition to the Present*. 2nd revised ed. Berkeley: University of California Press, 2005.

Pinney, Thomas. *A History of Wine in America: from the Beginnings to Prohibition*. Berkeley: University of California Press, 1989.

Pinney, Thomas. "The Language of Wine in English." *Wayward Tendrils Quarterly* 21, no. 1 (January, 2011).

Pogue, Kevin. "Influence of Basalt On The Terroir of The Columbia Valley American Viticultural Area." Presentation at the 8th International Terroir Conference, Soave, Italy, 2010.

Pollan, Michael. *Botany of Desire*. New York: Random House, 2001.

Ravaz, Louis. "Sur la brunissure de la vigne." Les Comptes Rendus de l'Académie des Sciences 136 (1903):1276–78.

Reeve, Jennifer R., Carpenter-Boggs, L., Reganold, John P., York, Alan L., McGourty, Glenn, and McCloskey, Leo P. "Soil and Winegrape Quality in Biodynamically and Organically

Managed Vineyards." *American Journal of Enology and Viticulture* 56, no. 4 (2005): 367–74.

Renouil, Yves (director), and de Traversay, Paul (collaborator). *Dictionnaire du Vin*. Bordeaux: Féret et Fils, 1962.

Rigaux, Jacky, ed. *Terroir and the Winegrower*. Translated by Catherine du Toit and Naòmi Morgan. Dijon: ICO, 2008.

Risvik, E., McEwan, J. A., and Redbotten, M. "Evaluation of sensory profiling and projective mapping data." *Food Quality and Preference* 8, no. 1 (January 1997).

Robinson, Anthony L., Adams, Douglas O., Boss, Paul K., Heymann, Hildegarde, Solomon, Peter S., and Trengove, Robert D. "Influence of Geographic Origin on the Sensory Characteristics and Wine Composition of *Vitis vinifera* cv. Cabernet Sauvignon Wines from Australia." *American Journal of Enology and Viticulture* 63, no. 4 (2012): 467.

Robinson, Jancis. *How to Taste*. New York: Simon & Schuster, 2000.

Robinson, Jancis, ed. *The Oxford Companion to Wine*. 3rd ed. New York: Oxford University Press, 2006.

Sacks, Gavin L., Acree, Terry E., Kwasniewswki, Misha T., Vanden Heuvel, Justine E., and Wilcox, Wayne F. "'Tell me about your childhood:' The role of the vineyard in determining wine flavour chemistry." *Proceedings of the Australian Wine Technical Conference* (June 2013): 39–46.

Schlachter, Kyle. "The Fallacy of Terroir." Colorado Wine Press. July 24, 2013. http://www.coloradowinepress.com.

Schlosser, James, Reynolds, Andrew G., King, Marjorie, and Cliff, Margaret. "Canadian terroir: sensory characterization of Chardonnay in the Niagara Peninsula." *Food Research International* 38, no. 1 (January 2005): 11–18.

Schoonmaker, Frank. *Frank Schoonmaker's Encyclopedia of Wine*. 5th ed. New York: Hastings House, 1973.

Schultz, Hans R., and Jones, Gregory V. "Climate Induced Historic and Future Changes in Viticulture." *Journal of Wine Research* 21, nos. 2 & 3 (2010): 137–145.

Seguin, Gerard. "Terroirs and the pedology of wine growing." *Experientia* 42, no. 8 (August 1986): 861–73.

Serres, Olivier. *The Theatre of Agriculture and the Management of Fields*. France, 1600.

Shaulis, Nelson J., Jordan, Trenholm D. and Tomkins, John P. *Cultural Practices for New York Vineyards*. Cornell Extension Bulletin 805, 1966.

Skinkis, Patty. "Basic Concept of Vine Balance." Extension. January 22, 2013. http://www.extension.org/pages/33109/basic-concept-of-vine-balance#.UoatkCemZMg.

Skinner, Paul. "Soils and Wine Grapes in the Napa Valley." Study conducted for Terra Spase, 2003.

Skinner, Paul. "Weather and Wine Grapes in the Napa Valley." Study conducted for Terra Spase, 2003.

Smart, Richard, and Robinson, Mike. *Sunlight Into Wine: A Handbook for Winegrape Canopy Management*. Underdale, South Australia: Winetitles, 1991.

Smith, Clark. *Postmodern Winemaking: Rethinking the Modern Science of an Ancient Craft*. Berkeley: University of California Press, 2014.

Smith, Rod. "Waiting for the Valley to Inhale." *Los Angeles Times*, May 7, 2002.

Steiman, Harvey. "Is It All in the Funk? How 'natural wines' can polarize wine drinkers." Wine Spectator. March 9, 2015. http://www.winespectator.com.

Stevenson, Robert Louis. *The Silverado Squatters*. New York: Scribner's & Sons, 1902.

Sullivan, Charles. *Napa Wine: A History from Mission Days to Present*. 2nd ed. San Francisco: Board and Bench Publishing, 2008.

Swinchatt, Jonathan. "The Foundations of Wine in the Napa Valley: Geology, Landscape and Climate of the Napa Valley AVA." Study for EarthVision in 2002.

Swinchatt, Jonathan, and Howell, David. *The Winemaker's Dance: Exploring Terroir in the Napa Valley*. Berkeley: University of California Press, 2004.

Theise, Terry. *Reading Between the Wines*. Berkeley: University of California Press, 2010.

Tomasino, Elizabeth, Harrison, Roland, Sedcole, Richard, and Frost, Andy. "Regional Differentiation of New Zealand Pinot noir Wine by Wine Professionals Using Canonical Variate Analysis." *American Journal of Enology and Viticulture* 64, no. 3 (2013): 357.

Trubek, Amy. *The Taste of Place: A Cultural Journey into Terroir*. Berkeley: University of California Press, 2008.

Ulin, Robert C. "Invention and Representation as Cultural Capital: Southwest French Wine-growing History." *American Anthropologist* 93, no. 3 (September 1995): 519–527.

Valentin, D., Chollet, S., Lelièvre, M., and Abdi, H. "Quick and dirty but still pretty good: a review of new descriptive methods in food science." *International Journal of Food Science and Technology* 47, no. 8 (August, 2012).

Varela, Paula, and Ares, Gastón. "Sensory profiling, the blurred line between sensory and consumer science. A review of novel methods for product characterization." *Food Research International* 48, no. 2 (October 2012).

Veinand, B., Godefroy, C., Adam, C., and Delarue, J. "Highlight of important product characteristics for consumers. Comparison of three sensory descriptive methods performed by consumers." *Food Quality and Preference* 22, no. 5 (July 2011).

Veseth, Mike. *Wine Wars: The Curse of the Blue Nun, the Miracle of Two Buck Chuck, and the Revenge of the Terroirists*. Lanham, Maryland: Rowman & Littlefield, 2011.

Viticultural Roundtable: all questions and answers from personal email communications, December 2013.

Wade, Nicholas. "Microbes May Add Special Something to Wines." *New York Times*, November 15, 2013.

Washam, Ron. "The HoseMaster's Comprehensive Guide to Wine, Chapter 8: The Proper Use of Wine Terminology." Hosemaster of Wine. June 5, 2014. http://www.hosemaster-ofwine.blogspot.com/2014/06/the-hosemasters-comprehensive-guide-to.html.

White, Robert E. *Soils for Fine Wines*. New York: Oxford University Press, 2003.

Williams, John. "Wine Quality as Influenced by Sustainable Practices." Presentation at session on Science of Sustainable Viticulture, 54th Annual Meeting of the American Society for Enology and Viticulture, Reno, Nevada, June, 2003.

Williams, Larry E. "Grape," from *Photoassimilate Distribution in Plants and Crops: Source-Sink Relationships*. Edited by Eli Zamski and Arthur A. Schaffer. New York: Marcel Dekker, 1996.

Williams, Larry E., Dokoozlian, Nick K., and Wample, Robert. "Grape," from *Handbook of Environmental Physiology of Fruit Crops*, Vol. 1: *Temperate Crops*. Edited by Bruce Schaffer and Peter C. Andersen. Boca Raton: CRC Press, 1994.

Wilson, James. *Terroir: The Role of Geology, Climate, and Culture in the Making of French Wines*. Berkeley: University of California Press, 1999.

Winkler, Albert J., Cook, James A., Kliewer, William, and Lider, Lloyd A. *General Viticulture*. Berkeley: University of California Press, 1974.

Wolikow, Serge, and Jacquet, Olivier. "A Victory of the Unions." *Tong*, no. 2 (Summer 2009): 12–27.

CREDITS FOR REPRINTED MATERIALS

Asher, Gerald. "Rìas Baixas—Albariño: A Fragrant Wine of the Sea," from *A Vineyard in My Glass*. Berkeley: University of California Press, 2011.

Asimov, Eric. "Champagne's Servants Join the Masters." *New York Times*, July 13, 2011. Copyright © 2011 The New York Times. All rights reserved. Used by permission.

Ballester, J., Mihnea, M., Peyron, D., and Valentin, D. "Exploring minerality of Burgundy Chardonnay wines: a sensory approach with wine experts and trained panelists." *Australian Journal of Grape and Wine Research* 19, no. 2 (June 2013): 140–51. Reprinted with permission from John Wiley and Sons.

Berger, Dan. "Regional Transparency—When Marketing Trumps Winemaking." http://www .appellationamerica.com (accessed September 4, 2007). Reprinted with permission from Appelation America (US) Inc.

Berkowitz, Natalie. *The Winemaker's Hand*. New York: Columbia University Press, 2014. Copyright © 2014 Natalie Berkowitz. Reprinted with permission of Columbia University Press.

Bird, David. *Understanding Wine Technology*. 3rd ed. San Francisco: Board and Bench Publishing, 2011. Copyright © David Bird 2010. Reprinted with permission of Board and Bench Publishing.

Bisson, Linda F. "Geographic Origin and Diversity of Wine Strains of Saccharomyces." *American Journal of Enology and Viticulture* 63, no. 2 (June 2012): 164–76.

Bokulich, Nicholas A., *et al.* "Microbial biogeography of wine grapes is conditioned by cultivar, vintage, and climate." *Proceedings of the National Academy of Sciences* (October 2013).

Bonné, Jon. "Drawing new lines for wine on the Sonoma Coast." *San Francisco Chronicle*, April 1, 2011. Republished with permission.

Boulton, Roger B., Singleton, Vernon L., Bisson, Linda F., and Kunkee, Ralph E. *Principles and Practices of Winemaking.* New York: Chapman & Hall, 1996.

Bourguignon, Claude and Bourguignon, Lydia. "Soil Searching." *Tong,* no. 2 (2010): 12–18. Reprinted with permission from the authors.

Boyd, Gerald D. *The Science Behind the Napa Valley Appellation.* Napa: Napa Valley Vintners Association, 2004. Reprinted with permission from the author.

Brook, Stephen. "Priorat." *Decanter,* no. 5 (May 30, 2007). Reprinted with permission from the publisher.

Busby, James. *Journal of a Tour Through some of the Vineyards of Spain and France.* Sydney: Stevens and Stokes, 1833.

Cadiau, Paul. *Lexivin (Français-Anglais).* Paris: Paul Cadiau, 1987.

Callahan, Sue. "The Illusive Matter of Terroir: Can it be duplicated in the New World?" *Proceedings of the Australian and New Zealand Marketing Academy (ANZMAC) Conference,* 2007.

Carey, Victoria, Archer, Ebe, and Saayman, Dawid. "Natural terroir units: What are they? How can they help the wine farmer?" *WineLand,* February 2002. Reprinted with permission of the publisher.

Cassell's French Dictionary. New York: Macmillan, 1962.

Charters, Stephen. "Marketing terroir: A conceptual approach." 5th International Academy of Wine Business Research Conference, Auckland, New Zealand, February, 2010. Reprinted with permission from the author.

Clarke, Oz with Rand, Margaret. Grapes & Wine. New York: Sterling Epicure, 2007. Text copyright © Oz Clarke and Margaret Rand 2001, 2003, 2007, 2008, 2009, 2010. New York: Sterling Publishing Co, Inc.

Colette. *Tendrils of the Vine,* Second Movement. 1908.

Comiskey, Patrick. "Back to Nature." *Bon Appetit,* February 2009, 58, 61. © Conde Nast. Reprinted with permission.

Dagueneau, Didier. "Man and Terroir," from *Terroir and the Winegrower.* Edited by Jacky Rigaux. Translated by Catherine du Toit and Naòmi Morgan. Dijon: ICO, 2008. Reprinted with permission.

Debuigne, Gerard. *Dictionnaire des vins.* Paris: Librairie Larousse, 1985.

Derenoncourt, Stephan. "Terroir," from *Terroir and the Winegrower.* Edited by Jacky Rigaux. Translated by Catherine du Toit and Naòmi Morgan. Dijon: ICO, 2008. Reprinted with permission.

Elliott-Fisk, Deborah. "Geography and the American Viticultural Process, Including a Case Study of Lodi, California," from *The Geography of Wine.* Edited by Percy H. Dougherty. Heidelberg: Springer, 2012. © Springer Science + Business Media B.V. 2012. Reprinted with permission from the publisher.

Estreicher, Stefan K. "Dark Ages, Light Wines," from *Wine: From Neolithic Times to the 21st Century.* Algora Publishing, New York, 2006. Reprinted with permission from the publisher.

Fanet, Jacques. *Great Wine Terroirs.* Berkeley: University of California Press, 2004.

Fischer, Ulrich, Roth, Dirk, and Christmann, Monika. "The impact of geographic origin, vintage and wine estate on sensory properties of Vitis vinifera cv. Riesling wines." *Food Quality and Preference* 10, nos. 4–5 (July 1999): 281–288.

Foulkes, Christopher, ed. *Larousse Encyclopedia of Wine.* Paris: Larousse, 1994.

Geiger, Rudolf. *The Climate Near the Ground.* Translated by Scripta Technica. Cambridge, MA: Harvard University Press, 1965. This publication is now in its 7th Edition, and reprinted with permission from Rowman and Littlefield, Inc.

Gladstones, John. *Wine, Terroir and Climate Change.* Copyright © John Gladstones 2011. Reprinted with permission from Wakefield Press.

Goode, Jamie and Harrop, Sam MW. *Authentic Wine: Toward Natural and Sustainable Winemaking.* Berkeley: University of California Press, 2011.

Goode, Jamie. *Wine Science: The application of science in winemaking.* Copyright © Octopus Publishing Group Ltd. 2005.

Goodwin, Ian. "Managing Water Stress in Grape Vines in Greater Victoria," published November, 2002 on website of the State Government of Victoria, Australia, Department of Environment and Primary Industries. © State of Victoria, Department of Economic Development, Jobs, Transport and Resources. Reproduced with permission.

Grahm, Randall. "A Meditation on Terroir: The Return." *World of Fine Wine,* edition 21 (September 2008). Reprinted with permission from the author.

Grahm, Randall. "The Search for a Great Growth in the New World," from *Been Doon So Long, A Randall Grahm Vinthology.* Berkeley: University of California Press, 2009. Originally published in installments in *World of Fine Wine* in 2006 and 2007. Reprinted with permission from the author.

Gray, W. Blake. "Demoting Terroir." Wine Review Online. October 19, 2010. http://www.winereviewonline.com. Reprinted with permission from the author.

Harrington, Robert J. *Food and Wine Pairing: A Sensory Experience.* New York: John Wiley & Sons, 2008.

Hellman, Edward W. *Oregon Viticulture,* edited by Edward W. Hellman. Copyright © 2003 Oregon Winegrowers' Association. Reprinted with the permission of Oregon State University Press.

Heymann, Hildegarde, Hopfer, Helene, and Bershaw, Dwayne. "An Exploration of the Perception of Minerality in White Wines by Projective Mapping and Descriptive Analysis." *Journal of Sensory Studies* 29, no. 1 (February, 2014).

Holt, Helen, Pearson, Wes, and Francis, Leigh. "Napping—A Rapid Method for Sensory Analysis of Wines." *Australian Wine Research Institute Technical Review* 208 (2014): 10–14. Reprinted with permission.

Iland, Patrick, Gago, Peter, Caillard, Andrew, and Dry, Peter. *A Taste of the World of Wine.* Adelaide, Australia: Patrick Iland Wine Promotions, 2009. Reprinted with permission. The authors and publisher invite you to read their recent publication, Iland, Patrick, Gago, Peter, Caillard, Andrew, and Dry, Peter. *Australian Wine: Styles and Tastes, People and Places.* Patrick Iland Wine Promotions, 2017.

Jackson, David, and Schuster, Danny. *The Production of Grapes and Wine in Cool Climates.* Christchurch, New Zealand: Gypsum Press, 1994.

Jackson, David. *Climate: Micrographs in Cool Climate Viticulture—2.* Wellington, New Zealand: Daphne Brasell Associates, 2001. Reprinted with permission from Dunmore Publishing Ltd.

Jayer, Henri. "I Have a Dream . . . " from *Terroir and the Winegrower.* Edited by Jacky Rigaux. Translated by Catherine du Toit and Naòmi Morgan. Dijon: ICO, 2008. Reprinted with permission.

Johnson, Hugh, and Halliday, James. *The Vintner's Art.* New York: Simon & Schuster, 1992.

Keller, Markus. *The Science of Grapevines: Anatomy and Physiology.* Copyright © 2010, Markus Keller. Reprinted with permission from Elsevier.

Kramer, Matt. "The Notion of Terroir," from *Making Sense of Burgundy.* Copyright © 1990 by Matt Kramer. Published by William Morrow and Company, Inc. Reprinted with permission from the author.

Kreydenweiss, Marc. "My philosophy as a man of the vines," from *Terroir and the Winegrower.* Edited by Jacky Rigaux. Translated by Catherine du Toit and Naòmi Morgan. Dijon: ICO, 2008. Reprinted with permission.

Lapsley, James. *Bottled Poetry.* Berkeley: University of California Press, 1996.

Laville, P. "Natural terroir units and terroir. A necessary distinction to give more coherence to the system of appellation of origin." Département Cartes et Synthèses Géologiques. B.P. 6009, 45060 Orléans Cedex, France: Bulletin de L'O.I.V., 1993.

Lewin, Benjamin. *What Price Bordeaux?* Dover: Vendange Press, 2009. Reprinted with permission from the publisher.

Librairie Larousse. *Wines and Vineyards of France.* New York and Paris: Arcade Press, 1991.

Lukacs, Paul. *Inventing Wine, A New History of One of the World's Most Ancient Pleasures.* Copyright © 2012 by Paul Lukacs. Used by permission of W. W. Norton & Company, Inc.

Lund, Cynthia M., Thompson, Michelle K., Benkwitz, Frank, Wohler, Mark W., Triggs, Chris M., Gardner, Richard, Heymann, Hildegarde, and Nicolau, Laura. "New Zealand Sauvignon blanc Distinct Flavor Characteristics: Sensory, Chemical, and Consumer Aspects." *American Journal of Enology and Viticulture* 60, no. 1 (January 2009).

Maltman, Alex. "Minerality in wine: a geological perspective." *Journal of Wine Research* 24, no. 3 (May 2013). Reprinted by permission of the publisher, Taylor and Francis Ltd.

Martini, Alessandro. "Origin and Domestication of the Wine Yeast Saccharomyces cerevisiae." *Journal of Wine Research* 1993 4, no. 3: 165–75. Reprinted by permission of the publisher, Taylor and Francis Ltd.

Matthews, Mark. "A Brief History of Plant Biology in Relation to Terroir," based on "The Terroir Explanation" in *Terroir and Other Myths of Winegrowing.* Berkeley: University of California Press, 2016. Reprinted with permission from the author.

Mayberry, Robert W. *Wines of the Rhône Valley: A Guide to Origins.* Copyright © 1987 Rowman & Littlefield.

McGee, Harold and Patterson, Daniel. "Talk Dirt to Me." *New York Times Magazine,* May 6, 2007. Copyright © 2007 The New York Times. All rights reserved. Used by permission.

McGovern, Patrick E. *Uncorking the Past: The Quest for Wine, Beer, and Other Alcoholic Beverages.* Berkeley: University of California Press, 2009.

Moran, Warren. "You said Terroir? Approaches, sciences, and explanations." Keynote address at Terroir 2006, a Symposium held at the University of California, Davis, March, 2006. Reprinted with permission from the author.

Napa Valley Destination Council, Preliminary Draft, FY2001–2015, TBID Destination Marketing Plan, V. 1.9, January 2010.

Noble, Ann C., Williams, Anthony A., and Langron, Stephen P. "Descriptive Analysis and Quality Ratings of 1976 Wines from Four Bordeaux Communes." *Journal of the Science of*

Food and Agriculture 35, no. 1 (January 1984): 88–98. Reprinted with permission from John Wiley and Sons.

Nowak, Linda, Wagner, Paul, and Arnold, Jean. "Marketing and Branding Wine," in *Wine: A Global Business*. Edited by Liz Thach and Tim Matz. Elmsford, NJ: Miranda Press, 2008. Copyright © Miranda Press 2008. Reprinted with permission of Cognizant Communication Corporation.

Office International de la Vigne et du Vin. *Lexique de la Vigne et du Vin*. Réimpression en facsimilé de l'édition de 1963. Paris: O.I.V., 1963.

Parker, Thomas. *Tasting French Terroir: The History of an Idea*. Berkeley: University of California Press, 2015.

Parr, Wendy V., Ballester, J., Valentin, D., Peyron, D., Sherlock, R., Robinson, B., Breitmeyer, J., Darriet, P., and Grose, C. "The nature of perceived minerality in white wine: preliminary sensory data." *New Zealand Winegrowers* 78 (Feb/March 2013): 71–75. Reprinted with permission from NZ Winegrower Magazine.

Patterson, Tim. "Rocks in my Head." *Wine Enthusiast* 20, no. 11 (November 1, 2007): 136. Reprinted with permission from Wine Enthusiast Media.

Pincus, Robert. "Wine, Place, and Identity in a Changing Climate." *Gastronomica* 3, no. 2 (Spring 2003). Reprinted with permission from The University of California Press, Journals Division.

Pinney, Thomas. *A History of Wine in America: from the Beginnings to Prohibition*. Berkeley: University of California Press, 1989.

Pinney, Thomas. *A History of Wine in America: from Prohibition to the Present*. 2nd revised ed. Berkeley: University of California Press, 2005.

Pinney, Thomas. "The Language of Wine in English." *Wayward Tendrils Quarterly* 21, no. 1 (January, 2011). Reprinted with permission from the publisher.

Pogue, Kevin. "Influence of Basalt On The Terroir of The Columbia Valley American Viticultural Area." Presentation at the 8th International Terroir Conference, Soave, Italy, 2010. Reprinted with permission from the author.

Reeve, Jennifer R., Carpenter-Boggs, L., Reganold, John P., York, Alan L., McGourty, Glenn, and McCloskey, Leo P. "Soil and Winegrape Quality in Biodynamically and Organically Managed Vineyards." *American Journal of Enology and Viticulture* 56, no. 4 (2005): 367–74.

Renouil, Yves (director), and de Traversay, Paul (collaborator). *Dictionnaire du Vin*. Bordeaux: Féret et Fils, 1962.

Robinson, Anthony L., Adams, Douglas O., Boss, Paul K., Heymann, Hildegarde, Solomon, Peter S., and Trengove, Robert D. "Influence of Geographic Origin on the Sensory Characteristics and Wine Composition of *Vitis vinifera* cv. Cabernet Sauvignon Wines from Australia." *American Journal of Enology and Viticulture* 63, no. 4 (2012): 467.

Robinson, Jancis, ed. *The Oxford Companion to Wine*. 3rd ed. New York: Oxford University Press, 2006.

Sacks, Gavin L., Acree, Terry E., Kwasniewswki, Misha T., Vanden Heuvel, Justine E., and Wilcox, Wayne F. "'Tell me about your childhood:' The role of the vineyard in determining wine flavour chemistry." *Proceedings of the Australian Wine Technical Conference* (June 2013): 39–46. Reprinted with permission from the Australian Wine Industry Technical Conference, Inc.

Schlachter, Kyle. "The Fallacy of Terroir." Colorado Wine Press. July 24, 2013. http://www
.coloradowinepress.com. Reprinted with permission from the author.

Schlosser, James, Reynolds, Andrew G., King, Marjorie, and Cliff, Margaret. "Canadian terroir:
sensory characterization of Chardonnay in the Niagara Peninsula." *Food Research International* 38, no. 1 (January 2005): 11–18.

Schoonmaker, Frank. *Frank Schoonmaker's Encyclopedia of Wine*. 5th ed. New York: Hastings
House, 1973.

Schultz, Hans R., and Jones, Gregory V. "Climate Induced Historic and Future Changes in
Viticulture." *Journal of Wine Research* 21, nos. 2 & 3 (2010): 137–145. Reprinted by permission of the publisher, Taylor and Francis Ltd.

Seguin, Gerard. "Terroirs and the pedology of wine growing." *Experientia* 42, no. 8 (August
1986): 861–73. With permission of Springer.

Skinkis, Patty. "Basic Concept of Vine Balance." Extension. January 22, 2013. http://www
.extension.org/pages/33109/basic-concept-of-vine-balance#.UoatkCemZMg. Reprinted
with permission of the author.

Smart, Richard, and Robinson, Mike. *Sunlight into Wine: A Handbook for Winegrape Canopy
Management*. Copyright © 1991 Ministry of Agriculture and Fisheries, new Zealand.
Reprinted with permission from Winetitles, www.winetitles.com.au

Smith, Clark. *Postmodern Winemaking: Rethinking the Modern Science of an Ancient Craft.*
Berkeley: University of California Press, 2014.

Smith, Rod. "Waiting for the Valley to Inhale," *Los Angeles Times,* May 7, 2002. Reprinted with
permission from the author.

Steiman, Harvey. "Is It All in the Funk? How 'natural wines' can polarize wine drinkers." Wine
Spectator. March 9, 2015. http://www.winespectator.com. Reprinted with permission from
Wine Spectator.

Stevenson, Robert Louis. *The Silverado Squatters.* New York: Scribner's & Sons, 1902.

Sullivan, Charles. *Napa Wine: A History from Mission Days to Present.* 2nd ed. San Francisco:
Board and Bench Publishing, 2008. Copyright © 1994, 2008 Napa Valley Wine Library
Association. Reprinted with permission of Board and Bench Publishing.

Swinchatt, Jonathan, and Howell, David. *The Winemaker's Dance: Exploring Terroir in the Napa
Valley.* Berkeley: University of California Press, 2004.

Theise, Terry. *Reading Between the Wines.* Berkeley: University of California Press, 2010.

Tomasino, Elizabeth, Harrison, Roland, Sedcole, Richard, and Frost, Andy. "Regional Differentiation of New Zealand Pinot noir Wine by Wine Professionals Using Canonical
Variate Analysis." *American Journal of Enology and Viticulture* 64, no. 3 (2013): 357.

Ulin, Robert C. "Invention and Representation as Cultural Capital: Southwest French Winegrowing History." *American Anthropologist* 93, no. 3 (September 1995): 519–527. Reprinted
by permission of American Anthropological Association and the author. Not for sale or
further reproduction.

Veseth, Mike. *Wine Wars: The Curse of the Blue Nun, the Miracle of Two Buck Chuck, and the
Revenge of the Terroirists.* Copyright © 2011 by Rowman & Littlefield, Publishers, Inc.
Reprinted with permission from the publisher.

Wade, Nicholas. "Microbes May Add Special Something to Wines." *New York Times,* November
15, 2013. Copyright © 2013 The New York Times. All rights reserved. Used by permission.

Washam, Ron. "The HoseMaster's Comprehensive Guide to Wine, Chapter 8: The Proper Use of Wine Terminology." Hosemaster of Wine. June 5, 2014. http://www .hosemasterofwine.blogspot.com/2014/06/the-hosemasters-comprehensive-guide-to. html. Reprinted with permission from the author.

White, Robert E. *Soils for Fine Wines*. New York: Oxford University Press, 2003.

Williams, John. "Wine Quality as Influenced by Sustainable Practices." Presentation at session on Science of Sustainable Viticulture, 54th Annual Meeting of the American Society for Enology and Viticulture, Reno, Nevada, June, 2003. Reprinted with permission from the author.

Wilson, James. *Terroir: The Role of Geology, Climate, and Culture in the Making of French Wines*. Berkeley: University of California Press, 1999.

Winkler, A.J., Cook, James A., Kliewer, W.M., and Lider, Lloyd A. *General Viticulture*. Berkeley: University of California Press, 1974.

Wolikow, Serge, and Jacquet, Olivier. "A Victory of the Unions." *Tong,* no. 2 (Summer 2009): 12–27. Reprinted with permission from the authors.

INDEX

Archer, Eben, 217
Argentina, 96, 103
Arnold, Jean, 246–47
aroma attributes, 194*fig.*, 196*tab.*
aroma compounds, grape-derived, 141*tab.*
aroma terms, 193*tab.*, 197, 210, 218, 219
aromatic molecules, 84, 173
Asens, Joan, 263
Asher, Gerald, 20–24
Asimov, Eric, 184, 263, 265–67
aspect, 47, 97, 99, 293
assemblage, 55, 169, 176
associative synesthesia, 207
Aube region (France), 265
Australia: basaltic soils in, 72; Cabernet
    Sauvignon wines from, 219–220; role of
    lakes and climate, 101; wine production in,
    1, 29–30, 60, 113, 165, 202, 237, 281; yeast
    flora in, 179
Australian geographical indications (GIs), 219
Austria, winegrowing in, 26, 96, 261, 286–88
Awatere Valley (New Zealand), 214
Azores, 72

Badascony region (Hungary), 72
Bakersfield (California), 108
Ballester, Jordi, 206, 208–11
Barbaresco (Italy), 78
Barbier, René, 262
Barbier, René Jr., 262
Barolo region (Italy), 78, 83, 139
Barossa Range (Australia), 117
Barossa Valley (Australia), 113, 139, 219
Barrett, Heidi Peterson, 31
Barsac-Sauternes wine, 169
basaltic bedrock, diagrammatic cross section,
    74*fig.*
basaltic soil: analysis of soil samples, 75*fig.*; in
    the Columbia River Valley, 72–76; effect on
    grape cluster temperature, 75*fig.*; effect on
    temperature, 76*fig.*; thermal properties of,
    73–74
Beaujolais wine, 78
Beaune Committee, 234
Beauroy (France), 66
bedrock: basaltic, 74*fig.*; in the Columbia Valley
    AVA, 72–76; and topsoil, 68–72
Beerenauslese wines, 285

beer industry, 186
Bench region (Ontario), 218
Benedictine monks, 34, 35
Bérard, Laurence, 238
Berger, Dan, 274–78
Bergerac region (France), 231
Berkowitz, Natalie, interviews with winemak-
    ers, 170–75
Bierzo wine, 261
biodynamic winegrowing, 147, 266
biogeography, 176
Bird, David, 155–58
Bisson, Linda, 180–81
bituminous soils, 100
Black Hamburg grapes, 121
Blaufränkisch grapes, 286
Bloch, Marc, 9
bloom (flowering, or anthesis), 127
Bokulich, Nicholas, 176, 177–78
Bonné, Jon, 256–58
Bonny Doon Vineyard, 24–26
Bordeaux (region), 34, 35, 36, 55, 77, 96, 125,
    139, 160, 164, 174, 179, 212, 214, 215, 283;
    classification system in, 227–29; invention
    of, 229–232; and the Judgement of Paris,
    189–190; soil types in, 136
Bordeaux wines: aroma terms selected for evalua-
    tion, 193*tab.*; canonical variates analysis of
    aroma attributes, 194*fig.*; commune or
    district of origin, 192*tab.*; descriptive
    analysis and quality ratings, 190–97; mean
    intensity ratings for wines, 196*tab.*
Bordelais area (France), 78, 79
Borden Ranch AVA (California), 242, 243, 244
Botrytis cineria. See noble rot (*Botrytis cineria*)
Bouchard, Charles, 234–35
Boulton, Roger, 182; on winemaking, 167–69
Bourgueil region (France), 78
Bourguignon, Claude: on climate, 112–13; on
    soil and *terroir*, 82–85
Bourguignon, Lydia: on climate, 112–13; on soil
    and *terroir*, 82–85
Bragato, Romeo, 267
Brazil, 96, 179
Brettanomyces, 160
Brillat-Savarin, 40
Brix (soluble solids), 138, 146, 147, 172–73,
    276–77

château (châteaux): and growth designation,
192*tab*.; as invention, 231–32

Château du Tertre, 174

Château Giscours, 174

Château Kirwan, 174

Château Latour, 232

Château Margaux, 164

Chateau Montelena, 255

Châteauneuf-du-Pape, 103

Château Pétrus, 227–28

chemical weeding, 85, 146

Chianti (region), 78, 179

Chianti (wine), 105

Chile, 1, 237, 269

Chiles Valley (California), 68, 70, 72, 96

Chimney Rock winery, 133

Chinon (France), 78

Christmann, Monika, 218

Cistercian abbeys, 36, 36–39

Cistercian monks, 21, 34, 35–36, 66

Cistercian wines, 35–36

Clarets, 105

Clare Valley (Australia), 219

clarification, 155

Clarke, Oz, 165–66

clay soils, 78, 79, 81, 265

Clements Hills AVA (California), 242, 243, 244

Cliff, Margaret, 218

climate: and air drainage, 9, 97, 289; and aspect
and slope, 97; and geology, 95–96; and
grape ripening, 95*fig*.; humidity, 22, 79–80,
81, 93, 102, 103, 106, 108, 113–15, 118, 120,
121, 205, 280; influence of, 92–93, 292,
293; and latitude, 96; in the Napa Valley,
250; and proximity to water bodies,
100–101; seasonal influences, 109; and
sunlight, 115–18; temperature adjustments
for site factors, 98*tab*.; and topography,
95–96; variations in, 153–54; wind, 113–14.
See also climate change; climate excerpts;
fog; heat summation; microclimate

climate change, 93–94, 154, 277, 278–79

climate change excerpts: Gladstones, 96–101,
288–89; Pincus, 284–88; Schulz and
Jones, 279–283

climate excerpts: Claude and Lydia Bourguignon, 112–13; Geiger, 119–121; Jackson,
109–12; Keller, 113–15; Mayberry, 102–3;

Schuster, 109; Smart and Robinson,
115–18; Winkler, 104–9

climate classification systems, 242

climate/maturity groupings, 95*fig*.

climats, 42, 44, 64, 153

clonal propagation, 154

clos, 42

Clos Eerasmus, 262

Clos Henri, 268

Clos Mogador, 262

Clos Vougeot, 31, 234

Cloudy Bay, 268

Colette, 42

Columbia Valley AVA (Washington), 72–76

Comiskey, Patrick, 158–160

Common Agricultural Policy (CAP), 216

Condado do Tea (Spain), 22

Confédération des Associations Viticoles de
Bourgogne (CGAVB), 235

Constantia wines, 105

Constellation Brands, 268

Cook, James, 104–9

Coonawarra region (Australia), 219

Cosumnes River AVA (California), 242, 243

Coteaux du Tricastin (France), 102

Côte Chalonnaise, 206

Côte de Beaune, 55, 84

Côte de Léchet, 66

Côte des Bar, 265, 266

Côte Dijonnaise, 235

Côte d'Or, 38, 77, 166, 237; and the evolution of
terroir, 233–37

Côtes de Beaune, 279

Côtes de Nuits, 55, 84, 234

Côtes du Rhône, 35, 102

Craggy Range, 268

Croatia, wines from, 202

crop load, 131; equations for, 134

Croser, Brian, 165, 185

cru (and *crus*), 9, 36, 38, 42

Cru Artisan, 228

Cru Bourgeois, 228

cultivars, 53, 55, 79, 135, 142, 149, 153, 154, 169,
170, 176, 217, 283

cuvées, 16

Dagueneau, Didier, 16–17

Dally plonk (junk wine), 268

d'Angerville (Marquis), 234–36

d'Aussy, Pierre Jean-Baptiste Legrand, 40

Debuigne, Gerard, 43

Degree Days (DD), 104, 131; in California, 106

Delfaut, Philippe, 174

Deronoroncourt, Stephan, 15–16

descriptive analysis, 218; and quality ratings, 211–12. *See also* sensory analysis methods

de Serres, Olivier, 39–40

Diamond Creek, 255

Dion, Roger, 53, 56–57, 126

Domaine Carneros, 170–71

Donkey and Goat Winery, 31

Dornfelder grapes, 139

Dosnon & Lepage Champagnes, 265

drainage: air, 97, 289; soil, 7, 13, 60, 64, 77, 79, 80, 82, 146, 149; water, 9, 80, 83, 166

Drappier, Michel, 265

Drappier winery, 265

dried-grape wines, 37

drought, 77, 81, 82, 137, 278, 287; resistance to, 80

Dry Creek Valley (California), 19

Dubourdieu, Denis, 55, 169

Due, Graham, 112

Dunn, 255

earthy smell, 204. *See also* minerality

Eiswein, 29, 285, 286, 288

elevation, 1, 9, 67, 72–73, 93, 101, 103, 116, 155, 239, 242, 244, 280

Elliott-Fisk, Deborah, on the Lodi AVA, 239–244, 256

Emperor grapes, 105

enhancements, 158, 159

enology, 112, 167, 183

erosion, 64, 67–69, 73, 77, 78, 99, 103, 242

Estreicher, Stefan, 35–36

evapotranspiration, 79, 244. *See also* transpiration

Fanet, Jacques, 48–49

Faux Chablis, 184–85

Feiring, Alice, 184

fermentation, 155, 160, 171; stuck, 144, 184, 276

fertilization and fertilizers, 88, 127, 159; nitrogen, 144–45

filtering, 155, 168–69, 200

Finca Dofí, 262

Finger Lakes region (NY), 96, 101, 139–42

fining, 155, 168, 200

First Annual Qvevri Symposium, 184

Fischer, Ulrich, 218

Fletcher, Doug, 133

flintiness, 203–4. *See also* minerality

flooding, 72, 286, 287

flowering, 42, 81, 110, 122, 127, 130, 134, 137–38

Flowers Vineyards, 258

fog, 19–20, 28, 93, 96, 103, 106, 113, 172, 244, 250

Folsom, California, 242

Fort Ross Vineyards, 258

Foulkes, Christopher, 60–61

Fourchaume (France), 66

France: basaltic soils in, 72; *la France profonde*, 10; *terroir* in, 24; white wines from, 212–14; winegrowing in, 105. *See also* French wine regions

Francis, Leigh, 220–21

Franciscan formation, 70

Frankland River region (Australia), 219

Freestone winery, 258

Freixenet, 262

French terminology, 13, 43–46

French wine regions: geographic and geologic map, 63*map*; geologic map of Chablis, 65*map*; in Southwest France, 230*map*. *See also* Bordeaux; Burgundy; Champagne; wine regions (France)

Fresno, California, 105, 108

Frog's Leap Winery, 173–74

Frost, Andy, 219

frosts, 99, 100, 101, 110, 119, 121, 284, 295, 296; advective, 97; radiative, 97

fruitful shoots, 127

fruit set, 23, 127, 134, 135, 136, 138

full bloom, 127. *See also* flowering

Galicia, 261; and *terroir*, 20–24

Gangjee, Dev, 126

Garnacha grapes, 262, 263

Gautherot, Bertrand, 266–67

Gayton, Don, 49

Geiger, Rudolf, 118–121

geographical information systems (GIS), 216

geography, 153–54

Greene, Joshua, 187–88
greenhouse gases, 288–89
Greenough, John, 86
Green Valley (California), 18, 257
Grenache grapes, 112–13, 139
Grieco, Paul, 87
Growth-Yield Relationship, 134
Grüner Veltliner grapes, 286, 287, 288
gun-flint aroma, 204. *See also* minerality
Guyot, Jules, 40

Halliday, James, 164–65
Hanzell Vineyards (California), 246–47
Harrington, Robert J., 50
Harrison, Roland, 219
Harrop, Sam: on authentic wine, 270–73; on
    natural wine, 185–86; on *terroir*, 50–51
Hautes Côtes (France), 235
Haut Pauillac, 232
Heathcote region (Australia), 101
heat summation, 105–6, 131; in California and
    various foreign locations, 107*tab.*; for
    California climate regions, 106, 108–9. *See
    also* temperature
Heat Summation Unit (HSU), 104
Heitz Cellar's Martha's Vineyard, 215, 255, 275
Hellman, Edward, 127–132
Henick-Kling, Thomas, 178
herbicides, 146. *See also* chemical weeding
Hess Collection, 70
Heymann, Hildegarde, 211–12
Hirsch, David, 258
Hirsch Vineyards, 258
Holt, Helen, 220–21
horizons of soil, 67
Howell, David, 66, 68–72
Hugel, Jean, 284–85
Hugel & Fils, 284
Huglin, Pierre, 104
humidity, 22, 79–80, 81, 93, 102, 103, 106, 108,
    113–15, 118, 120, 121, 205, 280
humus, 78, 83, 124, 149, 199, 206
Hungary, 72, 83–84, 96
Hunter, Ernie, 268

Ice wine. *See Eiswein*
identity: and marketing, 226*fig.*; national, 57
identity *terroir*, 53, 56–57

igneous rock, 61, 67
Iland, Patrick, 47–48
India, 72, 112, 179
indoor cultivation, 121
Inglenook's cask wines, 255
inoculation, 155, 177
Institut National de l'Origine et de la Qualité, 41
Institut National des Appellations de L'Origine
    (INAO), 41, 56
Interconnected Spatial System (ISS), 237–38
Intergovernmental Panel on Climate Change
    (IPCC), 288–89
international style, 27, 269–72
iron, in soil samples, 75*fig.*
irrigation: in California, 106, 108; and canopy
    temperature, 115; controlled, 83; decreas-
    ing, 142; deficit (regulated deficit irrigation,
    RDI), 123, 136–38; drip, 5; dripline, 278;
    eliminating, 146; with fertilizer (fertiga-
    tion), 144; growing grapes without, 158; as
    negation of *terroir*, 25–26; systems for, 249,
    278; and vineyard management, 45, 48, 61,
    79, 88, 132, 135, 154, 170, 249
Italy, 72, 105, 139, 179

Jackson, David, 104, 109–12
Jacquet, Olivier, 233–37
Jahant Hills AVA (California), 243, 244
Jamestown (California), 108
Japan, 179
Jayer, Henri, 13–14
Jefford, Andrew, 226
Jenny, Hans, 154
Jerez (Spain), 78, 79
J.H. Selbalch, 286
Johnson, Hugh, 164–65, 225
Jones, Graham, 86
Jones, Greg, 94; on climate change, 279–283
Jordan, Tony, 165
Judgment of Paris, 125, 189–190, 273

Kamptal region (Austria), 286–87
Keesing, Roger, 232
Keller, Markus, 124; on wind and humidity,
    113–15
Kemp, Carroll, 257
Kern county (California), 105
Khakheti region (Georgia), 184

Médoc region (France), 77, 78, 79, 228; chateau system in, 231–32; High Médoc gravel-sand soils, 79–80

Memstar, 277

Mendocino County (California), 146–47, 256

Mendoza region (Argentina), 103

Mentzelopoulos, André, 164

Mercedes Effect, 250

Merlot grapes, 146–47, 262, 283

Merlot wine, 267, 275

mesoclimate, 44, 116–17

meso-*terroir*, 61

metallic smell, 205. *See also* minerality

metamorphic rock, 67, 83

Meuse Valley (France), 100

microbes, 175–78

microbial *terroir* excerpts: Bisson 180–81; Bokulich et al., 176; Martini, 178–180; Wade, 176–78

microclimate, 44, 99–100, 117, 121; of the grapevine canopy, 116–18

microflora, 175–76

micro-oxygenation, 163

micro-*terroir*, 61

Mills, David A., 176, 177, 178

mimetic discourse, 232

minerality, 87, 185, 188, 197–98, 206–7, 263; in Burgundy Chardonnay wines, 208–11; earthiness, 87–88, 204; as the flavor of minerals, 198–99; flintiness, 203–4; gun-flint aroma, 204; and the inability to taste minerals, 201–3, 222; metallic smell, 205; and nutrient minerals, 199–201; in the Priorat, 263; proportions of ions in geologic minerals and wine, 200; seashells and fossilized shells, 205; sensory perception of, 207; smell of warm/wet stones, 204–5, 212, 222; and *terroir*, 16; in white wines, 211–14

minerality excerpts: Ballester et al., 206, 208–11; Heymann et al., 211–12; Maltman, 197–206; New Zealand Winegrowers study, 212–14; Patterson, 222

Mokelumne Hill (California), 108

Mokelumne River AVA (California), 242, 243, 244

Mondavi, Robert, 247

Monkey Bay, 268

Mont de Milieu (France), 66

Montée de Tonnerre (France), 66

Monterey County (California), 113

Montgolfier, Gislain de, 169

Montgueux region (France), 266

Montmains (Monts Mains; France), 66

Montrachet wine, 12

Moran, Warren: on *terroir*, 51–53, 52–57; on winemaking, 169–170

Moscato wine, 31

Moselle wine, 83

Mosel Valley (Germany), 26–29, 36, 77, 100, 101, 116, 179, 283; climate change in, 285–86

Moueix, 228

Mount Barker region (Australia), 219

Mount Veeder winery, 70

Mourvèdre grapes, 113

Mt. Etna (Sicily), 72

Müller-Thurgau grapes, 139

Muscat grapes, 121, 284

Muscat wines, 15, 140

mycorrhizae, 83

Nagambie Lakes (Australia), 101

Napa Valley (California) AVA, 68–72, 113, 117, 133, 139, 171–72, 177, 240, 246, 247–50; brand promotion in, 244–46; history of, 253–56; international style in, 273; and the Judgment of Paris, 190; marketing in, 250–53; rise of, 244–46; soils and geology, 247–250; soil types in 66; welcome sign, 251*fig*.

Napa Valley Destination Council, 252–53

Napa Valley Vintners' Association (NVVA), 244–46; research reports sponsored by, 247–250

napping, 220–21

Nashik region (India), 112

Natural Terroir units (NTUs), 215–17

natural wine movement, 152, 184–86; David Bird on, 155–58; Harvey Steiman on, 160–61; Patrick Comiskey on, 158–160

Nebbiolo d'Alba wine, 78

Nebbiolo grapes, 139, 149, 283

New World: *terroir* in, 1, 24–26, 46, 237–44; winemaking in, 45–46, 52, 56, 60–61, 112. *See also* Australia; California; New Zealand

tilth, 83

Tokaji (Hungary), 96

To Kalon vineyard, 31

Tokay grapes, 105, 108

Tokay wine, 83

Tomasino, Elizabeth, 219

topo climate. *See* mesoclimate

topography: effect on soil, 67; variations in,
153–54

transpiration, 100, 114–15, 118, 129. *See also*
evapotranspiration

Traversay, Paul de, 40

Trebbiano grapes, 139

trellising techniques, 243, 249, 275, 278;
California Sprawl, 275

Trockenbeerenauslese wines, 285

Tulare (California), 105

Turkey, and the origin of viticulture, 148–49

Tuscany, 139, 160

Uco Valley (Argentina), 103

Ulin, Robert, 229–232

United Kingdom, 121, 279, 283

United States: role of lakes and climate, 101;
winemaking in, 84. *See also* California;
New York; Oregon; Washington

UTN (*l'unité de terroir naturel*), 216–17

Vaca Mountains (California), 69

Vaillons (Les Lys), 66

Val do Salnés (Spain), 22

van Beek, Alexander, 174–75

vapour pressure, 201–2

veraison, 80, 130, 138, 281

vertical shoot positioning (VSP), 275

Veseth, Mike, 267–69

Vidal de la Blache, Paul, 40, 53

villages, as wine classification, 84

vine balance, 131, 133, 146; unbalanced vines,
136; vine-dictated, 135; viticulturist-dictated, 135–36

vine leaf area, 134

vineyard management practices, 48; and wine
flavor chemistry, 139–142

vineyard preparation, 154

viniculture, 148, 286. *See also* winemaking

vinification, 14, 55, 82, 89, 169, 170, 175, 200,
202, 229, 234, 236. *See also* winemaking

vini-*terroir*, 53, 55, 169

Vinovation, 277

vins d'effort, 24

vins de terroir, 24, 25, 293

Vintners Quality Alliance (VQA), 218

viticulture: biodynamic, 146; and climate
change, 261, 279–283, 288–89; in the
Columbia Valley AVA, 54; cool climate, 54,
112; degree day standards in, 242;
development of, 147–48, 217; in France, 13,
234–35; and geology, 82; growing season
average temperatures, 280*map*; high-
altitude, 103; limitations of, 99; and
microclimate, 121; modern approaches to,
123, 183, 243, 269; in monasteries, 34–36;
in the Mosel, 286; organic, 142–46; and
RDI, 137; restrictions on, 238; and soil
geology, 99–100; and soil types, 243–44; in
Southern Europe, 279; on steep slopes,
101; and *terroir*, 3, 123, 132, 166, 170; use of
chemical weeding in, 85; use of sulphur in,
202; variations in, 6, 44, 139, 238; and vine
balance, 133; warm, 112; world areas,
94*map*; zones of, 126. *See also* winegrowing

volcanic rock, 68, 69–70, 71, 83–84

volcanic soil, 69, 71, 84, 243, 254

Vosne-Romanée, 153

Vouette & Sorbée, 266

Wachau region (Austria), 96, 286

Wade, Nicholas, 176–78

Wagner, Paul, 246–47

Wahgunyah region (Australia), 101

Waipara region (New Zealand), 219

Wairau Lowlands (Australia), 214

Washam, Ron, 57–58

Washington (state), 72–76

water balance (*bilan hydrique*), 54

water deficit, 129

water management, 133

water status, 129

water stress, 115, 133, 137–38

water supply: absorption of, 129; availability and
drainage, 83; in clay soils, 81; and climate

change, 280–81; in gravel-sand soils, 79–80; in limestone soils, 81; in New World vineyards, 61; subterranean, 262. *See also* irrigation

water table, 80

Weingut Bründlmayer, 287

West Sonoma Coast Vintners, 258

White, Robert, 49–50

Whitehaven, 268

white wines: aromatic compounds in, 140, 142; from Burgundy, 12, 38; cultivation of, 108; effect of temperature and climate on, 111, 281; Heymann's study of minerality in, 211–12; less trend toward internationalization, 271–72; minerality in, 87; New Zealand study of minerality in, 212–14; from New Zealand, 268; soil requirements of, 85; sulphur dioxide requirements of, 182; from the United States, 84; use of carbon treatment for, 42; zinc content of, 202. *See also* wines; *specific white wines by name*

Wibberly, Leonard, 245

wildfires, 215, 278

Willamette Valley (Oregon), 72, 99

Williams, John, 173–74; on sustainable practices in winegrowing, 143–46

Williams Selyem winery, 258

Wilson, James E., 62–66

wind and wind speed, 113–14, 118

wine chemistry: Goode's thoughts on, 181–83; mineral content, 84; role of the vineyard in, 139–142

wine grapes. *See* grapes

winegrowing: average growing season temperatures, 280*map*; biodynamic, 266; carbon cycle, 157–58, 157*fig.*; modern scientific practices, 132–33; organic, 142–46; regions in Southern France, 230*map*; and the *terroir* mindset, 132; in the United Kingdom, 121; world viticultural areas, 94*map*. *See also* viticulture

winegrowing excerpts: Reeve et al., 147; Williams, 143–46

winemakers: interviews with, 170–75; involvement of, 155, 161–64, 184–85; role of, 5, 11, 14, 15, 17, 25, 30–31. *See also* winemaking

winemaking: acidulation, 155; amelioration, 155; amplified outline scheme, 168; anthropic influence, 153–54; assessment of, 270; chaptalization, 155, 166; clarification, 155; elevation, 155; fermentation practices, 155, 156; filtering, 155; fining, 155; inoculation, 155; and marketplace consistency, 159; natural wine movement, 152, 154–61, 184–86; outline scheme, 168*fig.*; reception, 155; stabilization, 155; sustainable, 185–86; technical control in, 11, 31; and *terroir*, 151–52, 166–69; trend toward style over substance, 269–278; in the United States, 84; use of oak in, 277; yeasts, 155. *See also* viniculture; vinification; winemakers; winemaking excerpts; winemaking history

winemaking excerpts: Bird, 155–58; Boulton, 167–69; Burt, 170–71; Clarke and Rand, 165–66; Clark Smith, 183–85; Comiskey, 158–160; Corison, 171–72; Delfaut, 174; Goode, 162–164; Goode and Harrop, 185–86; Johnson and Halliday, 165; Moran, 169–170; Prüm, 175; Schlachter, 153–54; Steiman, 160–61; Stevens, 172–73; van Beek, 174–75; Williams, 173–74;

winemaking history: Egyptian winemaking, 32, 37, 156, 217; Estreicher's analysis of, 35–36; Lukacs's analysis of, 36–39; McGovern's analysis of, 33–34; Neolithic, 32; Noah hypothesis, 148, 150; paleolithic hypothesis, 32–33; Roman winemaking, 34, 37, 156; winemaking in ancient Greece, 37, 156; winemaking in monasteries, 9, 34, 35–36, 36–39, 66

wine regions: Andes Mountains (Chile), 96; Azores, 72; Canary Islands, 72; Columbia Valley AVA (Washington), 72–76; Finger Lakes (New York), 96, 101, 139–42; Khakheti (Georgia), 184; Nashik region (India), 112; Willamette Valley (Oregon), 72

wine regions (Argentina): Andes Mountains, 96; Mendoza, 103; Salta, 103; Uco Valley, 103;

wine regions (Australia): Barossa Valley, 113, 117, 139, 219; Clare Valley, 219; Coonawarra, 219; Frankland River, 219; Goulburn Valley, 101; Greater Victoria, 137–38;

Heathcote, 101; Lake Eppalock, 101; Langhorne Creek, 219; Margaret River, 219; McLaren Vale, 219; Mount Barker, 219; Nagambie Lakes, 101; Padthaway, 219; Rapaura, 214; Rutherglen, 101; Wahgunyah, 101; Wairau Lowlands, 214; Wrattonbully, 219

wine regions (Austria): Kamptal, 286–87; Kremstal, 286; Wachau, 96, 286;

wine regions (California): Alexander Valley, 19, 241; Alta Mesa AVA, 242, 243; Amador County, 31; Bakersfield, 108; Borden Ranch AVA, 242, 243, 244; Camp Pardee, 242; Carneros District, 20, 69, 257; Central Coast, 177, 254, 255, 256; Central Valley, 108, 177, 239, 269; Chalk Hill, 20, 246; Chiles Valley, 68, 70, 72, 96; Clements Hills AVA, 242, 243, 244; Cosumnes River AVA, 242, 243; Dry Creek Valley, 19; Green Valley, 18, 257; Jahant Hills AVA, 243, 244; Kern county, 105; Lodi AVA, 240*map*, 243, 244, 256; Mendocino County, 146–47, 256; Mokelumne River AVA, 108, 242, 243, 244; Monterey County, 113; North Coast, 275; Paso Robles, 256; Pope Valley, 68; Redding, 243; Russian River Valley, 18, 19, 257; Rutherford AVA, 274–78; Rutherford Bench, 171–72; Sacramento Valley, 105, 108, 240, 242; Salinas Valley, 20, 113, San Diego County, 108; San Joaquin Valley, 105, 108, 177, 240, 243; San Luis Obispo, 20; Santa Barbara, 96; Santa Cruz Mountains, 103; Santa Maria Valley, 20; Santa Ynez Valley, 20; Sierra Foothills AVA, 242; Sloughhouse AVA, 242, 243, 244; Sonoma Coast, 257; Sonoma Valley, 117, 177, 246, 247, 253–57; St. Helena, 108, 171–72; Wooden Valley, 68. *See also* Napa Valley (California) AVA

wine regions (Canada): Bench, 218; Lake Okanagan, 101; Lakeshore, 218; Lakeshore plain, 218; Niagara peninsula, 218

wine regions (France): Aloxe Corton, 55; Alsace, 38, 61, 83, 99, 261, 272, 284–85; Anjou, 83; Aube, 265; Beauroy, 66; Bergerac, 231; Bordelais, 78, 79; Bourgueil, 78; Chablisien, 206; Chambertin, 11; Coteaux du Tricastin, 102; Côte Chalonnaise, 206;

Chinon, 78; Coteaux du Tricastin, 102; Côte de Beaune, 55, 84; Côte de Léchet, 66; Côte des Bar, 265, 266; Côte Dijonnaise, 235; Côte d'Or, 38, 77, 166, 233–37; Côtes de Beaune, 279; Côtes de Nuits, 55, 84, 234; Côtes du Rhône, 35, 102; Fourchaume, 66; Gevrey-Chambertin,, 55–56; Graves, 78, 79, 80; Hautes Côtes, 235; Languedoc, 113; Loire Valley, 38, 41, 62, 215; Médoc, 77, 78, 79–80, 228, 231–32; Meuse Valley, 100; Mont de Milieu, 66; Montée de Tonnerre, 66; Montgueux, 266; Montmains (Monts Mains), 66; Nuit-St-George, 55; Pomerol, 78, 81, 227; Pouilly, 16–17, 62, 213; Rhine river region, 36, 101; Rhône Valley, 38, 102–103, 160; Richebourg, 153, 234; Romanée, 153; Romanée-Conti, 153; Saint Bris, 214; Saint-Vivant. 234; Sancerre/Loire, 212, 214; Sauternais, 78; St. Emilion, 77, 78, 81. *See also* Bordeaux; Burgundy; Champagne; French wine regions

wine regions (Germany): Mosel Valley, 26–29, 36, 77, 100, 101, 116, 179, 283, 285–86; Palatinate, 119; Rheingau, 78, 281–82, 283; Würtzburg, 121

wine regions (Hungary): Badascony, 72; Tokaji, 96

wine regions (Italy): Alto Adige, 96; Barbaresco, 78; Barolo, 78, 83, 139; Chianti, 78, 179; Piedmont, 283; Tuscany, 139, 160

wine regions (New Zealand): Awatere Valley, 214; Central Otago, 219; Marlborough, 96, 139, 170, 212–14, 219, 268; Martinborough, 56, 219; Rapaura, 214; Southern Valley, 214; Waipara, 219; Wairu Lowlands, 214

wine regions (Sicily): Marsala, 78; Mt. Etna, 72

wine regions (Spain): Condado do Tea, 22; Galicia, 20–24, 261; Jerez, 78, 79; O Rosal, 22; Priorat, 27, 109, 139, 261; Rías Baixas, 22–23, 261; Rioja, 139, 226; Val do Salnés, 22

wines: aging of, 182, 200, 232; artisanal, 183–85; assessment of, 271, 272; authentication of, 86; bulk, 262; Cistercian, 35–36; as commodity, 272–73; diversity in, 273; from dried grapes, 37; grape-derived aroma compounds in, 141*tab.*; "international" style, 27, 269–72; literary references to, 39;